ESSAYS ON GALILEO AND 7
AND PHILOSOPHY OF S

VOLUME II

Stillman Drake

Selected and introduced by N.M. Swerdlow and T.H. Levere

For forty years, beginning with the publication of the first modern English translation of the *Dialogue Concerning the Two Chief World Systems*, Stillman Drake
was the most original and productive scholar of Galileo's scientific work. During
those years, Drake published sixteen books on Galileo, including translations of
almost all the major writings, and *Galileo at Work*, the most comprehensive
study of Galileo's life and works ever written. Drake also published about 130
papers, of which nearly 100 are on Galileo and the rest on related aspects of the
history and philosophy of science. The three-volume collection *Essays on Galileo and the History and Philosophy of Science* includes 80 of those papers.

The papers included in Volume II elaborate on some often misunderstood
aspects of Galileo's *Dialogue Concernng the Two Chief World Systems*, and
serve as an extended annotation to Drake's monumental translation of the work.
The papers cover still-controversial aspects of the *Dialogue* such as the notori-
ous Tower Argument, which Drake examines by addressing its background in
Aristotle, Ptolemy, and Copernicus.

One of Drake's most significant scholarly discoveries was that Galileo, early
in his studies, based his theory of tides upon the diurnal and annual motions of
the earth, proving that Galileo was a Copernican many years earlier than was
previously supposed. Drake considered the principal purpose of the *Dialogue* to
be the exposition of the theory of the tides in the Fourth Day, rather than direct
proof of the Copernican theory.

Volume II of Stillman Drake's *Essays on Galileo and the History and Philosophy of Science* showcases the papers that made Drake the touchstone of international Galileo scholarship.

N.M. SWERDLOW is Professor, Department of Astronomy and Astrophysics,
University of Chicago.
T.H. LEVERE is Professor and Director, Institute for the History and Philoso-
phy of Science and Technology, University of Toronto.

Stillman Drake. Photo courtesy of Mrs Florence Drake.

Essays on Galileo and the History and Philosophy of Science

VOLUME II

Selected and introduced by
N.M. Swerdlow and T.H. Levere

STILLMAN DRAKE

UNIVERSITY OF TORONTO PRESS
Toronto Buffalo London

© University of Toronto Press Incorporated 1999
Toronto Buffalo London
Printed in Canada

ISBN 0-8020-4343-7 (cloth)
ISBN 0-8020-8164-9 (paper)

Printed on acid-free paper

Canadian Cataloguing in Publication Data

Drake, Stillman
Essays on Galileo and the history and philosophy of science

Includes bibliographical references and index.
ISBN 0-8020-0626-4 (v. 1 : bound)   ISBN 0-8020-7585-1 (v. 1 : pbk.)
ISBN 0-8020-4343-7 (v. 2 : bound)   ISBN 0-8020-8164-9 (v. 2 : pbk.)

1. Galilei, Galileo, 1564–1642.   I. Swerdlow, N.M. (Noel M.).
II. Levere, Trevor H. (Trevor Harvey).   III. Title

QB36.G2D667 1999   520'.92   c98-932667-5

University of Toronto Press acknowledges the financial assistance to its publishing
program of the Canada Council for the Arts and the Ontario Arts Council.

Contents

Illustrations

The plates in this volume are all facsimiles or transcriptions of folios from volume 72 of the Manuscripti Galileiani. The plates originally appeared and are referred to in many papers but are reproduced here only once, with the most detailed caption. As a guide to finding each folio where the plate number is not given in the text, we give the folio numbers in order followed by the plate on which each is found. We also list the plates in order followed by folio numbers.

PART V

GALILEO: *DIALOGUE CONCERNING THE TWO CHIEF WORLD SYSTEMS*

Drake's translation of the *Dialogue Concerning the Two Chief World Systems*, his second publication on Galileo, has been the standard English version since it appeared in 1953. He was able to provide some revision and additional notes for an edition of 1962, but his wish to revise the translation and expand the annotation considerably was repeatedly refused by the publisher, which is a pity since over the years he gave the work a great deal of thought, only part of which appeared in published papers. Aspects of the *Dialogue* are considered in various papers of this collection, and those in this section concern both general and specific topics. The first three, "The Title Page and Preface of Galileo's *Dialogue*" (1), "The Organizing Theme of the Dialogue" (2), and "Reexamining Galileo's *Dialogue*" (3), are quite broad, considering, among much else, the history of the composition of the *Dialogue* and the types of hypothetical arguments used by Galileo, but a theme that runs through all of them is that the original intent of the *Dialogue* was the explanation of the cause of the tides, to which its original title referred, rather than the proof of the Copernican theory, which Galileo was explicitly forbidden to "hold or defend." Drake believed that this misunderstanding was the foundation of much unwarranted criticism of the *Dialogue* and of Galileo's conduct before the Holy Office in 1633.

One of the most criticized passages in the *Dialogue* is Galileo's *bizzaria* (curiosity or caprice) that the path of a body falling from a tower on the rotating earth is probably an arc of a semicircle extending from the top of the tower to the centre of the earth. The speculation is incorrect and the path a second degree spiral, as Fermat later pointed out, and as Galileo seems to have known all along as he told Fermat it was meant as a "rather audacious joke." "Semicircular Fall in the Dialogue" (4) considers various misunderstandings of this passage, such as that Galileo intended to describe the path of the falling body all the way to the centre of the earth; rather, Galileo's intention was to describe only the approximate path of the body to the surface of the earth, to show that its motion was uniform, circular, and passed through an equal distance whether it fell or remained at the top of the tower, and that its acceleration in falling was only a projection of its circular motion into a straight line along the tower. The semicircular fall raises another disputed question, that of "circular inertia" as it has been called, whether bodies have a natural tendency to follow circular paths, which Alexander Koyré considered fundamental to Galileo's physics. This is taken up in "The Question of Circular Inertia" (5), which concerns Galileo's treatment of uniform rectilinear and circular motion, neither of which can strictly be characterized as inertial although in different kinds of motion, as of projectiles or planets, each was taken to be natural and requiring no cause. Finally, "The Tower Argument in the *Dialogue*" (6) takes up the rather puzzling and indirect background to the argument in Aristotle, Ptolemy, and Copernicus.

We have noted that Drake considered the principal purpose of the *Dialogue* to be the exposition of the theory of the tides in the Fourth Day. One of Drake's most interesting and important discoveries was that Galileo had devised his theory of the tides, based upon the diurnal and annual motions of the earth, many years before its first appearance in his letter to Cardinal Alessandro Orsini of 1616. "Origin and Fate of Galileo's Theory of the Tides" (7) is concerned with three paragraphs written in Paolo Sarpi's notebook on physical problems in about 1595 that give a brief summary of an early stage of Galileo's theory, which he must have learned from Galileo himself, and show that Galileo, perhaps with the tidal theory as motivation, was already a Copernican. The paper goes on to consider later criticisms, and in some cases misunderstandings, of Galileo's theory – that of Ernst Mach is the most well known – a subject taken up at greater length in "History of Science and Tide Theories" (8), and its resemblance in part to Laplace's hydrodynamical theory of the tides, in which the periodicity and height of tides depends upon the length and depth of the sea, is pointed out.

1

The Title Page and Preface of Galileo's *Dialogue*

For a century it has been universally assumed that Galileo wrote his celebrated *Dialogue* in a deliberate attempt to defeat the purpose of an edict issued by the Church to prohibit Copernican books. Scholarly opinions on matters of great importance in our cultural history, to say nothing of opinions about Galileo's character and his scientific views, have been profoundly affected by that assumption. Yet in the form just stated, the assumption is certainly defective. The Church edict did not prohibit even the *De revolutionibus* of Copernicus himself, from which were removed only a few passages of no importance to its scientific content. Moreover, the basis of Galileo's summons to Rome for trial was not the edict, but an alleged earlier command applying only to Galileo personally. It would be closer to historical truth to assume that Galileo wrote his *Dialogue* in an attempt to aid the Church in overcoming contemporary misunderstanding of the edict – precisely as he said in his preface *Al discreto lettore*.

Such a complete reversal of a long-accepted belief is here set forth only after long study of the surviving documents relating to Galileo, his *Dialogue*, and his trial. To adduce all the evidence in favour of it and reply to all objections would be far more than can be done in the present paper. My principal purpose here is to throw light on Galileo's preface to the *Dialogue*, which has never been taken seriously because of its strange and artificial appearance. Also to be explained is a conspicuous disparity between the ending of that preface and the opening speech in the text which immediately follows it. No one seems to have commented on that, or to have perceived the relation of both to the printed title page of the *Dialogue*, with an examination of which I think it best to begin.

Although Italian writers now customarily refer to Galileo's book as *Dialogo dei* (or *sui*) *Massimi Sistemi*, no such precise phrase appeared on the long title

Reprinted from *Quaderni d'Italianistica* I (1980), 139–56, by permission.

page of 1632. The first English translator, probably Joseph Webbe of London, was content in 1634 to put the printed title page word for word into English, but his translation, left unpublished, remained without influence on others.[1] In 1635 the widely circulated Latin translation by Matthias Bernegger supplied a misleading short title,[2] followed by an accurate rendering of the Italian subtitle: *Systema Cosmicum authore Galileo Galilei ... in quo quatuor dialogis, de duobus Maximis Mundi Systematibus ... disseritur.* The first published vernacular translation, by Thomas Salusbury, appeared at London in 1661; though he worked from both the original and the Latin version, it was the latter which principally determined his title: *The Systeme of the World: in Four Dialogues. Wherein the Two Grand Systemes of Ptolomy and Copernicus are largely discoursed of ... By Galileus Galileus.* The next translator, Emil Strauss in 1892, avoided the previously established error of imputing to Galileo a book on *the* system of the world: *Dialog über die beiden haptsächlichen Weltsysteme.* In 1953, after pondering long over the original printed title, I decided to call my translation *Dialogue Concerning the Two Chief World Systems – Ptolemaic and Copernican.*

Thus, neglecting Joseph Webbe at the beginning, common Italian usage and all translators have supplied Galileo's book with some principal subject, though none appeared on the printed title page of 1632. Omitting only the author's employment and affiliations, that read:

Dialogo di Galileo Galilei ... Dove ne i congressi di quattro giornate si discorre sopra i due Massimi Sistemi del Mondo, Tolemaico, e Copernicano, proponendo indeterminatamente le ragioni Filosofiche, e Naturali tanto per l'una, quanto per l'altra parte.

Many precedents existed among books in this popular didactic form for a *dialogo nel quale ...*, but Galileo's *Dialogo Dove ...* may be unique. Usually a *dialogo* was *sopra*, or *intorno a*, or *del* some clearly named topic, after which various other things touched on in it were often recited in a subtitle to attract buyers. The printed title of Galileo's *Dialogue* occasions a certain discomfort because it is so easy for us to rearrange the same words more tastefully. Since the book was the production not of a novice but of a master of Italian style, we are entitled to suspect some compelling reason behind its curiously phrased title. Or we might suspect that Galileo had had a principal subject in mind while writing his book, to which discussion of the two world systems was subsidiary or incidental, and then that at the last moment he decided not to name it. However absurd that may sound now, it is what actually happened, though Galileo's strange last-minute decision was not, strictly speaking, entirely his, or really strange, or literally last-minute. The circumstances are easily documented.

In 1630 Galileo carried to Rome the manuscript of a completed book on which he had laboured since 1624. Because he intended it to be published at Rome under the auspices of the Lincean Academy, it needed the Roman imprimatur. The Master of the Holy Palace required various revisions before licensing it, and while Galileo was revising the book at Florence the head of the Lincean Academy died at Rome; moreover, an outbreak of plague not only made a return to Rome inconvenient but would have required fumigation on all pages of any manuscript sent there. Galileo decided to have his dialogue printed at Florence, which entailed his getting a new license there. No mystery remains about his original title for the *Dialogue* in the light of a letter dated 24 May 1631 from the Roman to the Florentine licenser:

Il Sig. Galilei pensa di stampar costì una sua opera, che già haveva il titolo *De fluxu et refluxu maris*, nella quale discorre probabilmente del sistema Copernicano secondo la mobilità della terra, e pretende d'agevolar l'intendimento di quel'arcano grande della natura con questa posizione, corroborandonla vicendevolmente con questa utilità. Venne qua a Roma a far veder l'opera, che fu da me sottoscritta, presupposti l'accommodamenti che dovevano farcisi, e riportatici ricever l'ultima approvazione per la stampa. Non potendo ciò farsi per gl'impedimenti delle strade e per lo pericolo degl'originali, desiderando l'autore di ultimare costì il negozio, V.P.M.R. potrà valersi della sua autorità, e spedire o non spedire il libro senz'altra dependenza dalla mia revisione; ricordando però, esser mente di Nostro Signore che il titolo e soggetto non si proponga del flusso e reflusso, ma assolutamente della matematica considerazione della posizione Copernicana intorno al moto della terra, con fine di provare che, rimossa la rivelazione di Dio a la dottrina sacra, si potrebbono salvare le apparenze in questa posizione, sciogliendo tutte le persuasioni contrarie che dall' esperienza e filosofia peripatetica si potessero addurre, sì che non mai conceda la verità assoluta, ma solamente la hipothetica e senza le Scritture, a questa opinione. Deve anco mostrarsi che quest'opera si faccia solamente per mostrare che si sanno tutte le ragioni che per questa parte si possono addurre, e che non per mancamento di saperle si sia in Roma bandita questa sentenza, conforme al principio e fine del libro, che di qua manderò aggiustati. Con questa cauzione il libro non haverà impedimento alcuno qui in Roma, e V.P.M.R. potrà compiacere l'autore e servir la Serenissima Altezza, che in questo mostra sì gran premura. Me le ricordo servitore, e la priego a favorirmi de'suoi commandamenti.[3]

Clearly the manuscript which Galileo took to Rome had carried the title *Dialogo del flusso e reflusso del mare* above the (printed) subtitle mentioning discussions of arguments for and against the two world systems, which latter has since been made to supply the absence of any specific title. If any doubt remains because the Roman licenser chose to put the title into Latin, it may be added that from 1624 to 1629, when Galileo was composing the book and had occasion to mention it in letters to friends, he invariably called it "my dialogues on the tides." Only once that I have noted, late in 1629, did he so much as mention the Copernican system in connection with it.[4] But once he knew the terms of the Roman license, in August 1630, he alluded in a letter to *i Dialogi che scrivo esaminando i 2 sistemi massimi Tolemaico e Copernicano in grazia del flusso e reflusso*.[5] Even while making the required revisions, then, Galileo himself continued to think of his book as on the subject of tides, discussion of the world systems being included only for the benefit of his explanation of that *arcano grande della natura*. And that is exactly how his manuscript was described in the first sentence of the letter of the licenser who had read it at Rome.

The *principio e fine* to be sent *aggiustati* from Rome are of great importance to the whole story of the *Dialogue*. By *principio* was meant Galileo's address *Al discreto lettore*, which needed rearrangement in order to conform with the altered title. The original form and content of this preface will be reconstructed below. By *fine* was meant what in other documents concerning the *Dialogue* was called "the medicine of the end"; it consisted of the pope's own argument designed to preclude any impression that Galileo's tide theory was, or could be, more than a mere scientific speculation. The basic position had been stated to Galileo by Urban VIII no later than 1624, and perhaps earlier, when he was still cardinal.[6] Of general applicability, and not confined to the tides, it held that motion of the earth could not be conclusively proved from any set of appearances, since it lay in God's power to produce the same effects by any number of means, of which many remain unimaginable by us. Galileo himself used that position, not only in the *Dialogue* but earlier, as a caution against overconfidence in scientific conclusions.[7]

Actual composition of the *Dialogue* was not commenced until late in 1624. In order to understand Galileo's preface, even before it was altered to comply with the conditions for publication imposed by the Roman licenser, it is necessary to consider some events of earlier years. We should in fact start with the Church edict of 1616 which regulated books dealing with motion of the earth, as Galileo himself did in the printed address *Al discreto lettore*, of which the opening sentence is:

Si promulgò a gli anni passati in Roma un salutifero editto, che, per ovviare a'pericolosi scandoli dell'età presente, imponeva opportuno silenzio all'opinione Pittagorica della mobilità della terra.[8]

The edict had come as a severe disappointment to Galileo at the time it was issued. For two years before that he had devoted most of his energies to a campaign against those who wished to see the Church make an article of faith out of a purely scientific question.[9] In 1613 he had gone so far as to write, in a widely circulated letter to a former pupil, that Scripture should be put in the last place in deciding such questions. In 1615 that letter was called to the attention of the Roman inquisitors, who submitted it for examination to a qualified theologian. His report stated that although some infelicitous expressions were found in the letter, Galileo's position was on the whole good Catholic doctrine. Galileo himself was confident throughout his campaign that if the matter were ever taken up by truly responsible Church authorities, they would follow the express warnings of St. Augustine and would not expose the faith to possible contempt on the part of heretics who were fully informed in scientific matters. In February 1616, at Rome, against the advice of Roberto Cardinal Bellarmino, Galileo forced the issue. The decision went against him and he was instructed by the cardinal to abandon the Copernican propositions. On 5 March the "salutary edict" was issued by the Congregation of the Index on order of Pope Paul V.[10]

Galileo had said for two years that he would be ruled by any official action taken, and he was as good as his word. Before 1616 he had vigorously opposed any intervention by the Church. After the edict was issued, he remained silent on Copernicanism for several years and occupied himself with other things. The only evidence of his disappointment that I have noted came in 1618, when he expressed regret that he could not publish his explanation of tides because that depended on ascribing the Copernican motions to the earth.[11]

Galileo's treatise on the tides had in fact been written out at Rome in January 1616, for Alessandro Cardinal Orsini, and it was immediately after Orsini's next meeting with the pope that Paul V conferred with Bellarmino and decided to initiate official action. He sent the two propositions of fixed sun and moving earth to the eleven Qualifiers of questioned doctrines, who unanimously held them to be "false and absurd in Philosophy" and therefore rash and erroneous in the faith. Galileo had supposed the question to be settled was whether the Bible had spoken to the sun's motion literally or only metaphorically; if the latter, then Copernicanism remained open to debate. The Qualifiers, on the other hand, treated all debate on the matter as settled by Aristotelian natural philosophy.[12] They even went so far as to classify belief in a fixed sun as heresy, that being contradicted literally by various biblical passages. The pope, however, wisely ignored that extreme opinion, and the edict was so worded as to prohibit only two things: attempted reconciliation of the Copernican proposition with the Bible, or assertion that motions of the earth were physically real.

That the edict was not intended to stop the use of terrestrial motion as a scientific hypothesis is evident from its express mention of Copernicus's own book, which was suspended until "corrected" by removal of a few sentences mentioning scriptural interpretation or calling the earth a "star"; that is, a planet. Only two other books were named, both by theologians, and though there were many astronomical books that contained Copernican diagrams and calculations, none were prohibited at the time.

In 1623 Maffeo Cardinal Barberini became Pope Urban VIII, and in 1624 Galileo journeyed to Rome to pay homage to this old friend and admirer of his. Cardinal Zollern, who was also in Rome, told Urban at this time that the 1616 edict was an embarrassment to the Church in Germany, where prospective intellectual converts were all Copernicans. The pope replied that Copernicanism had never been declared heretical and never would be, though neither could it be conclusively proved. Galileo knew of this conversation and reported it in a letter at the time.[13] The intellectual climate of Rome had greatly changed with the accession or Urban VIII, to whom Galileo's *Il Saggiatore* had been dedicated by the Lincean Academy in 1623. Several Linceans were appointed to positions in the Vatican.

When Galileo left Rome in June 1624, he had had six audiences with the pope, in the course of which it appears that he gained permission, and perhaps encouragement, to proceed with the book that was eventually published as the *Dialogue*. It is difficult to be sure how nearly Galileo and Urban were in agreement on the plan of that book. Certainly the pope had no intention of repealing the edict, though there is later evidence that he had never approved of its issuance. An event in 1630 which confirmed the situation reported by Cardinal Zollern suggests that Urban may have sympathized with any plan that would neutralize effects of the edict that were adverse to the interests of the Church, particularly among intellectuals. This event was reported to Galileo in March 1630 by Benedetto Castelli, a former pupil and close friend of Galileo's who had been called to Rome by Urban to superintend hydraulic engineering projects there:

Il Padre [Tommaso] Campanella, parlando a'giorni passati con Nostro Signore, li hebbe a dire che haveva hauti certi gentilhuomini Tedeschi alle mani per convertirli alla fede Catolica, e che erano assai ben disposti; ma che havendo intesa la prohibizione del Copernico etc., che erano restati in modo scandalizati, che non haveva potuto far altro: e Nostro Signore li ripose le parole precise seguenti: "Non fu mai nostra intenzione; e se fosse toccato a noi, non si sarebbe fatto quel decreto."[14]

It seems safe to infer that although Barberini did not actively oppose the edict in 1616, his view had always been similar to Galileo's; that is, that outright prohibition had not been intended by the Church, and that official intervention on its part could only work to its eventual disadvantage. In any case it is probable that in 1624 he agreed with Galileo that the edict did not prevent Catholic use of motion of the earth as a scientific hypothesis.[15] Now, Galileo's tide theory employed that hypothesis for solution of a scientific problem, and he had long wished to publish it. He could do so, if permitted, in a way that would show everyone that existence of the 1616 edict did not impede scientific progress, but only forbade overconfident assertions about physical reality or the mingling of science with scriptural interpretation. He would in fact show scientifically that no experiment practicable on earth could prove or disprove its motion, which he knew to support the pope's theological argument. If he raised and answered traditional arguments and added new ones, it would be clear to all that the edict was based not on ignorance of science, but on reasons of a totally different kind, wholly in the province of theology. Italy would be seen to retain its scientific leadership in Europe without contravening the true intent and scope of the edict.

Galileo's actions from 1624 to 1630 conformed to this plan, though the extent to which he outlined it to Urban and received his approval of it is debatable. His first move was to reply at length to the weak anti-Copernican arguments that had been submitted to him in 1616 by Francesco Ingoli, head of the Propaganda Fidei. In his long reply he stressed the need of showing that Catholics were fully informed in science, and developed the idea of relativity of motion which was incorporated in the *Dialogue*, begun in that form by the end of 1624. The pope was shown the reply to Ingoli, which appeared to have his approval.

Let us now return to the preface *Al discreto lettore*, which Galileo knew would be read by the pope and must therefore have been written in accordance with whatever plan Galileo had outlined to him. It continued as follows:

Non mancò chi temeriamente asserì, quel decreto essere stato parto non di giudizioso esame, ma di passione troppo poco informata, e si udirono querele che consultori totalmente inesperti delle osservazioni astronomiche non dovevano con proibizione repentina tarpar l'ale a gl'intelletti speculativi.

That sentence has seemed to all modern commentators heavy-handed irony bordering on hypocrisy, since it was primarily Galileo himself who had expressed such sentiments during the years 1613–16. But the above sentence in context referred to events *after* passage of the edict, not before any action had been taken. The rash critics meant were German Protestants such as those reported by Cardinal Zollern in 1624 and Thomas Campanella in 1630, not Italian Catholics who had questioned the advisability of Church intervention in the first place. Galileo himself had gone to Rome in 1615–16 to make sure that Church officials were fully informed, as he went on to say:

Non potè tacer il mio zelo in udir la temerità di sì fatti lamenti.[16] Guidicai, come pienamente instrutto di quella prudentissima determinazione, comparir publicamente nel teatro del mondo, come testimonio di sincera verità. Mi trovai allora presente in Roma; ebbi non solo udienze, ma ancora applausi de i più eminenti prelati di quella Corte; nè senza qualche mia antecedente informazione seguì poi la publicazione di quel decreto.

Here there was indeed a vein of irony, but not bitter; on the contrary it would much amuse those few who knew the story of Galileo's "antecedent information." Perhaps it was misleading to the public to imply that the Church had sought and received Galileo's comments before issuing the edict, but nothing could now be gained by greater precision. And since Galileo had never questioned the supreme authority of the Church in all matters of scriptural interpretation, neither could any harm come of this very incomplete but literally quite true account. Next:

Per tanto è mio consiglio nella presente fatica mostrare alle nazioni forestiere, che di questa materia se ne sa tanto in Italia, e particolarmente in Roma, quanto possa mai averne immaginato la diligenza oltramontana; e raccogliendo insieme tutte le speculazioni proprie intorno al sistema Copernicano, far sapere che precedette la notizia di tutte alla censura Romana, e che escono da questo clima non solo i dogmi per la salute dell'anima, ma ancora gl'ingegnosi trovati per delizie degl'ingegni.

Before proceeding further into the preface, I call attention to its printed opening sentence. That was extraordinarily abrupt, since the raising at once of the matter of the edict cannot possibly have been appropriate for a book on the subject of tides. No one but Galileo himself associated tides with motions of the earth at that time. Something else must originally have come first, alluding to tides; and this, or part of it, was then transferred to become the third of *tre capi principali* to be treated in the *Dialogue*:

Nel terzo luogo proporrò una fantasia ingegnosa. Mi trovavo aver detto, molti anni sono, che l'ignoto problema del flusso del mare potrebbe ricever qualche luce, ammesso il moto terrestre. Questo mio detto, volando per le bocche degli uomini, aveva trovato padri caritativi che se l'adottavano per prole di proprio ingegno. Ora, perchè non possa mai comparire alcuno straniero che, fortificandosi con l'armi nostre, ci rinfacci la poca avvertenza in uno accidente cosi principale, ho giudicato palesare quella probabilità che lo renderebbero persuasibile, dato che la Terra si movesse. Spero che da queste considerazioni il mondo conoscerà, che se altre nazioni hanno navigato più, noi non abbiamo speculato meno, e che il rimettersi ad asserir la fermezza della Terra, e prender il contrario per capriccio matematico, non nasce da non aver contezza di quant'altri ci abbia pensato, ma, quando altro non fusse, da quelle ragioni che la pietà, la religione, il conoscimento della divina onnipotenza, e la conscienza della debolezza dell'ingegno umano, ci somministrano.

Placed thus third among three topics to be considered, the tides could not be called more than an *accidente principale*; but when they had been the main subject of the whole book, Galileo must have begun by stressing the importance of the tide problem. Also a trace remains above of some previous mention of navigation and speculation, as well as of Italy and foreign nations. Though it is hazardous to put words into Galileo's mouth, even in English, I suppose that his original preface to the dialogue on tides began along these lines:

The problem of the cause of ocean tides has baffled natural philosophers from antiquity to our day. Aristotle himself could not explain them, and some say that in despair of this he hurled himself into the sea and so met with his death. The ignorant attribute these great movements of huge bodies of water to the influence of the moon, while the learned have been unable to improve on that. Yet now that Columbus and Vespucci have discovered new worlds across great oceans, to the glory of Italian navigation, the unsolved problem of the tides has acquired practical as well as speculative importance to mankind, and still no physical explanation has seemed possible.

If we now read the authentic paragraph previously cited (except the sentence *Nel terzo luogo* ...), the whole leads quite easily from the original topic of the book to Galileo's account of the 1616 edict which so abruptly opens the printed preface. Suggestion of the earth's motion as a possible key to solution of the tide problem would alert the *discreto lettore* to an attendant problem of a different kind, while the concluding reference to piety and religion at once allayed his fears and ushered in consideration of the edict. Not only Galileo's plan for the *Dialogue*, but his usual style, suggests this reconstruction of an earlier preface in place of the usual account of the printed version as having been written tongue-in-check by an ironical hypocrite.

Without the sentence *Nel terzo luogo* ... there were only two heads of discourse, so I have changed the first word of the paragraph that in the printed preface includes the allusion to the tides:

Due capi principali si tratteranno. Prima cercherò di mostrare, tutte l'esperienze fattibili nella Terra essere mezi insufficienti a concluder la sua mobilità, ma indifferentemente potersi adattare cosi alla Terra mobile, come anco quiescente; e spero che in questo caso si paleseranno molte osservazioni ignote all'antichità.

That was effectively done in the Second Day, the First Day being a general preparation for the wide range of discussions to follow. In the Second Day Galileo developed his new physics of relative motion, composition of motions, and conservation of motion. Galileo perhaps cared less that it answered traditional objections against Copernicus than that it was a creation of his own which was capable of resolving many problems of terrestrial motions. The idea that Galileo created a new physics in order to support Copernicus is mistaken, though widespread. It is now certain that Galileo's physics was virtually completed before the telescope diverted his attention to astronomy, and his first unequivocal statement in favour of Copernicanism came three years after that.

Secondariamente, si esamineranno li fenomeni celesti, rinforzando l'ipotesi Copernicana come se assolutamente dovesse rimaner vittoriosa, aggiugendo nuove speculazioni, le quali però servano per l'utilità d'astronomia, non per necessità di natura.

These things occupy the Third Day, in which commentators often point out the conspicuous lack of planetary astronomy of the kind to which Ptolemy, Copernicus, Tycho Brahe, and Kepler devoted their lives. This has appeared to them a very puzzling omission in a book purporting to discuss the two chief world systems. Indeed it would have been, had Galileo written his book for the purpose implied by the censored title page as printed. In a book on tides, however, it was physics and not astronomy that needed to be explained to readers. There was no previous book from which a useful knowledge of the elementary principles of physics could be obtained. On the other hand there was an abundance of books at any level of understanding from which astronomy could be learned. Ptolemaic astronomy had nothing to do with tides, and Galileo did not explain it or even describe it. Even the Copernican astronomy was described in the *Dialogue* only to the extent of an elementary diagram based on one in the introductory chapter of *De revolutionibus*. Traditional devices of planetary astronomy – eccentric circles, epicycles, and equants – were barely mentioned and not explained at all. The Third Day took up novel astronomical matters such as the location of supernovas, the motions of sunspots, and ways in which stellar parallax might be detected, but ignored the usual questions. Had the book retained the title *Dialogue on the Tides* ..., no one would now be surprised at its neglect of planetary astronomy, and that was the title Galileo had in mind when he wrote the book.[17]

Alteration of the title and preface of the *Dialogue* naturally entailed a change also of the opening speeches. A passage near the end of the Third Day, overlooked by Galileo and the censors when the book was revised for printing, betrays the way in which it had originally opened. There Salviati, who served as Galileo's spokesman, said:

E perchè mi pare che assai a lungo si sia questi tre giorni discorso circa il sistema dell'universo, sarà ormai tempo che venghiamo all'accidente massimo, dai quale presero origine i nostri ragionamenti; parlo del flusso e reflusso del mare, la cagione del quale pare che assai probabilmente si possa referire a i movimenti della Terra: ma ciò, quando vi piaccia, riserberemo al seguente giorno.[18]

This speech is found on p. 406 of the 1632 edition, where the first mention of the tides is on p. 205 and in no sense gave rise to the interlocutors' discussions. The First Day, as printed, begins no less abruptly than the printed preface, and inconsistently with the end of the preface:

Fu la conclusione e l'appuntamento di ieri, che noi dovessimo in questo giorno discorrere, quanto più distintamente e particolarmente per noi si potesse, intorno alle ragioni naturali e loro efficacia, che per l'una parte e per l'altra sin qui sono state prodotte da i fautori della posizione Aristotelica e Tolemaica e da i seguaci del sistema Copernicano. E perchè. ...[19]

Yet on the previous page, at the end of the preface, it was no well-defined topic which the gathering was to discuss, but various wonders of God in heaven and on earth:

Erano casualmente occorsi (come interviene) vari discorsi alla spezzata tra questi Signori, i quali avevano più tosto ne i loro ingegni accesa, che consolata, la sete dell'imparare: però fecero saggia risoluzione di trovarsi alcune giornate insieme, nelle quali, bandito ogni altro negozio, si attendesse a vagheggiare con più ordinate speculazioni le maraviglie di Dio nel cielo e nella terra. Fatta la radunanza nel palazzo dell'Illustrissimo Sagredo, dopo i debiti, ma però brevi, complimenti, il Sig. Salviati in questa maniera incomminciò.

The reader is not prepared by this for Salviati's opening statement that agreement had already been reached among the interlocutors to consider the two chief world systems, though of course the printed title page made Salviati's statement seem natural. Before the title was altered, the opening speeches must have been different, and we know from the allusion to tides as the origin of the discussions, near the end of the Third Day, that that topic was brought up originally before discussion turned to the system of the universe.

From Galileo's later *Two New Sciences*, written in dialogue form and using the same interlocutors, it is possible to judge with some confidence the probable form of the manuscript *Dialogue*. The speakers probably began, in accordance with the ending of the preface, by proposing for a first topic of discussion various wonders of God in heaven and on earth for which orderly explanations were desired. Among those one would expect some of the recently discovered celestial phenomena for which explanations were not found in the texts of Aristotle and on which natural philosophers were in disagreement. Then, because the setting of the dialogues had been placed in Venice, where the cyclic rise and fall of the sea is more conspicuous than elsewhere in Italy, the cause of that motion would afford a natural question. Sagredo could recall that their common friend the Academician (as Galileo was called in the *Dialogue*) had once proposed motion of the earth as a probable solution, to which Simplicio would object that Aristotle had proved the earth to be fixed at the center of the universe. Salviati would reply that nonetheless, the hypothesis of motion of the earth had been useful to astronomers and might be able to aid in other scientific speculations as well, though it could be neither proved nor disproved. There was also a question as to the validity of Aristotle's assumptions and demonstrations. The lively support of Simplicio for Aristotle would then lead naturally to a decision to put aside the problem of tides until arguments for and against the earth's motion had been duly examined, at which point Salviati's printed opening speech – except for the word "yesterday" – would be appropriate, and the rest could be left as in the manuscript.

The probable history of the title, preface, and opening speeches of the *Dialogue* has now been set forth. Galileo had spent five years organizing and writing the book in accordance with the plan accommodating his personal wish to expound his tide theory with the exigencies of the Church under conditions created by the 1616 edict. He was therefore fully conscious of the need to preserve the form of his manuscript as nearly intact as possible. Most of the objections he encountered at Rome were easy to meet because Galileo had anticipated them. His basic tactical principle in doing this was that of Cardinal Bellarmino, who had meanwhile died, but who when living had continually urged that as long as motions of the earth were treated as purely hypothetical, no conflict with scriptural passages could ever arise.[20] Since Galileo had taken pains throughout the *Dialogue* to comply with Bellarmino's principle, he was ready and willing to remedy any oversights in that regard, and nearly every objection could be met in that way.

Nevertheless one difficulty arose that was nearly fatal to the book, and that was papal opposition to Galileo's tide theory. Even though Galileo had taken care to present that as a purely hypothetical consequence of the Copernican motions, Urban was not satisfied. His special concern can be guessed from the compromise he finally accepted, which was removal of any mention of the tides from the title and subject of the *Dialogue*, and inclusion at its end of his own argument based on God's power to have produced the same phenomena in an unlimited number of ways. From this it appears to me that Urban VIII, who for strategic reasons wanted the Church to license a book which freely discussed the Copernican arguments, to counter the prevailing interpretation of the 1616 edict abroad, would under no circumstances permit it to appear that endorsement of that discussion constituted endorsement by the Church of a tide theory linked to the Copernican motions of the earth.

It was probably Galileo himself who proposed removal of the tides from the printed title page, though he was fully aware that it would be impossible to replace that with any other principal subject. This was toward the end of his stay at Rome in 1630, when he was not dealing directly with the pope, or even with the Master of the Holy Palace, but with Raffaello Visconti, who had been appointed to carry out the detailed review of the book. At any rate, on 3 June 1630 Orso d'Elci wrote to Galileo:

Mi rallegro che V.S. trovi il compagno [Visconti] del Maestro del Sacro Palazzo capace della verità della sua dottrina, et che egli speri di persuadervi anche il Papa per rimuonverlo dalla noia che dà a S.B. ne la dimostrazione che V.S. vuol fare, che il flusso proceda dal moto della terra.[21]

The *dottrina* meant was of course not Copernicanism, but the principle of Cardinal Bellarmino that hypothetical treatment obviated any conflict with scripture. That it was the title page as such which remained the last seemingly insuperable problem is clear from the letter written by Visconti to Galileo on 16 June:

Il Padre Maestro gli bacia le mani, et dice che l'opera gli piace, et che dommattina parlerà con il Papa per il frontispizio dell'opera, et che del resto, accommodando alcune poche cosette, simili a quelle che accommodammo insieme, gli darà il libro. Et io gli resto servitore.[22]

Ten days later Galileo left Rome for Florence, entirely satisfied. The pope granted him a parting audience at which he showed great affection for him, while Urban's nephew, Francesco Cardinal Barberini, gave Galileo a farewell dinner. So reported the Tuscan ambassador, who added that the entire Roman court honoured Galileo as was his due.

As mentioned above, removal of the tides as subject of the *Dialogue* did not permit substitution of another subject. It would have been absurd for the Church to licence a book on the Ptolemaic and Copernican systems exclusively, since the issue between them had been officially decided in 1616. It would not have been impossible to license a book discussing the merits of a hypothesis to be applied to a specific unsolved scientific problem such as the tides, and Galileo had designed his title page in that way. All that could be done was to delete the line referring to the tides, leaving everything else untouched; and that is what accounts for the amateurish appearance of the printed title page of a book written by a master of Italian style.

How the reordering of Galileo's preface and the truncation of the opening speeches in his *Dialogue* followed inexorably from papal interference with the title of a very astutely organized book has been amply explained. The book was destined to become a symbol of the struggle for freedom of inquiry and of a supposed inherent conflict between religion and science. As a result, the seemingly trifling problem of giving it an acceptable title has had consequences of great magnitude, because Galileo's *Dialogue* has been read by cultural historians as intended to accomplish purposes quite different from those envisioned by its author. To comment on even the most serious of the misunderstandings that now prevail would take me far beyond my present purpose, which is only to call attention to a neglected story of no little importance to larger issues. It is necessary, however, to explain briefly in closing how it came about that a book whose publication was approved by a friendly pope became the source, a few months later, of Urban's boundless rage and vindictive persecution of its author.[23]

The defective assumption set forth at the beginning of this paper associates the trial and condemnation of Galileo with his supposed deliberate violation of the 1616 edict by publishing the *Dialogue*. There was no such violation; in fact, it is a bit ridiculous to suppose that Galileo would set out to evade an old edict by reminding readers of it in the first sentence of his preface. No less absurd is the idea that the Roman licenser of books could fail, in view of that sentence, to examine the book most scrupulously for any possible evasion or violation of the edict. The principle of Bellarmino was known to, and accepted by, the theologians who were asked to re-examine the *Dialogue* after Galileo was first summoned to Rome, just as it had been known to and accepted by the licensers. Their report was concerned mainly with passages in which purely hypothetical treatment of the earth's motion might be overlooked; these, they said, could easily be amended if the book was judged to have sufficient merit in other ways to continue in circulation.[24]

Galileo was brought to trial at Rome in 1633 on the charge of "vehement suspicion of heresy" because he was believed to have violated not the 1616 edict, but a personal command given to him before the edict was issued. The pope was shown a memorandum to the effect that Galileo had been ordered never to teach the Copernican propositions in any way, orally or in writing, lest the Holy

Office proceed against him. Disobedience of such a command was *prima facie* evidence for the charge, and any information about the Copernican system disseminated by Galileo in the *Dialogue* was evidence of guilt under it. This had nothing to do with the edict as such, or with any other Catholic than Galileo, and the pre-trial documents as well as the hearings of the first day of his trial show that if the memorandum which so angered the pope had not existed, there would probably have been no trial of Galileo.

Urban's indignation against Galileo is understandable; he felt that his old friend had deliberately concealed from him information that alone would have sufficed in 1624 to have prevented his approval of Galileo's project, no matter how much it appealed to him otherwise. Now, the memorandum was unsigned by anyone named in it, though it purported to set forth events which had taken place in the presence of a notary and witnesses. To have any legal value, such a document would have had to be signed by all persons involved, one of whom was Cardinal Bellarmino. At his trial, Galileo produced an affidavit, in Bellarmino's own hand, that he had been told only of the decision applying to all Catholics as set forth in the edict, and had agreed to be ruled by it. In order to save its reputation and maintain its authority, the Roman Inquisition then obtained from Galileo a confession to some lesser offences, which became the legal basis of his conviction and sentencing. In that way the edict, as well as the legally worthless memorandum, came to be incorporated into Galileo's sentence and abjuration, to the everlasting shame of his judges and at the cost of misapplication of the edict for the next two centuries, until it was finally repealed.[25]

Perhaps nothing will remove from our folklore, or even from scholarly tradition, all the misapprehensions that currently exist about Galileo and his *Dialogue*. They arose, in my opinion, largely because the trial documents were not published in full until the 1870's, in a context of warfare between theologians and scientists over Darwinian evolution theory. Traditionalists and freethinkers formed anti-Galileo and anti-Church factions that did not provide the best milieu for objective research. Such factions are easily mistaken for pro-Galileo and pro-Church camps, but those would be very different adversary positions indeed. In my opinion they are just beginning to form, and are not adversary positions at all. From their future dialogue, I believe, will emerge not a picture of heroes and villains, but one of tragic human misunderstandings among those who, in Plato's words as paraphrased by Kepler, seek truth on one path and many paths.

NOTES

1 The bound manuscript, beautifully written, is preserved at the British Library, Harleian ms. 6320. Thomas Hobbes, who left England in mid-1634 to tour Europe, told Galileo of its existence when visiting him late in 1635.

2 The short title "system of the world" probably came to be applied to the *Dialogue* because of Galileo's promise in his *Sidereus Nuncius* of 1610 to publish such a book, described also in a letter of that year as 2 *libri De sistemate seu constitutione universi, concetto immenso e pieno di filosofia, astronomia, et geometria* (to Belisario Vinta, 7 May 1610; *Opere*, X, 351). In the same letter he mentioned a work *De maris estu* separately, and it was that, not *De sistemate*, which eventually became the *Dialogue*. He again mentioned his promised systematic work in a book published in 1612, but after the edict of 1616 when he received inquiries from abroad about it, he replied that it had been "stayed by a higher hand." Foreign translators of the *Dialogue* understandably assumed it to be the long-promised treatise, and doubtless some parts of that survive in the *Dialogue*, which is, however, rather a treatise on preponderance of evidence than a book full of philosophy, astronomy, and geometry.

3 *Opere di Galileo Galilei, Edizione Nazionale*, XIX, 327. Above, and hereinafter, referred to simply as *Opere*.

4 To Elia Diodati, 29 October 1629: "... ho ripreso i miei *Dialogi* intorno al flusso e reflusso ... e subito li publicherò; dove, oltre a quello che s'aspetta alla materia del flusso, saranno inserti molti altri problemi et una amplissima confermazione del sistema Copernicano ..." (*Opere*, XIV, 49). Galileo alluded to a newly discovered argument based on annual variations in the paths of sunspots which had induced him to take up again and complete the book which he had laid aside for some time.

5 To G.B. Baliani, 6 August 1630, only a month after Galileo's return from Rome (*Opere*, XIV, 130).

6 The event was described by Augustino Oregio in his *De Deo uno* (Rome, 1629). Scholars have disagreed on the date, his phrase *adhuc Cardinalis* being of somewhat ambiguous reference in the original text. I am inclined to believe that the event involving Galileo belonged to 1624 and that Oregio meant only that Barberini when still a cardinal was noted for his doctrine, since in thereafter telling the specific story he began: *Sanctissimus dixit ...* . No personal meeting between Galileo and Barberini between 1611 and 1624 is recorded.

7 Thus in 1613, before any of the public controversies involving theology, Galileo had published in his *Sunspot Letters* reasons for regarding as merely probable all the *controverse condizioni delle sustanze naturali; le quali poi finalmente sollevandoci all'ultimo scopo di nostre fatiche, cioè all'amore del divino Artefice, ci conservino la speranza di poter apprender in Lui, fonte di luce e di verità, ogn'altro vero* (*Opere*, V, 188). In the First Day, considering whether light and dark parts of the moon might indicate the presence of land and water, he wrote: "But because there are more ways known to us that could produce the same effect, and perhaps others that we do not know of, I shall not make bold to affirm one rather than another to exist on the moon," and, in reply to a speculation whether plants and animals might exist on the moon: "... regarding the production there of things similar to or different

from ours, I should always reply, 'Very different and unimaginable by us'; for this seems to me to fit with the richness of nature and the omnipotence of the Creator and Ruler." The arrogance of scientific opinion ascribed to Galileo by writers like Arthur Koestler is not detectable in his books or letters; what was at issue in Galileo's day was not his firmness of conviction, but his challenge of the firm convictions of professors of philosophy.

8 *Opere*, VII, 29. Continuations of the preface cited below will not be further referenced, as being also on this or the following page.

9 This move was originally sponsored by professors of philosophy at the University of Pisa, as Galileo clearly stated in his letters. At least a year elapsed before they succeeded in bringing a priest into the attack. When the Roman inquisitors had questioned the priest and others whom he named, they dropped the inquiry.

10 The only book prohibited by it was by a theologian; Galileo's *Sunspot Letters*, in which he had expressed confidence that the Copernican system would triumph, was not mentioned.

11 Galileo's tide theory had been linked with his preference for the Copernican system from the outset, in 1595, and appears to have been the basis of his first real interest in the new astronomy.

12 The phrase cited above had been standard for centuries in judging philosophical opinions to be erroneous in the Catholic faith. It appears not to have been previously applied to a question of fact capable of decision by scientific inquiry.

13 To Federico Cesi, 8 June 1624 (*Opere*, XIII, 182).

14 16 March 1630; *Opere*, XIV, 87–88.

15 This is what I shall call "Bellarmino's principle"; see below.

16 Galileo used the word *zelo* frequently to describe his Catholicism in the letters of 1613–16, and elsewhere later in life, but so far as I know he never employed the word in any other connection. It is evident that many cardinals, some other highly placed ecclesiastics, and the rulers of Tuscany never doubted Galileo's Catholic zeal or ever suspected him of the Copernican zeal now commonly ascribed to him. I can find no evidence of the latter, and much counter-evidence, such as his refusal to speak out in favour of Copernicanism when urged by Kepler to do so in 1597, his advice to Castelli not to teach the new astronomy at Pisa coupled with the statements that he himself had never done so in eighteen years as professor at Pisa and Padua, and the paucity of references to Copernicanism in letters, even to close friends, except during 1613–16 when its official suppression was being rumoured.

17 It is only the baseless claim that Galileo was a Copernican zealot, mentioned in the preceding note, coupled with inattention to the circumstances under which the *Dialogue* was begun and to Galileo's original title for it, that has given rise to the notion that he concerned himself with planetary astronomy. Galileo's lifelong interest was in questions of motion, which included the cosmological question of deciding which

heavenly bodies were at rest and which were moving. His only detailed investigations of celestial motions related to Jupiter's satellites, concerning which his working papers include voluminous observations and calculations. Hardly a page survives which reflects work on planetary motions.

18   *Opere*, VII, 439.

19   *Opere*, VII, 33.

20   Cf. n. 15, above. From 1613 to 1616 Galileo vigorously opposed the treatment of the earth's motion as merely hypothetical, or the ascription of such an intention to Copernicus. After passage of the edict in 1616 he accepted Bellarmino's principle as the best choice open to Catholics. Its only deleterious effect on Catholic scientists would be in time to make them look foolish by reason of an obligation to go on treating as hypothetical some propositions that would come to be regarded by other scientists as established as firmly as anything in physical science ever is.

21   *Opere*, XIV, 113.

22   *Opere*, XIV, 120.

23   My reconstruction of the events leading up to the trial of Galileo, and translations of some crucial trial documents, may be read in my *Galileo At Work* (Chicago, 1978), pp. 222–30, 238–56, 289–96, 312–14, 336–52. See also Ludovico Geymonat, *Galileo Galilei* (New York, 1965), Appendix A, and Jerome J. Langford, *Galileo, Science, and the Church* (Ann Arbor, 1966), pp. 93–97.

24   *Opere*, XIX, 326–27. Items 3 and 4 of the report are amusingly counterposed: 3. *Mancarsi nell'opera molte volte e recedere dall'hipothesi. ... 4. Tratta la cosa come non decisa.* Later on some other examiners were extremely hostile to the book and its author, but that was after it had become a focus of internal Church politics, and does not concern us here.

25   Actual repeal, or rather the dropping of Copernican books from the Index, awaited deliberate defiance of it by a Catholic astronomer in 1818, but in fact the *Dialogue* had been printed in 1744 with Church permission, together with an approved explanatory preface, and after 1757 the absurd precedent established by Galileo's condemnation was tacitly ignored under liberalizing rules decreed by Pope Benedict XIV for the Index of Prohibited Books.

2

The Organizing Theme of the *Dialogue*

It is a singular privilege for me to speak of the organizing theme of Galileo's immortal *Dialogue* at Rome on invitation of the Accademia Nazionale dei Lincei. In 1630 Galileo brought to Rome a manuscript intended to be published under the auspices of the Lincean Academy. Here, a papal command changed the title of the manuscript and thereby concealed, though it did not alter, the organizing theme of the book whose 350th anniversary we are celebrating. A subsequent event at Rome, the untimely death of the founder and leader of the Academy, Prince Federico Cesi, deprived the city of the honor of publication of the *Dialogue* and the Academy of its official sponsorship. Yet is to the fate of its transfer to Florence that we owe our certain knowledge that the original title was *Dialogo del flusso e reflusso del mare* and that not Galileo, but Urban VIII, removed those words, with consequences that have ramified down to the present.

Galileo's original manuscript is lost to us, but it had been read and provisionally licensed at Rome by Niccolò Riccardi, Master of the Holy Palace, after his careful revision to assure strict compliance with the edict of 1616 regulating publication of Copernican books. Only then did the pope decide to impose the further condition that tides be not made the title and subject of the printed book. Writing in 1631 to the chief inquisitor at Florence, Riccardi accurately described the *Dialogue* in a letter that opened with these words:

Il Sig. Galilei pensa di stampar costì una sua opera, che già aveva il titolo *De fluxu et refluxu maris*, nella quale discorre probabilmente del sistema Copernicano secondo la mobilità della Terra, e pretende d'agevolar l'intendimento di quel'arcano grande della natura con questa posizione, corroborandola vicendevolmente con questa utilità.[1]

Reprinted from *Atti dei convegni Lincei* 55 (1983), 101–14, by permission.

Three centuries later the most eminent and influential modern analyst of the Scientific Revolution and critic of Galileo's place in it, Alexandre Koyré, described it very differently:

Le *Dialogue sur les deux plus grands systèmes du monde* prétend exposer deux systèmes astronomiques rivaux.[2]

Such was the ultimate effect of a change in title and a related adjustment of Galileo's preface and opening speeches of the interlocutors. In his long commentary on the *Dialogue* in his celebrated *Études Galiléennes* of 1939, Koyré did not even touch on Galileo's tide theory, nor do I recall his having discussed it elsewhere. In contrast, Riccardi's letter did not speak of rival astronomical systems; rather, he said that the *Dialogue* was written:

... con fine di provare che, rimossa la rivelazione di Dio e la dottrina sacra, si potrebbero salvare le apparenze in questa posizione, sciogliendo tutte le persuasioni contrarie che dall'esperienza e filosofia peripatetica si potessero addure, sì che non mai conceda la verità assoluta, ma solamente hipothetica e senza le scritture, a questa posizione.[3]

From Ricardi's words it is easy to recognize the *Dialogue* we have all read, and also its role in a crisis that the Church had generated by its own action in 1616, as well as its bearing on a very ancient philosophical tradition that confined astronomy to mathematical fictions. But its hardly possible to recognize that same *Dialogue* in the next sentence written by Koyré:

Mais, en fait, ce n'est pas un livre d'astronomie, ni même de physique.[4]

In Koyré's view the *Dialogue*, though it claimed to expound two rival astronomical systems, was not an astronomical book. But where was the alleged claim supposed to have been made? The title page of the 1632 *Dialogue* is as innocent of the word "astronomical" as Ricardi's letter had been; all that was promised was discussion of physical and philosophical reasoning about two systems of the world. That is what Riccardi had implied in speaking of arguments from experience and the Peripatetic philosophy. Astronomical systems as such, which are complex mathematical schemes, were of little interest to Galileo, of even less interest to prospective readers of his *Dialogue*, and discussion of them could be no value to the Church. The central issue was not planetary astronomy, but motion of the earth – a topic less astronomical than physical and philosophical; that is, metaphysical with regard to propositions used in or implied by physics.

It is not hard to understand why Koyré, reading instead of the Florentine title page of 1632 the misleading paraphrase that since 1744 has been universally adopted for the *Dialogue*, assumed an astronomical claim to have been made; for the names of Ptolemy and Copernicus have come to symbolize for modern minds two rival astronomical systems. On the other hand it is not easy to see why Koyré went on to deny that the *Dialogue* is a book of physics. Its entire Second Day is mainly devoted to presentation of the first viable physics in the modern sense of the term – a physics of relative motion, conservation of motion, and composition of independent tendencies to motion. Koyré did not regard Galileo's physics as modern or viable, although he discussed it at length. Not recognizing that it was necessary for Galileo to include physics if his readers were to understand the explanation of tides presented in the Fourth Day as the climax of his *Dialogue*, Koyré supposed the physics of the Second Day to be governed by metaphysical propositions found early in the First Day. Thus it appeared to him incorrect and contradictory, not viable and modern. Had Galileo's physics been so governed, Koyré's conclusions would be valid and his ensuing remarks would be useful in interpreting the *Dialogue*:

C'est avant tout un livre de critique; un oeuvre de polémique et de combat; c'est en même temps un oeuvre philosophique; et c'est enfin un livre d'histoire: "l'histoire de l'esprit de M. Galilée."[5]

I fear that by now you may feel my own remarks to be polemic and combative, rather than constructive or even objectively critical, but I hope that the reason for them will become clear as I proceed. To explain Koyré's puzzling denial that the *Dialogue* is a book of physics, I note first that he did affirm it to be a philosophical work. For Koyré, the physics of the *Dialogue* remained natural philosophy, as it had been since Aristotle; he did not see that it was science, founded on sensate experience and necessary demonstrations, to borrow Galileo's own definition of the scope of science in letters written to Benedetto Castelli and the Grand Duchess Christina before issuance of the edict of 1616. The organization of the *Dialogue* goes against Koyré; the physics of the Second Day prepared readers for explanation of the tides in the Fourth Day. Libero Sosio recognized this in 1970, when he wrote in introduction to a new edition of the *Dialogue*:

Se consideriamo nel loro insieme i *Massimi Sistemi*, le prime tre giornate si presentano come una preparazione lenta e paziente, un lavoro assiduo di fondazione su cui dovrà innalzarsi il fastigio della quarta.[6]

Libero Sosio wrote in opposition to the long tradition among modern critics of the *Dialogue* which regards the Fourth Day as a sort of appendix to the others, not firmly attached to them and even ill-advised and not entirely relevant. Now, it is true that one of the four Days is not firmly attached to the rest, but this is not the Fourth Day; it is the First Day. That may be seen from Salviati's opening words in the Second Day:

Le diversioni di ieri, che si torsero dal dritto filo de' nostri principali discorsi, furon tante e tali, ch'io non so se potrò senza l'aiuto vostro rimettermi su la traccia, per procedere avanti.[7]

Entirely different is the relation of the Fourth Day to the Third, near the end of which Salviati says:

E perchè mi pare che assai a longo si sia in questi tre giorni discorso circa il sistema dell'universo, sarà ormai tempo che venghiamo all'accidente massimo, dal quale presero origine i nostri ragionamenti; parlo del flusso e reflusso del mare, la cagione del quale pare che assai probabilmente si possa referire a i movimenti della Terra; ma ciò, quando vi piaccia, riserberemo al seguente giorno.[8]

And just as the Third Day led to the Fourth, so it had been led to in the Second, when Salviati said:

Ricordiamoci in grazia che il cercar la costitutzione del mondo è de' più nobili problemi che sieno in natura, e tanto maggiore poi, quanto viene indirizzato allo scioglimento dell'altro, dico della causa del flusso e reflusso del mare, cercata da tutti i grand'uomini che sono stati sin qui e forse da niun ritrovata; però, quando altro non ci resti da produrre per l'assoluto scioglimento dell'istanza presa dalla vertigine della Terra, che fu l'ultima portata per argomento della sua immobilità circa il proprio centro, potremo passare allo scrutinio delle cose che sono in pro e contro del movimento annuo.[9]

Thus on indisputable internal evidence in the text it is seen that explanation of tides remained the organizing theme of the *Dialogue*, which was not altered when Riccardi cautioned the Florentine licenser about the title:

... ricordando però, esser mente di Nostro Signore che il titolo e soggetto non si proponga del flusso e reflusso ... .[10]

Removal of tides from the title of the printed book did entail adjustment of Galileo's prefatory address *Al discreto lettore*, which would originally have begun with a statement about the importance of that *arcano grande della natura*. Riccardi himself undertook that alteration, writing:

Deve ancora mostrarsi che quest'opera si faccia solamente per mostrare che si sanno tutte le ragioni che per questa parte si possono addurre, e che non per mancamento di saperle si sia in Roma bandita questa sentenza, conforme al principio e fine del libro, che di qua manderò aggiustati.[11]

The adjusted preface was duly sent, but only after the opening speeches as revised by Galileo at Florence had already been printed. Originally those had introduced the tides as the subject of discussion. That is evident from the statement near the end of the Third Day, cited above, that long discussion of the system of the universe had originated from an inquiry into the cause of tides (though in the printed book that subject was not mentioned until midway in the Second Day). Galileo had to remove his original introductory speeches to comply with the papal command that tides must not appear as the subject of the book. When the adjusted preface arrived from Rome it was printed without substantial change, as Riccardi had instructed, with the result that any *discreto lettore*, then or now, could detect intervention by the censors. The preface, after identifying the three interlocutors, ended:

Erano casualmente occorsi (come interviene) vari discorsi alla spezzata tra questi signori, i quali avevano più tosto ne i loro ingegni accesa, che consolato, la sete dell'imparare: però fecero saggia risoluzione di trovarsi alcune giornate insiemi, nelle quali, bandito ogni altro negozio, si attendesse a vagheggiare con più ordinate speculazione le meraviglie di Dio nel cielo e nella terra. Fatta la radunanza nel palazzo dell'illustrissimo Sagredo, dopo i debiti, ma però brevi, complimenti, il Signor Salviati in questa maniera incomminciò.[12]

But Salviati's first speech no longer alluded to natural marvels in the amended text as already printed, which began:

Fu la conclusione e l'appuntamento di ieri, che noi doveremmo in questo giorno discorrere, quanto più distintamente e particolarmente per noi si potesse, intorno alle ragioni naturali e loro efficacia, che per l'una parte e per l'altra sin qui sono state prodotte da i fautori della posizione Aristotelica e Tolemaica e da i seguaci del sistema Copernicano.[13]

From the evidence already in hand it is not hard to reconstruct the probable tenor of the original introductory section of Galileo's manuscript carried to

Rome with the title *Dialogo del flusso e reflusso del mare*, and with a subtitle promising discussion of reasons for and against the systems of Ptolemy and of Copernicus. The three friends, having agreed to consider marvels in the heavens and on earth, would begin by selecting one in particular. The scene of the *Dialogue* was Venice, where tides are conspicuous and would naturally be proposed by Sagredo. Salviati would offer to throw light on their cause by assuming motions of the earth. Simplicio would object that it was idle to examine the consequences of an impossibility, Aristotle having proved the earth to be motionless. Nevertheless, Salviati would say, the Copernican hypothesis had advanced astronomy, and similar assumptions might likewise advance other sciences. The only philosophically valid reason not to explore their possible bearing on tides would be conclusive proof against the assumption, whereas Aristotle's demonstrations remained inconclusive. The friends would then agree to put aside the problem of tides until all arguments for and against motion of the earth, apart from divine revelation, had been considered and weighed. Then, if nothing demonstratively disproved such motions, probable arguments relating them to tides would be taken up.

With such a beginning, the printed *Dialogue* would have been quite orderly and comprehensible to educated laymen at its time. It has been less orderly and comprehensible to modern critics for several more or less independent, but not entirely unrelated, reasons. First among those is the supposed title – *sopra i due massimi sistemi* – under which it has long been read, that being a loose and misleading paraphrase of the original subtitle. Those words had been separated from the 1632 title *Dialogo di Galileo Galilei Linceo* by five lines of type reciting his official university and court positions, and two more lines of description also came before the word *sopra*. Then, following the phrase now universally regarded as the title of the book, were four more lines of print that described the entire treatment as discussion without decision. It is rash to suppose, as many scholars now do, that that was mere pretense designed to conceal an ulterior purpose of propagandizing for the Copernican system.

I recall that thirty years ago, when I translated the *Dialogue* into English, I long hesitated over accepting the traditional abbreviated title. That certainly does not convey the true sense of the 1632 title page, whether nor not Galileo was completely candid in his wording of it. Translators are more concerned with conveying the true sense of an original than about its possible historical and philosophical implications. In retrospect, I believe that those were not entirely clear to me simply because agreement among previous students of the *Dialogue* had created preconceptions about Galileo that obscured possible alternative implications.

All persons studious of Galileo have long known that he originally intended to title his book *Dialogo del Flusso e Reflusso del Mare*. From 1624, when he began its composition, to 1630, when he submitted it for licensing at Rome, he invariably referred to his work in progress as *i miei dialoghi del flusso e reflusso* in letters to friends. The text grew in size and in variety of topics, but Galileo's original conception of his book remained unchanged. That fact alone might reasonably have suggested that explanation of tides was the organizing theme of the *Dialogue* throughout its six years of composition, and that it remained so even after the title was altered; indeed, it must have done so unless the whole book was rewritten at Florence after 1630. It certainly was not; yet the nature of preconceptions is such as to inhibit any suggestions that sharply conflict with them. Galileo was well aware of that fact himself, from the responses of philosophers against motion of the earth. Simplicio's greatest difficulty was that of putting aside, even for a moment and for purposes of argument, the inveterate idea that the earth is fixed. That preconception was grounded both in Simplicio's own experience and in his faith in Aristotle. Any preconception left immune from examination can reduce us to the state of persons described by Salviati early in the Third Day:

Dove io finalmente, osservando, mi sono accertato esser tra gli uomini alcuni i quali, preposteramente discorrendo, prima si stabiliscono nel cervello la conclusione, e quella, o perchè sia proprio loro o di persona ad essi molto accreditato, sì fissamente s'impronono, che del tutto è impossibile l'eradicarla giammai.[14]

To extricate oneself from the effects of preconceptions linked together by long and wide agreement among scholars is not easy when they are numerous and independent. Thus even Libero Sosio, when he perceived the first three Days to be a preparation for the Fourth Day, did not question the propriety of the modern printed title that has so long supported common preconceptions when he wrote:

Del valore conclusivo della sua teoria delle maree Galileo era così certo che pensava appunto di intitolare *De fluxu et refluxu maris* il suo *Dialogo*. Solo l'intervento di Urbano VIII potè indurlo a modificare il titolo in quello, meno impegnativo nella sua ambiguità, sotto il quale l'opera ci è pervenuto.[15]

Along with the preconception that Galileo's tide theory was detached from the rest of the book and hardly deserved attention today, challenged by Libero Sosio, went a second preconception that Galileo was simply mistaken in supposing tides to be conclusive of the earth's motions. Perhaps that preconception now exists only because tide theory seems to historians of science out of place in a book about the systems of Ptolemy and Copernicus, who never discussed tides. At any rate it was accepted by Albert Einstein when he mentioned Galileo's explanation in his foreword to my English translation of the *Dialogue* and accounted for its presence:

It was Galileo's longing for a mechanical proof of the motion of the earth which misled him into formulating an incorrect theory of tides. The fascinating arguments of the last Day would hardly have been accepted as proofs by Galileo if his temperament had not got the better of him. It is hard for me to resist the temptation to deal with this subject more fully.[16]

When Einstein wrote that in 1953, there seemed to be no doubt that Galileo had been a Copernican long before he hit upon his physical theory of tides while searching for evidence in support of motions of the earth. That the theory was not correct as an explanation of observed tidal phenomena was known long ago; I shall speak later of the way in which it is truly fascinating from the standpoint of physics, which did not escape Einstein as it has escaped historians of science. But a few years later, in 1961, I discovered in the notebooks of Fra Paolo Sarpi a succinct statement of reasoning fundamental to Galileo's tide theory, written in 1595.[17] That invalidated the belief that the content of the Fourth Day represented a product of Galileo's later reflections as a physicist after his telescopic discoveries in 1610 had made him a convinced Copernican. Implications of this deserve our attention.

It is a curious fact that, except for two letters in 1597, Galileo's letters and working papers contain no trace of interest in Copernicanism before 1610. No one but Johann Kepler ever consulted Galileo, during those dozen years, on anything relating to Copernican astronomy or motion of the earth. Even Kepler's letters soon stopped because Galileo did not reply. Thee is no documentary evidence that Galileo maintained interest in Copernicanism after 1597, when he told Kepler that he had favored the new system for several years and had been able to explain by it some natural phenomena that could perhaps not be explained under the old astronomy. Kepler correctly surmised that Galileo must refer to the tides, the only observable natural phenomenon that could possibly be explained by motions of the earth. But Kepler was certain that tides could not be so explained, and he did not inquire further.

Next, it is certain that Galileo was not a Copernican when he moved to Padua in 1592. He had then recently completed his *Sermones de motu* in which the earth was treated as the center of the universe. In it he said that he had written a commentary on Ptolemy's *Almagest*: he mentioned Copernicus only in connection with a purely mathematical theorem, and he treated the daily revolution of the fixed stars as a real motion. Galileo's work at Pisa and during his first years at Padua related to motion and mechanics, not to astronomy. The "several years" of his Copernican preference before 1597 appear not to have begun before 1593, and no reason for it has direct or indirect documentary support except his theory of tides. Reasons for my belief that Sarpi recorded from conversations in 1595 a theory propounded by Galileo were set forth in my 1961 paper. Since Libero Sosio holds the opposite view, I may add that Fulgenzio Micanzio, Sarpi's biographer and friend, assures us that Sarpi accepted but a single motion of the earth,[18] whereas Galileo's tide theory required two terrestrial motions.

The tides at Venice could not fail to attract Galileo's interest as a student of motion and mechanics when he moved to Padua. On the other hand nothing is known that would have awakened his interest in astronomy for its own sake during his first ten years as professor of mathematics at Padua, before 1604. It is accordingly possible, and I believe it overwhelmingly probable, that Galileo's first preference for the Copernican motions arose from his tide theory, not the other way round. As a physicist, seeking a mechanical explanation for the rise and fall of water in the Venetian Gulf, Galileo hit upon an analogy with water in a container carried at uneven speed; and, seeking a reason for which sea-basins might move with varying speeds, he found this in the doubly circular Copernican motions of the earth. Sea-basins moved epicyclically around the sun, and every astronomer recognized (and some were scandalized by) the non-uniformity of epicyclic motion with respect to a second center. Galileo was thus led, I believe, to formulate fascinating but inconclusive proofs for the Copernican motions not by prior conviction that the earth moved, but because they fitted with a mechanical explanation of vast observed motions of the seas and could thus be argued probably, not demonstratively.

That view of the matter is strikingly in accord with Galileo's own account, early in the Fourth Day, of motions of water in a barge carrying it from Fusina to Venice. Long-continued oscillations, he wrote, occur after change of speed by the barge, the water rising and falling alternatively at the ends but not at the middle. It is plausible that a physicist, riding to Venice on such a barge, should see the analogy between motion of the water contained in it and that of water beneath the barge, held in a much larger container. For Galileo to propound this idea to his friend Sarpi after such a voyage would be quite natural, for Galileo and Sarpi frequently discussed problems of motion over many years.

In any case, the terrestrial problem of tides was associated in Galileo's mind with the Copernican astronomy in 1597 when he wrote to thank Kepler for a book. By 1610 Galileo had written a treatise on tides, now lost but mentioned in a letter of that year – a year crucial in the story of Galileo's Copernicanism. His earliest treatise on tides surviving in his own handwriting is dated in January 1616, on the eve of the edict regulating Copernican books. In 1618 he sent a copy to Archduke Leopold of Austria, expressing fear that unless it were published, if only as an ingenious fantasy, some foreigner would gain credit for a scientific explanation that was rightfully Italian and Galileo's own. Conditions in 1624 made it possible for Galileo to publish it himself as an application of the Copernican hypothesis supported by probable reasoning. In that year a combination of circumstances caused Urban VIII to perceive value to the Church in a book published by a Catholic scientist, with official permission, showing that the 1616 edict did not hamper advance in science and correcting current misconceptions among intellectuals about the reason for issuance of the 1616 edict, its scope, and its proper interpretation. Galileo began to organize such a book that same year, as a *Dialogo del flusso e reflusso del mare* based on three decades of reflection about the tides as linked with Copernican motions of the earth.

Galileo's explanation of tides as organizing theme of the *Dialogue* is of great significance with respect to the course historically taken by the Scientific Revolution inaugurated by Galileo in physics and Kepler in astronomy. That culminated in 1687 with the Newtonian system of the world, in which physical and astronomical phenomena were forever linked in a unified cosmology. Since antiquity, physics and astronomy has been kept rigorously apart for philosophical reasons. A few words about that are in order here, even though the facts are already known to you.

All philosophers, Platonist and Aristotelian alike, agreed that the earth was at absolute rest at the exact center of the various celestial spheres, whose true motions we uniformly circular. That conclusion was cast in doubt when Hipparchus found, two centuries after Aristotle's death, that recorded eclipses required displacement of the earth from the exact center of the sun's orbit. In order that astronomers might continue their useful work while philosophers remained in sole custody of true knowledge, Geminus proposed that astronomers refrain from considering physical causes, leaving that to the philosophers who knew the true celestial motions and could ignore mathematical fictions grounded in phenomena.

This admirable compromise served for centuries because, at any rate as I see things, Aristotelian physics applicable to terrestrial events remained undisturbed even if the earth's position were not central, provided that it was absolutely fixed. Motion of the earth was, however, incompatible with Aristotelian terrestrial physics. The Scientific Revolution thus became inevitable after the *De revolutionibus* of Copernicus in 1543, and it was set in motion by two books which proposed physical explanations compatible with astronomical knowledge. The first was Kepler's *Astronomia Nova* of 1609, proclaiming on its very title page a celestial physics in defiance of the ancient tradition. Kepler's astronomy was truly modern (though he called it "Copernican" in an excess of courtesy), but his proposed physics was ad hoc and scientifically worthless. The second book was Galileo's *Dialogue*, which would have proclaimed defiance of the ancient separation of physics from astronomy if its title page had been printed as Galileo wrote it. The physics of the *Dialogue* was modern and of great scientific value, while its planetary astronomy remained merely schematic and Copernican, not at all modern except as to Galileo's own telescopic discoveries. The astronomy of Kepler and the physics of Galileo, greatly advanced by Newton, were brought together in his system of the world.

Now, it happened that Kepler included an explanation of tides in the preliminary pages of his *Astronomia Nova*, and that Galileo rejected this in his later *Dialogue* as no more than immature appeal to occult properties, not a truly physical explanation. Modern critics defend Kepler and disparage Galileo in this matter. The fascination found by Einstein in Galileo's reasoning about his incorrect theory is replaced in their critiques by fascination with Kepler's seemingly correct anticipation of Newtonian tide theory. In fact, neither Galileo's explanation nor Kepler's – nor, for that matter, Newton's – gave anything like an adequate account of observed tides. But since Galileo's explanation was the organizing theme of the *Dialogue*, both that and his objection against Kepler's suggestion deserve our attention. Both throw light on Galileo's conception of scientific method and on the physics of the *Dialogue*, which is scientifically its most important contribution to knowledge.

Galileo's tide theory fell into neglect as soon as Newton applied his law of universal gravitation to the problem. The first modern comment on Galileo's theory was offered by Emil Strauss in his German translation of the *Dialogue* in 1891. Strauss believed that disturbance of large seas by doubly circular motion of the earth, quite apart from universal gravitation, must occur but is too small to account for the observed tides.[19] Soon afterward, Ernst Mach refuted a tidal explanation he called "Galileo's," in which the earth was assigned a rotation and a uniform straight motion. The absurdity of this is obvious, since no Copernican (and least of all Galileo) believed the earth to move forever farther from the sun. Yet Mach's reputation as a physicist induced historians to dismiss Galileo's tide theory as refuted. Emil Wohlwill, as Galileo's biographer, consulted persons trained in science or engineering and found that in 1909 they generally agreed with Strauss.[20] In 1954 the Linecan Academician Vittorio Nobile published a detailed mathematical analysis of Galileo's "primary cause of tides" and related it to the fundamental problem of geodesy.[21] Since Nobile's analysis was highly technical, and Strauss did not publish his own reasoning, I shall offer a simple Newtonian analysis without universal gravitation to show you that the "Galileo effect" is not imaginary. First, however, I shall appeal to your physical intuition, which is probably the same as Galileo's when he devised a mechanical model to exhibit the effect that he derived in the *Dialogue*.[22]

Many entertainment parks have a device consisting of a circular seat that rotates about its center while being carried around a central column at the end of a long beam. Physical intuition tells us that persons seated in it feel pushed and pulled by the doubly circular motion. If the seat were filled with water and a barrier were placed across the middle of the seat, water would rise and fall against the barrier as the seat rotated and revolved. Such was Galileo's picture of the water in wide east-west seas having long north-south coasts, their basins rotating around the earth's axis while revolving about the sun. No force to lift water was necessary; changing speeds sufficed to make it flow, and barriers to make it rise.

In Newtonian physics, neglecting universal gravitation and assuming only the natural tendency of water as a heavy body to move downward, continual disturbance of water in large seas is also assured. Merely by circular motion around the sun, every point on the earth's surface is subjected to a centripetal acceleration toward the sun. This acceleration is greatest at the point nearest to the sun, and ever less at points more distant from it. Water levels in large seas must differ at points, as a result of such differing accelerations. If the earth also rotates on its axis, the surface of such a sea at each point must rise and fall cyclically throughout each day. The undulations may be small, even immeasurably small, but Galileo's conception of continual flows in large seas was not absurd.

Of course this was not Galileo's reasoning, and of course Newton's tide theory is based entirely differently, on attractive forces. Also, Galileo's was a two-cause theory, continual disturbance of seas being explained by doubly circular motion, while periodicity of tides was explained by reciprocations dependent on the lengths and depths of sea-basins. What is fascinating about his basic argument is that by reasoning from differences of speeds at widely separated points in a sea, and changing speeds at any point, Galileo reached a conclusion that follows also from considerations of acceleration, and hence of force. In Newtonian dynamics, that was possible because the uniform circular motions of Galileo's kinematics are uniformly accelerated motions by definition in later dynamics, and because Galileo's conception of relative and absolute motion in the *Dialogue* anticipated Newton's famed discussion of that subject in the scholium to his definitions in the immortal *Mathematical Principles of Natural Philosophy*. Once we see that the Second Day was preparatory to the tide theory of the Fourth Day, and was not derived from metaphysical propositions found in the First Day, the many alleged contradictions in the *Dialogue* vanish, all being based on a confusion of Galileo's physics with previous natural philosophy.

Kepler's celestial physics remained natural philosophy, not science, both as to its magnetic chains and polarities in the heavens and as to a supposed power of lunar attraction to lift the seas without budging the earth; that is, their waters but not their basins. Three principal phenomena for which Galileo offered kinematic explanations could not be accounted for by Kepler's lunar bulge of water – absence of tides at island shores, spring and neap tides, and the occurrence of two high tides each day. Galileo's characterization of lunar influence as an occult property remained appropriate until Newton created the science of dynamics. Galileo's discovery of the law of falling bodies, set forth in the *Dialogue* and utilized to demonstrate that whirling of the earth could not cast off resting heavy bodies from its surface, made possible Newton's creation of dynamics, while Galileo's physics remained devoid of appeals to forces. His kinematics of heavy bodies was inadequate to explain tides, but his attempt to do that expelled occult causes from science in favor of sensate experiences and necessary demonstrations.

To see that attempt as the organizing theme of the *Dialogue*, making the Third Day necessary after the Second as joint preludes to the Fourth, makes understandable the detachment of the First Day from the rest, and the new beginning signaled by Salviati at the opening of the Second Day. What was to be discussed there was physics, not as natural philosophy founded on metaphysics, but as science concerned with the insensible world and not merely *un mondo di carta*.

The first thirty or forty pages of the *Dialogue* examined the foundations of Peripatetic natural philosophy, which were metaphysical. That served a necessary purpose, of exposing defects in Aristotle's analysis of natural motions. Starting anew in the Second Day, Galileo did not then set forth new metaphysical principles, but appealed to observed motions as his basis for a fresh analysis. It is understandable that modern commentators look back to the First Day for supposedly missing metaphysical principles in order to interpret the physics of the Second Day as ancient and medieval natural philosophy. Obtaining inconsistent results, they then impute to Galileo contradictions that have arisen only in this century, and from misapprehension of Galileo's view of metaphysical propositions. To Galileo, as I believe also to Aristotle, those came *after* physics, as distillations from it, and not as sources from which to deduce physics. That aberration of medieval scholasticism was prevalent in the universities of Galileo's time, and is finding favor again in them today among historians of science. I believe that the assumption that Galileo's concept of science was that of traditional natural philosophy is misleading.

The *Dialogue* is now often said to have been Copernican propaganda written by an enthusiast who lacked a scientific basis. To me it appears that the *Dialogue* was a sustained lesson in weighing all the known arguments, putting aside preconceptions and abiding by the preponderance of evidence at its time. That is a conception of science abandoned by philosophers at the time of Geminus and resumed by Kepler and Galileo. In their time preponderance of evidence favored the Copernican system, thanks in no small part to their own serious investigations. Philosophers attacked them then, and some disparage their work even now. For my part, I cannot read the *Dialogue* without recalling the words of Luca Holste, written in 1642 from Rome, at the palazzo of Francesco Cardinal Barberini, to Giovanni Battista Doni at Florence:

Oggi poi si è aggiunta anco la nuova della perdità del Signor Galilei, che già non riguarda solamente Firenze, ma il mondo universo e tutto il secolo nostro, che da questo divin uomo ha ricevuto più splendore che quasi da tutto il resto de' filosofi ordinarii. Ora, cessata l'invidia, si comincerà a conoscer la sublimità di quell'ingegno, che a tutta la posterità servirà per scorta nel ricercar il vero, tanto astruso e seppellito tra il buoio dell'opinioni.[23]

NOTES

1 *Opere di Galileo Galilei*, Edizione Nazionale, xix, 327.
2 Alexandre Koyré, *Études Galiléennes* (Paris, 1966), 212.
3 *Opere*, xix, 327.
4 Koyré, 212.
5 Ibid.
6 Galileo, *Dialogo*, a cura di Libero Sosio (Torino, 1970), lxxii.
7 *Opere*, vii, 132.
8 Ibid., 439.
9 Ibid., 236–7.
10 *Opere*, xix, 327.
11 Ibid.
12 *Opere*, vii, 31.
13 Ibid., 33.
14 Ibid., 299–300.
15 *Dialogo*, ed. Sosio, loc. cit.
16 Galileo, *Dialogue*, tr. S. Drake (Berkeley, 1953), xvii.
17 S. Drake, "Origin and Fate of Galileo's Theory of Tides," *Physis*, 3 (1961), 185–94.
18 F. Micanzio, *Vita del Padre Paolo* (1646), cited by L. Sosio, *Dialogo*, lxxix: ... *egli ha trovato di salvare tutti i fenomeni con un unico moto* ...

19 Galileo, *Dialog*, tr. E. Strauss (Leipzig, 1891), 566.
20 Emil Wohlwill, *Galilei und sein Kampf* ... (Hamburg & Leipzig, 1909), vol. 1, 600, n. 2.
21 Vittorio Nobile, "Sull'argomento galileiano della quarta giornata ... ," *Atti dell' Acc. Naz. dei Lincei, Classe di Scienze Fisiche*, 6 (1954), 426–33.
22 *Opere*, vii, 456: E benchè impossibil possa parer a molti che in machine e vasi artifiziali noi possiamo esperimentare gli effetti di un tale accidente, nulla dimeno non è però del tutto impossibile; ed io ho la costruzione d'una machina, nella quale particolarmente si può scorgere l'effetto di queste meravigliose composizioni di movimenti. See also my paper cited in n. 17 above, and my *Galileo Studies* (Ann Arbor, 1970), 260–7.
23 *Opere*, xviii, 378.

3

Reexamining Galileo's *Dialogue*

Reinterpreting Galileo with an eye to his significance for the history and the philosophy of science has occupied many able scholars during the present century. It is tempting to review the succession of fashionable interpretations and reasons for which thoughtful persons have moved from one to another, but I must limit myself to some brief generalizations and then proceed to a particular topic, the reexamination of Galileo's *Dialogue* as printed at Florence in 1632.

In 1891 Emil Strauss set the stage for modern interpretations of Galileo by presenting the first vernacular translation of the *Dialogue* to appear in 230 years.[1] A scholarly introduction and copious notes written by Strauss related Galileo's most famous text to all his other books, assessing its cultural context and scientific contributions with remarkable breadth and accuracy. While his book was in press, Strauss died. Full documentation of Galileo's trial, then only recently published, has subsequently been subject to more profound analysis. Antonio Favaro had just commenced publication of his twenty-volume edition of Galileo's works, which during the next two decades brought to light for the first time much unpublished material. Researches into medieval science have since transformed the history of physics. Evidences of Galileo's Platonist leanings in the *Dialogue*, noted by Strauss and later developed by Ernst Cassirer and E.A. Burtt, came to be seen by Alexandre Koyré as the key to the entire scientific revolution, with profound consequences for the interpretation of Galileo during the past forty years.[2] Renewed studies of Renaissance Aristotelianism by John H. Randall[3] and others challenged the Platonist thesis that had become

Reprinted from *Reinterpreting Galileo*, ed. W.A. Wallace, vol. 15 of *Studies in Philosophy and the History of Philosophy* (Washington: Catholic University of America Press, 1986), 155–75, by permission.

the majority view of historians and philosophers of science and offered an opposing philosophical interpretation. Most recently a long neglected aspect of sixteenth-century Aristotelianism has been shown by Father William A. Wallace to have strongly influenced Galileo during his formative years.[4]

In short, scholarly interpretations of Galileo began a century ago with close examination of his own books, then shifted to the writings of ancient, medieval, and Renaissance natural philosophers, and now consist largely of competing views founded on a vast philosophical literature that scholars believe determined Galileo's goals and ideas. Critics holding the most divergent viewpoints have easily found passages in Galileo's writings that support varied interpretations of his science. Selected passages do support the interpretation urged by the scholar who searched them out and sometimes also fit conflicting interpretations, since even otherwise cautious scholars may become overzealous to prove a thesis. The overall result of this trend has been growing uncertainty about the proper interpretation of Galileo – the very problem that so much labor was undertaken to eliminate. If I may be allowed a general comment, what we need now may be not so much *re*interpretation as *de*interpretation of Galileo.

In all studies during the present century one may detect an underlying assumption almost universally accepted – that Galileo wrote his *Dialogue* as an ardent Copernican who allowed that burning conviction to affect his scientific judgment, while attempting to cloak it under an insincere and merely formal compliance with the Church edict of 1616 regulating Copernican books. Archbishop Richard Whateley amusingly pointed out long ago that things everyone takes for granted are not necessarily things that they have most closely examined.[5] As Galileo remarked, concerning the universal belief at his time that a weight dropped from the mast of a ship alights differently when the ship is at rest or in motion, everyone supposed that someone else must have observed this phenomenon. Yet no one could have observed it because it does not happen (144). Ardent Copernican zeal of Galileo may similarly exist only in modern books. Josh Billings used to say that the trouble is not what we don't know; it is that we know so much that ain't so.

About three years ago I reexamined Galileo's physics in the second day of the *Dialogue* and came finally to understand it, thirty years after translating the book. There is a standpoint from which Galileo's mature physics was consistent and correct. Having seen this, I reexamined Galileo's metaphysics in the first day and his application of physics in the fourth day. Finally, I completed a thorough reexamination of the *Dialogue* and many documents relating to it. As a result, I no longer accept the usual assumptions of overzealous Copernicanism, defective scientific judgment, and insincerity of purpose on Galileo's part. On the contrary, it is my considered opinion that we should regard those as common preconceptions rooted in events that followed *after* publication of the *Dialogue*. They do not reflect the circumstances of its composition. The great explanatory power of those preconceptions does not prove their truth. Indeed, if I understand modern logic correctly, false propositions have greater explanatory power than true ones because they imply all propositions, true or false, whereas true propositions imply only other true ones. The common preconceptions suggest that the very severe action taken against Galileo by the Church was his own fault. They conveniently explain any scientific error that has been found (or alleged) in Galileo's writings, without even troubling us to understand Galileo's physics. But since there is a standpoint from which Galileo's physics was correct and consistent, the elementary blunders in science now charged to him are largely illusory. That casts doubt, for me at least, on Galileo's overzealous Copernicanism, on which depends the idea that he wrote his *Dialogue* insincerely or in defiance of his Church.

I do not expect you to abandon assumptions hallowed by a century of scholarly opinion just because I doubt them. But no irreparable harm can be done by putting preconceptions aside for purposes of argument, and I do ask you to do that. We can then examine anew some striking oddities about the printed *Dialogue* of 1632 that I noticed thirty years ago when I started translating the book but could not rationally explain because I accepted the usual preconceptions.

First among these oddities is the printed title page of 1632.[6] Bibliographically speaking, it bore no proper title beyond the words *Dialogue of Galileo Galilei, Lincean.* That is how the prosecutor at Galileo's trial in 1633 legally identified the book. The one-word proper title was followed by seven lines of print identifying the author and ending with a full stop. Then, beginning with a capital letter, came a very long subtitle forming a new complete sentence. When the Church first permitted a reprinting, in 1744, the editors took from that long subtitle, out of context, the phrase that has ever since masqueraded as Galileo's title and subject for the *Dialogue*: "concerning the two chief world systems, Ptolemaic and Copernican." Later scholars thus came to imagine that Galileo promised a book about planetary astronomies, though he certainly did not present one. Ptolemaic astronomy is not even described in the *Dialogue*. Description of Copernican astronomy hardly goes beyond one schematic diagram (323), adapted to include Jupiter's satellites but otherwise identical with a most unastronomical picture in the introductory chapter of Copernicus's *De revolutionibus*. The title page of the *Dialogue* promised no astronomical reasoning whatever, but only philosophical and "natural" – which is to say, metaphysical and physical – arguments, presented without positive determination. Scholars knew that Galileo's original title for his book had been deleted by censors before the *Dialogue* went to press, but they did not attach great importance to that. (In context, the phrase taken from the subtitle had originally made *reasonings*, not astronomical *systems*, the subject of discussion.)

The second curiosity is that the printed preface opened with a vigorous defense of the 1616 edict regulating Copernican books. In view of Galileo's battle before 1616 against *any* official Church regulation, the preface has appeared to be ironical almost to the point of hypocrisy. Such a preface presents a difficult problem under the common assumption that Galileo was a zealous Copernican, intent on evading censorship. No author calls the attention of censors to a regulation he does not want them to enforce. It would also be most unusual for censors to leave such a preface in place if they considered it not in compliance with the regulation.

The third striking oddity in the printed *Dialogue* is an opening breach of continuity. Galileo introduced the three interlocutors in his preface and then concluded it thus: "They very wisely resolved to meet together on certain days during which, setting aside all other business, they might apply themselves more methodically to contemplation of God's wonders in the heavens and on earth. They met in the palazzo of the illustrious Sagredo; and, after the customary but brief interchange of compliments, Salviati commenced as follows." (7) But Salviati opened the first day with these words: "Yesterday we resolved to meet today and to discuss as clearly and in as much detail as possible the character and efficacy of those physical reasonings which up to the present time have been put forth by partisans of the Aristotelian and the Ptolemaic position on the one hand, and by followers of the Copernican system on the other." (9)

Salviati could not have forgotten overnight an agreement to discuss wonders in heaven and on earth, which hardly extend to reasonings of a philosopher and two astronomers. Yet the other interlocutors did not correct him. What was missing at the very opening of the printed text becomes apparent some four hundred pages later, near the end of the third day, when Salviati says: "It seems to me that in these three days the system of the universe has been discussed at great length, so it is now time for us to take up that principal event from which our discussions *took their rise* – I mean the ebb and flow of the oceans, whose cause may be very probably assigned to movement of the earth" (413; italics added).

Tides had not even been mentioned until the middle of the second day. What had been said there was in a sense only incidental, but I shall quote it because it is important to something that the *Dialogue* is now often said to lack – rational organization. It is also of great interest with regard to Galileo's conception of science, for what Salviati said was this:

Please, gentlemen, it seems to me that we have gone off woolgathering. Since our argument should continue to be about serious and important things, let us waste no more time on frivolous and quite trivial altercations. Let us remember that to investigate the constitution of the universe is one of the greatest and noblest problems in nature, and that it becomes still grander when directed toward another discovery; I refer to the cause of the flow and ebb of the sea, which has been sought by the greatest men and has perhaps been revealed by none. [210]

Galileo's view that study of the heavens could be further ennobled by a mere terrestrial phenomenon is neither in the Platonist nor in the Aristotelian tradition. It was not philosophical but scientific. It was motivated by the same reason for which Galileo laid the scene of the *Dialogue* in Venice, where tides are conspicuous as they are nowhere else in Italy. Hence tides could plausibly be named at the outset as a "wonder on earth" deserving methodical discussion. The point will be pursued later, after we consider explanations of the oddities already pointed out.

The first, which was absence of a specific title for the printed *Dialogue*, is easily explained. From 1624 through 1629, when it was being written, Galileo invariably referred to his work in progress as "my dialogues on the flow and ebb of seas."[7] Even at the end of 1629, when the book was complete except for the metaphysical introduction that Galileo composed last, to give his book some elegance beyond dry science, corroboration of the Copernican system was mentioned only incidentally in one letter to a friend. The manuscript Galileo carried to Rome in 1630 for licensing to be published there bore his full title, "Dialogue on the Flow and Ebb of the Seas." That is shown by the document presented below (which further implies that the subtitle later printed was already present). An outbreak of plague had closed the roads after Galileo returned to Florence to make various required revisions, so that any manuscript sent to Rome would be taken apart and fumigated page by page. In May 1631 Niccolò Riccardi, Master of the Holy Palace, wrote as follows to the chief inquisitor at Florence:

Signor Galilei is thinking of printing there a work of his that formerly had the title "On the Flow and Ebb of Seas," in which probable reasoning[8] is given concerning the Copernican system according to mobility of the earth, and he claims to facilitate understanding of that great mystery of Nature with this position, reciprocally corroborating that by such utilization of it. He came here to Rome to show the work, which I subscribed assuming the accommodations that had to be made in it and his bringing it back to receive final approval for printing. That being impossible to be done because of hindrances on the roads and danger to the originals, and the author wishing to finish the business there, Your Reverence may exercise your own authority and send the book forth, or not, without any dependence on my review – but keeping it in mind that it is the will of His Holiness that the title and subject may not propose the tides, but absolutely mathematical consideration of the Copernican position about motion of the earth, to the end of proving that except for God's revelation and sacred doctrine, all the appearances could be saved in that position, collecting together all the contrary arguments that can be adduced from experience and the Peripatetic philosophy, so that absolute truth is never conceded to that position, but only hypothetical, and without the Bible. It must also be shown that this work is done solely to show that all the reasons are known, and that it was not from lack of knowledge that this opinion was banned at Rome – in accordance with the beginning and ending of the book which I shall send adjusted, from here. With this precaution the book will have no impediment here at Rome, and Your Reverence will be able to satisfy His Serene Highness who is showing such pressure in this. Remember me as your servant, and favor me with your commands. [*Opere* 19:327, trans. mine]

Galileo's explicit title had been deleted by the Roman censor at the pope's personal order, as was also reflected in some letters at Rome in 1630. But the pope had not ordered removal of Galileo's explanation of tides because that was still promised in the preface as adjusted by Riccardi, duly sent to Florence with instructions against Galileo's making any substantial change in it. Why Urban VIII did not want tides to appear as title and subject, though they could be discussed, was never officially explained.[9] Galileo removed the opening speeches that had been appropriate under his original title. Some fifty pages of his revised text had been printed before July, when the adjusted preface arrived. That was printed last, unaltered, creating the inconsistency between it and the beginning of the text.[10]

There is also a reasonable explanation for the seemingly strange opening of the printed preface. Under the original title, Galileo's preface would naturally have begun with reasons for writing a book on tides. As printed, those reasons appear in third place instead of at the very beginning (6). Their transfer of place was, in all probability, Riccardi's only major adjustment to the preface Galileo had left with him.[11] Galileo's own wording had led naturally to his discussion of the 1616 edict, beginning thus:

I shall propose an ingenious speculation. It happens that long ago I said that the unsolved problem of the ocean tides might receive some light from assuming motion of the earth. This assertion of mine, passing by word of mouth, found loving fathers who adopted it as the child of their own ingenuity. Now, so that no foreigner may ever appear who, arming himself with our [Italian] weapons, shall charge us with want of attention to so important a matter, I have thought it good to reveal those probabilities which might render this plausible, given that the earth moves. I hope that from these considerations the world will come to know that if other nations have navigated more, we have not theorized less. It is not from our failing to take account of what others have thought that we have yielded to asserting the earth to be motionless, and holding the contrary to be a mere mathematical caprice; but – if for nothing else – for those reasons that are supplied by piety, religion, knowledge of divine omnipotence, and awareness of the limitations of the human mind. [6]

Galileo's reference to his intended assumption of motion of the earth (which could be dealt with only as a hypothesis) made it appropriate for him next to discuss the 1616 edict. That was indeed not only appropriate but absolutely necessary, for reasons recited above in Riccardi's letter. Accordingly, Galileo went on to clarify the legitimacy of the assumption by explaining why the edict had been issued, which was primarily to stop controversial scriptural interpretations by unqualified persons, creating scandals: "Several years ago there was published at Rome a salutary edict which, in order to put an end to dangerous scandals of our present age, imposed seasonable silence on the Pythagorean opinion of mobility of the earth" (5).

That sentence opened the printed preface abruptly and for no apparent reason. Following it was a reproach against hostile critics of the edict and Galileo's own account of the facts based on his presence in Rome at the time. In its original position, his account would not have appeared cynical.[12] Riccardi, being fully informed, probably did not even notice how it might be misconstrued in the absence of any preliminary context.

It is clear that the organizing theme of the *Dialogue* had been Galileo's explanation of tides as presented in the fourth day, where it constitutes the climax and conclusion of the work. As Galileo's own scientific theory, elaborated over many years, he was justly proud of it. Once we recognize this, and realize that neither the title page nor the opening speeches of the printed *Dialogue* fully disclose Galileo's purposes in writing it, the preconception that he wrote as a Copernican zealot, or intended a primarily astronomical work, becomes gratuitous.

Galileo had very good reasons for organizing his book as an explanation of tides and for subordinating his discussion of Aristotelian, Ptolemaic, and Copernican cosmologies to that. In 1616 he had to abandon his project of publishing a book on *the* system of the world, promised to his readers in 1610 and again in 1612.[13] A Copernican book on that subject could no longer be licensed. Even comparison of the two chief systems could be approved only if incidental to some other valid purpose, unless the old system were made to appear clearly superior to the new. In 1620, replying to an inquiry from abroad, Galileo wrote that his promised book on the system of the world had been "stayed by a higher hand." But in 1624, after six audiences with the recently elected Pope Urban VIII, Galileo saw how he could modify it so as to be of service both to the Church and to Italy. By that time the 1616 edict had come to be widely misunderstood among intellectuals, particularly in Germany, as an outright prohibition of any discussion among Catholics of the Copernican position. Urban, himself an intellectual and no less proud than Galileo of Italy's traditional primacy in science, was concerned about this. Although Urban, like Galileo, believe in 1616 that the Church should not take any official action on Copernicanism, he would not repeal the edict then issued. In that, I think the pope was wise, for a reason that appears to be neglected and deserves mention.

From 1613 to 1616 Galileo had battled against *any* official action, while his enemies, professors of natural philosophy, urged that motion of the earth be declared a heresy. Neither side won. The edict was carefully worded, probably by Cardinal Bellarmine, to leave open the traditional right of astronomers to employ any hypothesis that advanced their mathematical understanding of celestial motions.[14] Only unauthorized biblical interpretation was forbidden and, of course, any *positive* assertion that the earth did move as a planet. If official action had to be taken, that was the best that could be done. Galileo never complained of it, as a Copernican zealot would surely have done. For Urban to repeal so moderate and traditional an edict, as things were then seen, would have been viewed within the Church as his siding against every professor of philosophy and nearly every astronomer in Italy, for no discernible reason except perhaps his long-standing personal friendship with Galileo. Although I am far from being an admirer of Urban VIII, I can see how unwise any such action by him would have been.

Whatever happened later, all scholars agree that Galileo won Urban's approval in 1624 of his project to write a book showing that Catholic scientists were still in the lead, unimpeded by the 1616 edict, and testifying that in Italy, and at Rome in particular, all scientific arguments for motion of the earth were known and were irrelevant to the scope and purpose of the edict. Galileo returned to Florence, bearing papal gifts and letters of recommendation, to begin composing his dialogues on the tides. In them he would advance science by hypothetical use of the Copernican motions to explain the existence of tides in very large seas, displaying along the way many other Italian and Catholic advances in both physics and astronomy. If Galileo had any reason to doubt that such a book would be licensed not only as permissible but as having value to the Church, he would hardly have invested five years in its composition, starting at the age of sixty with an already established reputation and a secure position in a very Catholic court.

The principal form of argument employed in the *Dialogue* lay at the heart of Galileo's mature conception of science. I have just identified this form as hypothetical – or as Galileo called it in the fourth day, reasoning *ex suppositione*. There he had Simplicio, representing the philosophers, attack his tidal explanation, beginning with these words:

> I do not think it can be denied that your argument goes along very plausibly, the reasoning being *ex suppositione* as we say, that is, assuming that the earth does move in the two motions assigned to it by Copernicus. But if we exclude those movements, all the rest is vain and invalid; and exclusion of this hypothesis is very clearly indicated to us by your own reasoning. Under the assumption of the two terrestrial movements, you give reasons for the ebb and flow, and then vice versa, reasoning circularly, you draw from the ebbing and flowing the sign and confirmation of those same two movements. Passing to a more particular argument, you say also that on account of the water's being a fluid body, not firmly attached to the earth, it is not rigidly constrained to obey all the earth's movements, and from that you deduce its ebbing and flowing. [436]

Simplicio went on to explain why Galileo's own reasoning, if valid, should apply also (and even more cogently) to the air, which is still less firmly attached to the earth than are the seas. Since tides do not occur in the air, Galileo's assumption was to be rejected. Salviati's reply was confined to observed phenomena of water and air and to proper application of reasoning *ex suppositione* about them. He did not answer Simplicio's imputation to him of circular reasoning about the tides.

Galileo deliberately placed the charge of circular reasoning in the mouth of the spokesman for philosophers, among whom that is a very damaging allegation. If Galileo had no answer, it seems that he might better have omitted this charge, thus sweeping under the rug an inherent flaw in his conception of science. Circularity of reasoning is in fact a charge commonly brought by philosophers today, not just against Galileo's conception of science but against the naive confidence placed by later scientists in their methods and results. I am no philosopher of science, nor of anything else, nor is it my task to defend scientists or modern science. My interest is limited to understanding the science of Galileo and answering the question why he introduced the charge of circular reasoning against himself but did not reply to it. If what I say about this is foolish and absurd in philosophy, as the Catholic Qualifiers in 1616 ruled motion of the earth to be, it may nevertheless interest other serious students of early modern science.

Galileo introduced the charge because if any basic fault existed in his conception of physical science, it was the presence of circular reasoning. He did not reply to that charge because his *Dialogue* was written in Italian for readers of good sense, not in Latin for professors of philosophy. Galileo was willing to submit his conception of science to the judgment of people of good sense, knowing that it could, and would, be faulted as circular by professors of philosophy. I cannot agree with Father Wallace that Galileo, following certain technicalities debated among philosophers, believed that reasoning *ex suppositione* could in some way establish *scientia* in the classic philosophical sense of absolute truth. Still less can I agree with Father Ernan McMullin that Galileo believed he could demonstrate motion of the earth from undeniable first principles. Reexamination of the *Dialogue* reveals no trace of any such illusions on Galileo's part. I believe they are now ascribed to him only under the mistaken preconception that Galileo was an overzealous Copernican and allowed that to cloud his reasoning powers. Later scientists may have had illusions about absolute truth, but not Galileo, who wrote that no event in nature would ever be completely understood in theory (101).

It is of interest that Riccardi had approved the line of argument described by Simplicio, but Riccardi did not call it "circular reasoning." He called it "reciprocal corroboration." Similar procedures have been used in physics ever since Galileo and had been used in astronomy long before his time. To claim corroboration of suppositions by observing phenomena deduced from them does not disturb people of good sense, and it did not upset the theologian Riccardi. It outraged only professors of philosophy, who held that true science requires rigorous proofs from undeniable first principles. In that view no astronomy was pure science because it was contaminated with practice and hence lacking in completeness and certainty. The true science of the heavens was for them cosmology, as in *De caelo*, when Aristotle proved the earth to be fixed at the precise center of all celestial revolutions. On that proof, from undeniable first principles, Aristotle also rested his analysis of the motions of heavy bodies, without which there would have been no science of physics. His scientific conclusions could not be shaken by circular reasoning in which motions of the earth were assumed, tides were deduced, and their existence was treated as corroborating the assumed motions.

It is true that parts of physics, though not many, can be derived from undeniable first principles. Archimedes found some, and Galileo added some others; but most of useful physics has grown out of skillfully chosen hypotheses corroborated by observational evidence and careful measurements. In those ways Galileo extended physics at the cost of absolute certainty, and he knew it. He derived the physics of the second day by repeated appeals to observation, not to undeniable first principles. Still less did Galileo appeal in the second day to his metaphysical statements in the first day, of which I will speak later. Indeed, Salviati sharply separated the second day from the first by opening it with these words: "Yesterday took us into so many and such great digressions, twisting away from the main thread of our principal argument, that I do not know whether I can proceed without your assistance in putting me back on the track" (106).

Reexamination of the second day reveals a standpoint from which its physics is consistent and correct. Probably Galileo adopted such a standpoint, knowing that hostile philosophers skilled in logic would examine his book. The best way to describe his physics is as a kinematics of natural motions of heavy bodies near the earth's surface. The common ascription of errors and inconsistencies to Galileo's physics arose from viewing it as a kind of dynamics – either medieval, with impetus as an impressed force, or Newtonian, with inertia as an inhering force.[15] Galileo, like Aristotle, regarded force as contrary to nature by definition, depriving it of any proper place in physics as the science of nature. The "natural" motions to which Galileo's physics was restricted were those undertaken spontaneously by heavy bodies when simply freed from support or from the action of external force.

Galileo retained the medieval term "impetus," but early in the *Dialogue* he redefined it as speed acquired (25, 27), omitting the idea of "impressed force" for which medieval writers had introduced it. In the second day Simplicio brought up that idea, only to have Salviati promptly remove it again by showing that all that need be conserved in a thrown ball is the motion it had shared with the hand, and not also a force (149–51, 156). Since motion is measured by speed, Galileo's kinematics assumed not inertia but simply conservation of speed. Together with composition of motions, that served for his tide theory, in which he introduced no concept of force, but only one of absolute motions (427). What is now incongruously styled "circular inertia" in Galileo's physics is better called "conservation of geocentric speed." His purely kinematic derivation of it is found in the argument that rotation of the earth could not cast off a resting heavy body (197–200). The same proposition is found in Newton's physics, where it first became dynamical.

These few hints should enable anyone to understand the physics of the second day, applied in the fourth day to explanation of tides, abandoning common confusions entailed by attempts to link Galileo's physics with medieval or Newtonian dynamics. It contains no elementary blunders or logical contradictions of the kind alleged by critics who cling to the preconception that Galileo wrote the *Dialogue* as an overzealous Copernican. Only the charge of circular reasoning that Galileo deliberately had Simplicio bring, and that Salviati left unanswered, can legitimately be brought against Galileo's procedures in scientific argument – as it can against any science attempting to encompass nature in a symbolic framework.

Galileo was well aware that circular reasoning proves nothing beyond doubt, and many of his refutations of Aristotle in the *Dialogue* exposed examples of question-begging by the founder of logic himself (e.g., 35–36). Conclusions derived from an undeniable principle cannot be used to support it, as if it could somehow be made more undeniable. The case is different with conclusions derived from a hypothesis adopted to fit known phenomena of nature. The inherent uncertainty of any assumption is reduced when conclusions derived from it point to other phenomena that were unknown, or not considered, when it was adopted. Such phenomena lend support to the hypothesis through what Galileo called "demonstrative advance" in his first book on physics, published in 1612 just before the Copernican issue began to be fought out.[16] In the second day Galileo advanced physics by showing how the Copernican motions fitted natural motions of heavy bodies near the earth's surface, contrary to conclusions reached by Aristotle's principles. If relative motion, conservation of motion, and compositions of motions had been undeniable first principles, Simplicio's charge of circular reasoning would be applicable here also; but Galileo supported those by appeals to observation and demonstrative advance. Galileo did not regard them as undeniable first principles; that would be foolish because philosophers continued to deny them. Still less was his tide theory derived from undeniable first principles. Hence Simplicio's charge did not require a reply; all Galileo's science stood or fell together.

Galileo's explanation of tides was a conspicuous example of his view of science advanced by reasoning *ex suppositione*, as may be seen from Salviati's first speech in the fourth day:

Let us see, then, how nature has allowed – whether the facts are actually such, or whether at a whim and as if to play on our fancies – has allowed, I say, movements that have long been attributed to the earth for every reason except in explanation of the ocean tides to be found now to serve also that purpose, with like precision; and how, reciprocally, this ebb and flow itself cooperates in confirming the earth's mobility.

... We have already shown at length that all terrestrial events from which it is ordinarily held that the earth stands still and the sun and fixed stars are moving would necessarily appear to us just the same if the earth moved while those others stood still. (416)

Here we may properly interpolate "necessarily, that is, if the physics of the second day is correct and consistent." That is what Galileo had shown at length by probable reasoning, hypothetically, on the basis of repeated appeals to observation and experience, leading to demonstrative advance in Galileo's terminology. And indeed, even in his preface he had declared: "First, I shall try to show that all experiences practicable on earth are insufficient means for proving its mobility, since they are indifferently adaptable to an earth in motion or at rest. I hope in so doing to reveal many observations unknown to the ancients" (6).

It was to that hypothetical demonstration in the second day that Galileo referred at his trial when he claimed that in his book he had shown the contrary of the Copernican opinion and that the reasoning of Copernicus was inconclusive. That starting statement did not sway his judges, and it has led a prominent historian of science to say that Galileo lied egregiously at his trial.[17] Its meaning was that the entire *Dialogue* was intended to show that absolute proof of the Copernican system lay beyond the power of science, which could do no more than reason about it *ex suppositione*. Undeniable first principles from which Copernicanism could be derived simply did not exist. In Galileo's conception of science the uncertainty of the Copernican assumptions, and of his own assumptions in physics, would be indefinitely reduced with the passage of time, but that would still not be irrefutable proof of them, and the procedure was legitimate for Catholics. Riccardi understood and agreed.

Galileo's metaphysical statements in the first day were not intended as undeniable first principles. He presented them not as grounds from which to derive his physics in the second day but as higher speculations grounded on that physics. Like Aristotle, Galileo regarded metaphysics as something coming after physics. A letter of his at the end of 1629 shows that the metaphysics of the first day still remained to be added. Galileo's kinematics of heavy bodies, though very limited in scope, was correct; it does not contradict Newtonian kinematics. Neither did Galileo's metaphysical statements contradict Newtonian dynamics, as they may appear to do when read inattentively. One of them was that indefinite straight motion cannot exist in nature, which seems to contradict Newtonian inertia (18). But that was a dynamic concept, and Newton's dynamics also forbids indefinite straight motion, by the law of universal gravitation. Many critics forget to read the second part of Newton's inertial law, which explicitly excludes the action of impressed force, including gravitation. Galileo's kinematics of heavy bodies included the effect of gravitation, which he regarded as a natural motion and not a force. Alleged contradiction of Newton by Galileo's physics reduces to a difference of classification, not to mistaken assertions about phenomena capable of measurement.

Again, Galileo said in the first day that only circular motion could be truly uniform, introducing that idea in connection with "integral world bodies" such as planets (32). In Newtonian dynamics, and in fact, planets move nonuniformly and in ellipses. They could move uniformly only if the two foci of such ellipses should coincide, in which event the orbits would be circular. Galileo did not assert that planets moved uniformly; on the contrary, he remarked in the fourth day on the inadequacy of existing planetary theory, and for his explanation of spring and neap tides he required nonuniform motion of the earth around the sun (453–55). Galileo did not confuse ideal motions with actual ones in scientific reasoning *ex suppositione*; he merely found ideal motions easier to analyze with the mathematics at his disposal.

But let us get back to Galileo's explanation of tides, which I have said is a conspicuous example of his view of science as advanced by noncircular reasoning *ex suppositione*. Tides are very complex phenomena that cannot be explained by any single cause. Galileo recognized that fact and listed half a dozen causes that bear on observed tidal phenomena. Among these, he singled out one as the primary cause of tides, without which tides would not occur; this was the doubly circular motion of the earth around two different centers, one of rotation and the other of revolution (428). And in fact it is true that without both the Copernican motions, the observed tides would not exist, though scientific proof was not forthcoming until Pierre Simon de Laplace produced it a century after Newton. After showing *ex suppositione* that his primary cause would indeed create daily cyclical disturbances in very large seas extending east and west, Galileo went on: "Now, secondly, I shall resolve the question why, though there is in this primary principle no cause of moving the waters except from one twelve-hour period to another ... the period of flow commonly appears to be from one six-hour period to another. Such a determination [of periodicity], I say, can in no way come from the primary cause alone." (432)

Despite this explicit statement early in the fourth day, you will find in nearly every modern discussion of the topic the simple falsehood that Galileo's theory of tides implied a single high tide each day, which goes against observation and could therefore have been inspired only by overzealous Copernicanism. What might indeed be said with some truth is that Galileo's primary cause of tides implied one high and one low tide daily, but that single cause was far from constituting Galileo's theory of tides. And strictly speaking, no single daily change of tide was entailed by Galileo's physics as applied in the fourth day, even if we consider only his primary cause. Disturbances as such are not tides, and all that Galileo's primary cause explained was the necessity of continual cyclical disturbances in large seas. To transform those into tides, Galileo required other causes, of which he wrote: "These, although they do not operate to *move* the waters ... are nevertheless the principal factors in fixing the *duration* of reciprocations, and they operate so powerfully that the primary cause must bow to them" (432).

Some say that Galileo invoked "minor" causes to explain away the observed periods. Galileo himself clearly declared them not minor, but more powerful than the primary cause. First among them he placed length and depth of the given sea, things that cannot move the water but do in fact powerfully influence its periodicity (much more than does the double motion of the earth). Galileo explained this by showing that water in a container can be set rocking by moving the container unevenly, but that the period of its then flowing back and forth cannot be regulated by motions given to the container, any more than the period of a pendulum can be altered by moving the hand that holds the string. In Galileo's tide theory, observed tide periods are determined mainly by length and depth of sea basin, which differ from one sea to another and are clearly independent of the daily cycle of disturbances. Again it was Laplace who first offered proof that seas like ours, of very different sizes and with most dissimilar connections between them, would, in many thousands of years, come to have tide periods related to the common cycle of disturbances. Galileo's modern critics take this as obvious, but physicists in Laplace's time argued that it was a very dubious assumption, steady-state conditions being then difficult to prove.

Experience has shown me that most people still suppose tides to be evidence of bulges in seas drawn up by the gravitational force of moon and sun. They are then often unable to account for a nearly equal tide diametrically opposite to those bulges. Kepler proposed a lunar-bulge explanation without accounting for the second daily high tide. Newton explained that problem and calculated a bulge of eleven feet in certain conditions; actually it is never more than three feet, even in the largest ocean at the equator, and much less in most places. Modern tide theories do not depend on a Keplerian bulge; they are flow-theories rather than bulge-theories. Galileo's was necessarily a flow-theory because his physics excluded forces, and a bulge in water could be created only by some force.

Space does not permit my outlining the arguments used by Galileo, which Albert Einstein found fascinating. It is true that Einstein also said that Galileo would not have accepted them had he not been seeking new support for Copernicans (xvii). But Einstein wrote that years ago, when it was generally believed that Galileo's interest in the tides began not much before 1610, when he first mentioned his treatise on tides shortly after his first telescopic discoveries. In fact, Galileo's basic concept dated back to 1595 at the latest. This puts the whole matter in a new light, as I shall explain in conclusion.

Before Galileo moved from Pisa to Padua in 1592 he had never seen appreciable tides. He was then not yet a Copernican, for his latest Pisan writings on motion treated the earth as the center of the universe and daily revolution of fixed stars as a real motion. In 1597 he told Kepler he had preferred the new system for several years (though it could not have been more than four years). His first preference for Copernicanism probably arose from his physical explanation of tides – exactly the reverse of the ordinary view that he hit on an inadequate and incorrect tide theory in search of new evidence for Copernican astronomy.

When Galileo moved to Padua, and for several years thereafter, his interests centered on motion and mechanics. His first known astronomical observation was made in 1604, up to which time his surviving papers show no astronomical calculations except for horoscopes, using Ptolemaic tables. The lectures on cosmography he gave until at least 1606 dismissed motion of the earth as needless to discuss, though accepted by some great mathematicians. There is no evidence that Galileo paid much attention to astronomy as such during his first years at Padua.

According to Galileo, huge volumes of water rise four or five feet at Venice twice a day, though tides are virtually absent along the Adriatic coast below Venice. This anomaly could not fail to puzzle a scientist concerned with mechanics. No moving bulge of water could explain the facts observed by Galileo because the Adriatic Sea rose only at its closed end. A flow theory best explained the observed facts, and Galileo's source for it is suggested early in the fourth day:

These effects can be very clearly explained and made evident to the senses by means of barges which are continually arriving from Fusina filled with water for the use of Venice. Let us imagine to ourselves such a barge coming along the lagoon with moderate speed, carrying placidly the water with which it is filled, when, either by scraping bottom or encountering some obstacle, it becomes greatly retarded. The water will not thereby lose its previously acquired impetus equally with the barge; keeping that impetus it will run forward toward the prow, where it will rise perceptibly, sinking at the stern. ... The parts in the middle rise and sink imperceptibly ... [doing so] the less according as they are nearer the center, and the more as they are farther from it. [424–25]

Actual observation probably suggested to Galileo an analogy here with the sea beneath the barge, and then all he needed for a physical explanation of the occurrence of tides was some reason for which any sea basin must move at varying speeds. Doubly circular motion of the earth supplied a reason, and over the next thirty years Galileo developed the first rational account of tides, including a cause for spring and neap tides associated with phases of the moon as well as explanation of two high tides daily in the Mediterranean. For these he added considerations other than the Copernican motions, creating perhaps the first multicause physical theory after the medieval impetus explanation of accelerated fall of bodies. Most scientific theories, like both of those, began as incorrect and inadequate to explain observed facts, but that does not reduce their interest to historians of later physics, once the sources of error are correctly perceived.

In the notebooks of Galileo's Venetian friend Fra Paolo Sarpi for 1595 there is a succinct summary of basic ideas behind Galileo's primary cause of tides. Sarpi himself accepted but a single motion of the earth and had a tide theory of his own, according to his biographer Fra Fulgenzio Micanzio (also a friend of Galileo from early days). It is therefore probable that what Sarpi recorded in 1595 came from Galileo's conversation during a visit to Venice, to which he had traveled on a barge carrying fresh water from the mainland.

Galileo wrote to Kepler in 1597 saying that he preferred the Copernican system because he had explained certain natural phenomena by it that could not be explained under the old assumptions. Kepler correctly surmised that Galileo must refer to tides, and he wrote a friend to that effect, but Kepler did not believe it possible to explain tides by motion of the earth, so he did not ask Galileo for further details. Instead he asked him to make astronomical observations that might reveal stellar parallax. Galileo appears to have believed that idea as unlikely to succeed as Kepler believed any attempt to explain tides by motion of the earth must be. At any rate, the correspondence lapsed, increasing the probability that Galileo first preferred the Copernican system not for astronomical reasons but through his interest in a problem of terrestrial physics.

One conclusion that, from my reexamination of the *Dialogue*, appears to me inescapable is that the fourth day cannot be properly regarded as detached from the others. That treatment has long been customary among scholars who seek the clue to Galileo's conception of science in the metaphysical parts of the first day, which Galileo himself explicitly detached from the rest. It is not at all surprising that he chose explanation of tides as the subject of his *Dialogue* in 1624, being in need of some topic other than planetary astronomy for a book that would be useful to the Church in correcting widespread misunderstanding of the scope and purpose of the 1616 edict. The seemingly rambling and poorly organized *Dialogue* as printed in 1632 is accordingly seen by me in a very different light – not as polemically astronomical but as introducing a new physics capable of enlarging man's understanding of his universe, always *ex suppositione* and without philosophical finality.

In presenting this interpretation, I have had occasion to quote the openings of the preface, the first day, the second, and the fourth. I shall close with Salviati's first speech in the third day, in which reasoning for and against motion of the earth around the sun was to be discussed: "Over the long run my observations have convinced me that some men, reasoning preposterously, first fix some conclusion in their minds which, either because it is their own or because of their having received it from some person who has their entire confidence, impresses them so deeply that one finds it impossible ever to get it out of their heads" (276–77).

Galileo here used the word "preposterously" in its literal sense; that is, of putting the cart before the horse. To understand nature, men's minds must be free, as they never are with any conclusion fixed in advance. In that way people merely collect evidence in support of some favorite idea, which may be wrong despite a wealth of evidence in its support. That is what was done for many centuries with the idea of a fixed earth. I believe it is being done now by scholars with preconceptions about Galileo's overzealous Copernicanism, lack of scientific judgment, and insincerity in religion. Scholars in his time resented Galileo's saying they reasoned preposterously, and they did him in. I think it sad that the Inquisitors favored scholars rather than men of knowledge and good sense – for over the long run my observations have convinced me that the two are not necessarily the same. When there is a conflict, I now usually banish inveterate scholarly conclusions from my mind, confident that in the end good sense will prevail in the interpretation of Galileo.

NOTES

Because citations of the *Dialogue* are frequent, they are indicated in the text by page numbers in my translation, Galileo, *Dialogue Concerning the Two Chief World Systems* (Berkeley and Los Angeles: University of California Press, 1953). The one reference to *Opere* is to the National Edition of Galileo's works, Antonio Favaro, ed., *Le Opere di Galileo Galilei*, 20 vols. in 21 (Florence: G. Barbèra, 1890–1909; repr. 1968).

1 Galileo Galilei, *Dialog über die beiden hauptsächlichsten Welt-systeme* (Leipzig: Teubner, 1891; repr. 1982).
2 See especially E.A. Burtt, *The Metaphysical Foundations of Modern Physical Science* (London: Routledge and Paul, 1932), and Alexandre Koyré, *Études Galiléennes* (Paris: Hermann, 1939), trans. J. Mepham, *Galileo Studies* (Atlantic Highlands, N.J.: Humanities, 1978).
3 J.H. Randall, Jr., *The School of Padua and the Emergence of Modern Science* (Padua: Editrice Antenore, 1961).
4 W.A. Wallace, *Galileo's Early Notebooks: The Physical Questions, A Translation from the Latin, with Historical and Paleographical Commentary* (Notre Dame: University of Notre Dame Press, 1977).
5 *Historic Doubts Relative to Napoleon Buonaparte*, published anonymously in 1819 and frequently reprinted with the author's name.
6 This title page is shown in facsimile but not translated in my English-language edition. Literally it reads:

Dialogue

of

Galileo Galilei, Lincean

Special mathematician
of the University of Pisa
And Philosopher and Chief Mathematician
of the Most Serene
Grand Duke of Tuscany.
Where, in the meetings of four days, there is discussion
concerning the two
Chief Systems of the World,
Ptolemaic and Copernican,
Propounding inconclusively the philosophical and physical reasons
as much for the one side as for the other,
with copyright
In Florence, by G. B. Landini, 1632.
Licensed by the authorities.

7 There are half a dozen such references in Galileo's letters to friends during the years named.

8 "Probable reasoning" was a technical term excluding absolute demonstration. No philosopher of that period would confuse the two: note the remark made below about absolute truth as distinguished from hypothetical.

9 Probably Urban did not want anyone to think that in licensing hypothetical use of motions of the earth, the Church in any way endorsed a tide theory depending on such motions.

10 It was printed in italic type, unlike the text, and that became one of the complaints against the book later. Critics charged that this dissociated the preface from the body of the book, weakening its effect.

11 Once the pope had forbidden tides as the subject, they could no longer be mentioned first in the preface. Some suppose that Riccardi rewrote Galileo's preface, but it contains too much personal information and stylistic integrity for that to be true.

12 Galileo had used similar arguments before 1616. Once the Church acted officially, he abandoned them, having said throughout that he would abide by any official ruling. Before 1616 he cautioned against any action as premature. After action was taken, Galileo did not criticize it, even in private letters.

13 The *Starry Messenger* of 1610 contained the promise, repeated in *Bodies in Water* of 1612. The matter is discussed in my *Telescopes, Tides and Tactics* (Chicago: University of Chicago Press, 1983).

14 Even before Ptolemy, this working agreement between astronomers and philosophers had been necessary. Astronomers could not invade physics, but they were free to assume any fictions that assisted successful prediction of planetary positions. Robert Cardinal Bellarmine had advised Galileo to avoid asking more, and he made sure that the Church did not grant astronomers less.

15 Newton wrote of *vis inertiae*, but we no longer think of inertia as a force because no acceleration occurs in inertial motions.

16 Concerning "demonstrative advance" see my *Cause, Experiment, and Science* (Chicago: University of Chicago Press, 1981), pp. 26, 128, 132.

17 R.S. Westfall, "Deification and Disillusionment," *Isis* 70 (1979): 274.

4

Galileo Gleanings – XVI
Semicircular Fall in the *Dialogue*

I

In the Second Day of the 1632 *Dialogo dei sistemi*,[1] Galileo introduced a speculation concerning the path of a body falling from the top of a tower to the surface of the earth, as seen by an imaginary stationary observer removed from the earth. The path, he said, would be approximately along a semicircle whose diameter was the line from the top of the tower to the center of the earth.

Labeled a digression at the beginning, the speculation is called a *bizzarria* at its end. Nevertheless, it was taken quite seriously by Marin Mersenne, who devoted a long section of his *Harmonie Universelle* to a criticism of it. Pierre Fermat offered a correction, which was sent to Galileo by Pierre Carcavy. Galileo acknowledged to him the merely approximate nature of his analysis and its obvious absurdity when applied to regions near the earth center, pointing out once more (as he had in the text) that it was not intended in all seriousness, and saying that it was rather a somewhat audacious caprice. Yet modern critics have been quite severe with Galileo, not for his having introduced a *jeu d'esprit* into a published book, but for his having tried to conceal his innermost convictions in such a guise; and even worse, for his having tried to worm out of a detected error by pretending that he did not really mean it.

If that were all, it would scarcely be worthwhile to reopen the matter merely for the purpose of vindicating Galileo's intelligence and his good faith. Other circumstances, however, render the passage in question of sufficient importance to Galilean scholarship to merit its complete re-examination. Chief among these is the fact that the passage in question constitutes the only one in all Galileo's extant writings to present us with a specific application of the concept of "circu-

Reprinted from *Physis* 10 (1968), 89–100, by permission.

lar inertia" as opposed to general suggestions of that concept. Moreover, it is now widely believed that this passage constitutes Galileo's first analysis of the trajectory of a projectile – a belief which I once accepted, and perhaps helped to advance, but which I now believe to be quite mistaken. Viewed in that way, the passage seems to bear on the chronology of Galileo's knowledge of the parabolic character of trajectories.

Both these matters are of central importance to students of Galileo's physics. But even that is not all. A re-examination of the passage in question will show on the one hand how carefully we must examine the precise words of an author if we wish to grasp his own thought, and on the other hand how far from a correct understanding we may wander when we go drawing one inference from another without returning constantly to the underlying basis of inference, whether or not it seems to have been settled once and for all.

The passage is one that has occupied my attention for many years, because in addition to the matters mentioned above, and others that have occupied its many critics, there is one that was called to my attention by Professor Einstein long ago. Why does Galileo in this passage make the falling body stop at the center of the earth, when it would be going most swiftly? I do not know that anyone has attempted to answer that question. But whether Galileo believed inertial motions to be circular or rectilinear in character, it seems he ought not to have overlooked their implications here, of all places. In two other passages of the same book, he discusses the continuance of motion of a body dropped through a tunneled earth, past the center to the opposite surface.[2]

I regret that a solution of all the difficulties of this passage had still not occurred to me until after the printing of a second revised edition of my translation of the *Dialogue*, though I reconsidered them again in revising my notes. Still more, I regret the survival in my revised text of the designation CIA for the correct CI of Galileo's text. The relevance of that seemingly trivial difference will shortly become apparent.

II

Let us recall first how the passage in question was introduced into the *Dialogue*. Salviati had begun to speak of the fact of accelerated motion. Sagredo, over Simplicio's protest, requested a discussion of that matter. Salviati replied that: "It would be too great a digression for us to interrupt with this our present discussion, which for that matter is a digression itself; it would make, so to speak, a play within a play." Thus the fact, but not the law, of acceleration is supposed to be known to the reader. The question discussed is: "what one may believe with regard to the line which is described by a heavy body falling naturally from the top of a tower to its base" on a rotating earth. A straight line is seen by a person sharing that rotation: what would be seen by one who could see the earth revolve and the body fall? Salviati attempts to combine the "straight movement toward the center of the earth ... and the circular motion toward the east," assuming the latter to be uniform, and the former to be accelerated. The actual law of acceleration is given only some sixty pages farther on.[3]

Now, it seems to me fairly evident that at this point of the *Dialogue*, Galileo had not the slightest intention of discussing the "trajectory of a projectile" in the ordinary sense of that term. He was concerned geometrically with the apparent path of a simply falling body, moving at every instant straight along a line directed toward the earth's center, while that line moved in uniform angular rotation toward the east. If we wish to say that this is the same thing as tracing the trajectory of a projectile fired on a stationary earth, we may do so; but Galileo certainly did not say this. Even though Galileo himself had established a principle of relativity of motion in the *Dialogue* that would imply such an analogy, we should not assume that he saw that implication in the discussion of absolute path here. Galileo does not say: "If the earth and tower were to stop suddenly, the stone would take such-and-such a path." He says that on a rotating earth the straight central motion of the falling body is accelerated, according to a law not yet set forth, and that its circular motion toward the east is uniform. These statements are made as assumptions, not as observed or demonstrable facts of the physical world.

Proceeding on the above assumptions, Galileo observes that the successive downward departures of the body from the top of the tower will be ever-increasing per unit of time, and that at the beginning they will be extremely small. Next he remarks that those conditions could be obtained by having the body move along a semicircle whose diameter is bounded by the top of the tower and the center of the earth. The argument is plausible, not rigorous; Galileo claims no more than that this curve will meet the conditions. Knowing perfectly well that the argument is not rigorous, he has Salviati say only: "I think it is very probable that a stone dropped from the top of the tower will move" along this semicircle. And concluding the entire discussion, he says: "But that the descent of heavy bodies does take place in exactly this way, I will not at present declare; I shall only say that if the line described by a falling body is not exactly this, it is very near to it."

Now, it was long ago pointed out by Mersenne that the versed-sine law of fall which would be followed by the stone under the conditions described by Galileo could not be distinguished by actual observation from the time-squared law of acceleration.[4] That is, for the few seconds during which free fall can actually be observed from any accessible experimental height, the versed-sine law is inappreciably different from $S = at^2$. We must not lose sight of that fact, or look upon the semicircular path as if it were easily contradicted by observation, or somehow absurd. So far as the points marked out in equal times along a high tower are concerned, the semicircular fall hypothesis of Galileo would, as he said, agree very nearly with the observed acceleration of free fall.

But Mersenne was struck by an absurdity which does seem to be implied, if Galileo's argument relates to the body's total fall to the center of the earth, and not just its fall from the top of the tower to the surface of the earth. If the earth revolves in 24 hours, the total fall to the earth's center, Mersenne pointed out, would occupy six hours exactly. But Galileo himself believed, as shown elsewhere in the same book, that a body would fall from the moon's orbit to the earth in less than four hours.[5] How, then, could it take the stone six hours to reach the earth's center from a point very close to its surface?

Unfortunately, this question was not specifically submitted to Galileo when his speculation was challenged in favor of a spiral path suggested by Pierre Fermat and forwarded to him by Pierre Carcavy. Had the six-hour objection been raised in that letter, Galileo's answer would have given us the key to his whole discussion of the absolute path of a body falling on a rotating earth.

But perhaps a painstaking inspection of his diagram, and of the text accompanying it, will throw sufficient new light on the problem to modify very considerably the received interpretation of it.

III

The problem which led to the digression was first stated by Sagredo thus: "What may one believe with regard to the line which is described by a heavy body falling naturally from the top of a tower to its base?" After a long preliminary discussion, he repeated the problem: "Meanwhile let us get back to the line described by the body falling from the top of a tower to its base." The problem is thus quite specific; it relates only to the actually observable part of the fall, stopping at the surface of the earth.

Salviati, in order to solve the problem, makes the two assumptions previously mentioned, and continues: "It being further true that the descending weight tends to end at the center of the earth, then the line of its compound motion must be such as to travel away from the top of the tower at an ever-increasing rate. ... This line must tend to terminate at the center of the earth." Note that he speaks here not of the body, but of the line of its motion. These considerations give him a basis for drawing such a line, which is the semicircle drawn from the top of the tower to the center of the earth, thus:

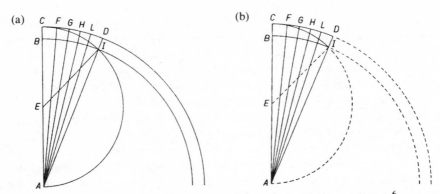

(a) Diagram shown in Galileo's *Dialogue*. The center of the earth is at *A*.[6]
(b) Diagram redrawn to distinguish construction lines from lines referred to in Galileo's demonstration.

The diagram has tempted everyone to assume that Galileo's discussion is intended to concern the entire path, through the solid earth, as if Galileo had assumed a tunnel from the base of the tower to the center of the earth. But nothing is said in the text of any tunnel. On the contrary, if we inspect the diagram, we see that the radii drawn to successive positions of the tower terminate precisely where the falling body reaches the earth's surface. That was the original problem, and the text goes on to say quite specifically that the stone will be situated at the intersections of the assumed semicircle with the extended radii, always above the surface of the earth. Nothing whatever is said about continued motion to the center of the earth. All that is said is that the motion would always be toward that center: "Finally, one may understand how such motion tends eventually to terminate at the center of the earth." Every specific reference in Galileo's text to the falling body deals with the arc CI only; that is, with the part of the CIA that lies above the earth's surface.[7] That is the only part about which Sagredo's original inquiry was concerned.

Now, if (as I believe) Galileo never intended to discuss any motion of the stone beyond that from the top of the tower to its foot, the puzzles that have plagued his critics simply vanish. The six-hour implication does not hold for a real stone that never reaches the center of the earth, but stops at the foot of the tower.[8] The apparent contradiction with other passages in the *Dialogue*, in which Galileo positively asserts that a stone falling through a tunneled earth would *not* stop at the center, vanishes likewise. The arc IA of the semicircle CIA is a construction line only, one which today we would draw as a dotted line; it shows how the arc is derived, but not how the body travels. Only the arc

CI represents actual motion; the body falls from the top of the tower to its base, and no further. Sagredo's question is answered. The motion of the stone along CI is essentially circular, Galileo says, just as it was when at rest on the top of the tower. And indeed, an imaginary observer on the moon would have some difficulty in distinguishing the absolute path of the stone from the circular path of any point on the tower.

In the *Assayer*, discussing a quite different matter, Galileo says: "Whether in fact the curvature has the shape of a circular arc, or an ellipse, or a parabolic line, or a spiral, or something else, I believe could not be determined in any way; as in an arc of at most two or three degrees the differences of such curvatures would be entirely imperceptible."[9]

For the falling body, assuming a tower 400 feet high, which would require five seconds for an object in free fall to reach its base, we would be talking about $1/48$ of $1°$ in the absolute path of which Galileo was speaking (the semicircle of fall). That is as much of the path as would be seen during the fall by the hypothetical observer removed from the earth and not sharing its motion. Such an observer could not tell that circular path from a parabola or from Fermat's spiral. An observer on earth, of course, would see the path of fall simply as a straight vertical line. It is a confusion of Galileo's purpose to speak of the parabolic trajectory at all in this connection. Those who do so should think, not of the ordinary parabolas we see in the fall of a gently propelled projectile, but of the parabola of a projectile moving uniformly about 1,000 m.p.h. horizontally while falling vertically 400 feet in accelerated motion. This path would be difficult to distinguish from the circle suggested by Galileo.[10]

IV

The semicircular path seen by a stationary outside observer, for a body which we see to fall plumb from a tower, has profound implications in any reconstruction of Galileo's own physical and cosmological thought. Even though it was presented as a *bizzarria*, and was later dismissed as a *scherzo* by Galileo, he did include it in a published book and he did not retreat from it as an approximation to truth when it was challenged in favor of Fermat's spiral path. Scholars have accordingly drawn extensive conclusions from it. Those conclusions in turn have been fitted into a general pattern that is now widely accepted concerning Galileo's fundamental conceptions.

It is my considered opinion that the accepted interpretation of the *bizzarria* is mistaken. A reassessment of it yields rather different implications. If these do not fit as easily into the received general pattern for Galileo's thought, it may mean that they are wrong; on the other hand, it may mean that the general pattern is defective. We are conscious of the grave errors that were bequeathed to the history of science by nineteenth-century students of Galileo, and of the great labors that were required to correct them and to revise the general pattern that had emerged from them. So much the more should we be conscious of the danger that we in turn may bequeath to our successors a pattern of Galileo's thought in which we place excessive confidence, putting them eventually to equal trouble in revising it. Our only danger lies in refusal to re-examine constantly some passages of fundamental importance, of which the *bizzarria* is one.

The excessive confidence of our nineteenth-century predecessors arose from their seeing, rightly enough, that the raw data could be coherently fitted together in a certain pattern, from which Galileo emerged as a completely original thinker who shook the foundations of philosophical speculation and introduced the methods of modern experimental science. What they failed to see was that the same data could be coherently fitted together in other ways also; hence their confident agreement in that erroneous conception. Without doing more violence to the raw data than they had done, it is possible to give an equally coherent picture of Galileo as a Platonic speculator, both inspired by and confined to the doctrine of the perfection of the circle. Fashionable as that view may be, it will in time be regarded as no less misleading (and no less plausible) than the view it superseded. To avoid this, it is important for us to see that the mutual reinforcement of interpretations of key passages in Galileo's works is always partly the work of the interpreter. Difficult as it may be to re-examine a single passage in the face of a prevailing pattern into which the received view of it fits, one must always keep in mind the possibility that the pattern itself, seemingly drawn from a safely large number of passages, has in fact colored our interpretation of each of them.

In the present instance, there has been unanimous agreement from Galileo's time to our own on the sense and intention of the *bizzarria*. That being the case, no competing view can expect to gain serious attention merely by equal plausibility. It is necessary also to account reasonably for the pre-existing unanimity. This I shall now attempt to do.

The passage in question is one of relatively few in the *Dialogue* to be accompanied by a diagram. The diagram is simple and striking; a glance at it practically forces upon the reader the concept of a heavy body descending from the top of a tower to the center of the earth. That is how Mersenne and Fermat looked at it; in that way they interpreted the text, and on that assumption they proposed in place of the semi-circle CIA a certain kind of spiral line. The question they attacked was this: "What would be the path of a body falling from a tower on a rotating transparent earth, tunneled to the center from the base of the tower, as seen by a stationary observer sufficiently distant to observe the entire descent?" To this question they applied the law of acceleration and derived a kind of spiral.

In the *Dialogue*, however, the question introduced was "what one may believe with regard to the line which is described by a heavy body falling from the top of a tower to its base." In reply, Salviati introduced the conception of acceleration, but postponed discussion of the specific law relating velocity to time on the grounds that here it would be a digression within a digression. Accepting the postponement, Sagredo then said: "In the meantime let us get back to the line described by the body falling from the top of the tower to its base." The words, "to its base," were specifically included in both statements of the question to be discussed, and at no point was any suggestion made that some further generalization of the problem be undertaken. In the *bizzarria* no phrase occurs such as "and it would continue, if free to do so, along this line." Galileo added three "little reflections" of a very general character, which in all probability constituted the real purpose of his inserting the whole passage in the text; but he did not add any generalization of the sort assumed by Mersenne and Fermat, and by all modern critics.

The suggestion of generalization to total fall beyond the surface of the earth exists strongly in the diagram. Nevertheless, if one reads the text with care, keeping in mind the specific question (line from top of tower to its base), a striking fact emerges. First, the semicircle CIA is constructed. Then the successive positions of the tower are marked off and connected to the center of the earth. Finally, the positions of the body are identified with the intersections, which stop at I. Looking once more at the diagram, this interruption of the path CIA becomes conspicuous. It is only the arc CI that Sagredo has asked about, and only that short arc of the semicircle is ever discussed. Thee is no "little reflection" about the body after it has reached I. The fall is over at that point, and so is Galileo's discussion.

But the text contains three phrases which could easily mislead any reader who had formed a contrary idea; one who, from seeing the diagram, was misled into thinking that total fall was implied. In setting forth his assumptions, Galileo supposes that "the descending weight tends to end at the center"; "the line of compound motion must tend to terminate at the center of the earth"; and in conclusion he remarks that "such motion tends eventually to terminate at the center of the earth." His words, *va per terminare, vadia a terminar*, and *andrebbe a terminar*, are by no means simple indicatives. Galileo does not say that the body goes to the center, or ends at the center, for it goes only to the base of the tower and stops there. But the body, the path, and the motion must tend towards the center and not just towards the ground. That specification by Galileo was necessary for the selection of any definite path of absolute motion.

Thus while the diagram and text are consistent with the answer to Sagredo's specific question, they also contain elements that easily and understandably suggested Mersenne's and Fermat's generalization. Once that generalization was made by them on Galileo's behalf, it is not surprising that it was never questioned by later critics – especially modern critics, who are more concerned with reconstructing Galileo's overall conceptions than with reviewing the basis of past critiques.

Now, since Galileo was not concerned in his *bizzarria* with the hypothetical behavior of the body in a tunneled earth, and was through with it when it hit the ground, he did not consider the possibility of its continued uniform circular motion along CIA, something that was physically impossible. Mersenne did, and saw that the stone would in that case take six hours to reach the center; moreover, the height of the tower would make no difference in that six-hour period. Either of those elementary absurdities should have brought Galileo up short. It has become customary to suppose either that he did not perceive them, or that he preferred to ignore them in order to save his circular fancy for all natural motions.

When suppositions of that sort seem necessary, they ought to warn us that we may be off the right track. The simplest supposition is that the absurdities seen by Mersenne were irrelevant to Galileo's problem. To suppose that Galileo did entertain in his own mind the hypothesis of total fall to the center at the time when he wrote the *bizzarria*, but that he failed to perceive its absurdities, is a strain on our credulity. A man capable of such elementary oversights as to the elementary implications of a simple proposition would have made frequent blunders in logic.

To suppose, on the other hand, that Galileo wanted to present a theory of fall to the center, and simultaneously tried to conceal from his readers a fatal flaw in it, is at best gratuitous. Such suppositions about Galileo's motives and procedures are not uncommon today, but a simpler and more creditable view can be found in every instance that I know of. In this case, the simplest view is that Galileo proposed and answered Sagredo's question, limiting it to observable fall to the surface of the earth, without considering the extension of that motion that has since been attributed to him. Galileo's real purpose in introducing the *bizzarria*, in my opinion, was to set forth the three "little reflections" at its end. These state that the natural motion of the body is the same, as to shape of path and total distance traversed, in either of two natural motions – that of diurnal rotation at rest on top of the tower, or that of free fall to its base – and that the acceleration we observe is only the projection of a uniform circular motion against a straight line. The first of these served his main purpose in the *Dialogue*, that of destroying Aristotle's exclusive attribution of circular motions to celestial and straight motions to terrestrial bodies. The last fitted in with Galileo's own determination to discourage any search for "causes" of acceleration, a position reiterated in the *Two New Sciences*.

Now if it is true that Galileo introduced his *bizzarria* not as a general theory of falling bodies but as a disgressionary pretext for his three "little reflections," then this passage must appear in a new light with regard to the much-vexed question of "circular inertia." I shall deal with that topic in another paper of this series; for the present it may merely be observed that the *bizzarria* contains the only unambiguous application of circular inertia to be found in all Galileo's writings. And here the concept makes its appearance not as a physical law, but as a geometrical implication of the observed fact that a falling stone grazes the side of a high tower from top to bottom, the tower being assumed to move eastward in uniform rotation. The postulated path is part of a purely geometrical exercise, and Galileo concludes by saying that he does not assert actual fall to take place exactly in the way assumed.

What Galileo had in mind may be gleaned from his reply to Pierre Carcavy in 1637, when Fermat's spiral derivation was sent to him. After remarking that his own *bizzarria* was not meant to be taken as anything but a somewhat audacious joke, he admits that the application of the law of uniform acceleration would indeed give Fermat's spiral for fall to the center. But "so long as we stay above the surface of the earthly globe, I do not hesitate to ascribe to that composition (of motions) a parabolic line, asserting such to be the lines described by projectiles. This assertion of mine might afford a much better basis for impugning me than the semicircular motion, which I did at least make to end at the center, where I am sure projectiles would go to end, whereas the parabolic line would go always departing from the axis ... ."[11] The letter goes on to speak of the assumption of Archimedes that lines of descent are parallel, and generally to distinguish mathematical from physical requirements.

Those who read into the *bizzarria* an attempted and mistaken analysis of trajectories (as I formerly did) are tempted to suppose that when Galileo wrote the *Dialogue*, he had not yet discovered their parabolic nature. Apart from abundant evidence to the contrary provided by Cavalieri's letter to Galileo of 21 September 1632,[12] Galileo's letter to Carcavy suggests that it was precisely Galileo's knowledge of parabolic trajectories at the earth's surface that induced him to emphasize, at the end of the *bizzarria*, that he did not assert actual fall to be precisely semicircular in character. Had Galileo believed literally in the privileged character of circular motions, no reason appears why he should have considered this a mere approximation.

NOTES

1 Original edition, pp. 158–160; Ed. Naz., VII. 190–193; English translation (Berkeley 1953, 1967) 164–167.

2 Ed. Naz. VII, 47, 161–162, 253, 262; Engl. trans., 22–23, 135–136, 227, 236.

3 Ed. Naz. VII, 253–256; Engl. trans., 227–229.

4 *Harmonie universelle* (Paris, 1637), Livre II, *Du mouvement des toutes sortes de corps*, pp. 93–96: Nous avions déja consideré cette meme ligne [i.e., a circular path] avant que d'avoir vû ses dialogues; mais puis qu'il met la cheute des poids en raison doublee des temps, comme nous avons fait, à laquelle la raison des sinus verses des arcs égaux est quasi semblable, principalement au commencement de la cheute lors qu'ils sont petits ... And : Le rayon estant 10.000, le sinus verse de 15, [of arc] qui se font dans 1′ [of time] par le mouvement journalier, est un, celuy de 30′ est 4. et celuy de 45′ est 9. Or nous ne sçaurions icy observer de cheute dans 1′ d'heure, et encore moins en 2! car en 1′ le poids seroi 7,200 toises, et 28,800 toises en 2′, et neantmoins la raison des espaces est exactement double de celle des temps en cette distance, encore que ce soit la raison des sinus aux arcs ... Finally: Or il est aisé à conclure de tout ce discours, que Galilee s'est contenté d'avoir une proportion de cheute qui luy sembloit s'accorder avec les apparences, et que pensant davantage aux belles corre-spondences et consequences qu'il en tiroit [i.e., the "three little reflections"], il n'a pas approfondi cette matiere, attendu qu'il n'est pas croyables qu'un tel homme se fust tellement mespris, s'il eust examiné de plus pres la cheute des poids, suivant les experiences qu'il a fait luy-meme. (The final remark alludes to Galileo's alleged belief that a cannon-ball falls 100 *braccia* in 5 seconds.)

5 Ed. Naz., VII, 249–252; Engl. trans., 223–226.

6 In the original edition of 1632, this diagram was printed in an inverted position. The line EI was not shown; E is midway between the center of the earth (A) and the top of the tower (C).

7 In the text of my English translation, the designation CIA was erroneously put in place of Galileo's CI, eight lines from the bottom of p. 165.

8 That implication would hold only if it were asserted that uniform circular motion would continue beneath the earth's surface. Galileo asserted that uniformity only for the arc CI. We know that a different gravitational law holds below the surface. Galileo did not know this, but he may well have suspected it; at any rate, he made no conjec-ture concerning the speed along IA, or even the possibility of continued motion there.

9 Ed. Naz., VI, p. 244.

10 The absolute path would closely approximate a circle about 50 kilometers in diame-ter; cf. Paul Mansion, *Mélanges Mathematiques* (Paris, 1898), p. 111.

11 Ed. Naz., XVII, 89–90; cf. VIII, 273–275 (*Discorsi*).

12 Ed. Naz., XIV, 394–395.

5

Galileo Gleanings – XVII
The Question of Circular Inertia

In writing notes to my translation of Galileo's *Dialogue* fifteen years ago, I accepted as substantially correct the idea that Galileo had believed inertial motions to include some essentially circular characteristic. That idea has been current among scholars for several decades, particularly since its development in 1939, in a general and philosophically interesting way, by Alexandre Koyré in his *Études Galiléennes*. Koyré's general thesis has now been widely adopted as a basis for further interpretations of Galileo's physics and his place in the overall history of science.

Over the past several years, however, I have become increasingly doubtful about the validity of the general thesis. In the course of revising the notes to successive printings of the *Dialogue*, I reconsidered the passages relevant to inertial motion and modified my support of the prevailing view. Ultimately only a single passage remained in which Galileo still seemed to me to have unequivocally employed the concept of "circular inertia" in a mathematical demonstration. Despite the fact that he had styled that passage a *bizzarria*, its presence deterred me from going against the considered opinion of the best Galilean scholars, even after I had put forth what I believe to be a new and correct account of the origin of the inertial concept in Galileo's own mind.[1]

Now, however, the last seemingly conclusively argument in favor of the prevailing view – the *bizzarria* of the *Dialogue* – has lost its force in my eyes. In the previous paper of this series, I set forth my reasons for believing that no inertial considerations are involved in that mathematical exercise. It therefore seemed appropriate to suggest that the entire received interpretation of "circular

Reprinted from *Physis* 10 (1968), 282–98, by permission.

inertia" in the work of Galileo be thoroughly re-examined.[2] The topic is fundamental not only to the understanding of Galileo and his outlook, but to the history of a concept that distinguishes medieval from modern physics.

We are all conscious of the enormous amount of work that was required to
overthrow the mythology in the history of physics that had arisen before 1900
as a result of the general but mistaken thesis that medieval science was an arid
desert. If we are to avoid the construction of a new mythology that will occasion an equal amount of work for our successors, we must be willing to re-
examine from time to time even the most widely accepted and apparently most
solidly established thesis. The underlying problem, as I see it, always arises
from an unnecessary positiveness in the presentation of what are at best tenable
hypotheses, so that they are made to appear as demonstrated conclusions. The
third and last of Koyré's *Études*, entitled *Galilée et la loi d'inertie*, illustrates
that point. I shall therefore begin by commenting on its structure and on the factual content of its point of departure.

The first 44 pages of Koyré's monograph are devoted to a review of the physical problems introduced by the Copernican theory, proposed solutions to them
offered by Copernicus, Bruno, Brahe and Kepler, and the arguments that
were brought against these. As an approach to Galileo's *Dialogue* and its anti-
Aristotelian polemic, Koyré devotes four or five pages to a statement of the thesis in the light of which he will examine the *Dialogue*. There follow some
eighty pages of interpretation and comment on selected passages of that book,
by which the said thesis is ably supported. The balance of the monograph,
except for two paradoxes presented immediately after the described above sections, will not concern us here.

Now, considering the structure of this work, it is important to be clear about
the basis on which Koyré's thesis is put forward, because the appearance of
invincible solidity given by the whole argument is largely achieved by the interpretation of suitably chosen passages, selected with the thesis already in mind.
Let us therefore examine the nature of the evidence offered by Koyré in support
of his general position. He has summarized Galileo's earliest work, *De Motu*, in
these words:

Le problème central qui préoccupe Galilée à Pisa est celui de la persistance du mouvement. Or, il est clair que lorsqu'il étudie le cas du monde, ainsi que celui d'une sphère
placée au centre du monde, ainsi que celui d'une sphère placée en dehors de ce centre, il
a en vue la situation créée par la doctrine copernicienne; la sphère marmoréenne dont il
analyse les mouvements représente, sans nul doute, la terre; et ses mouvements sont ceux
de la terre.

Mais le résultat auquel il aboutit – en contradiction, d'ailleurs, avec les prémisses essentielles de la physique de l'impetus – nous révèle d'une manière éclatante les diffi-cultés, et la source des difficultés, que rencontraient sur leur chemin la physique et l'astronomie nouvelles.

En effet, le résultat auquel aboutit l'analyse galiléenne, c'est la persistance naturelle, ou, plus exactement, la situation privilégiée du mouvement *circulaire*.[3]

"The central problem with which Galileo was preoccupied at Pisa is that of the persistence of movement."

We can judge Galileo's interests at Pisa objectively only by his own writing while there – the theorems on centers of gravity, *La bilancetta*, and *De motu*. In the first two, his preoccupation was with extending the work of Archimedes in mechanics. The first half of *De motu* is devoted to the application of the principle of Archimedes in refutation of Aristotle's laws of falling bodies. That constitutes the most important part of *De Motu*. Chapters 8, 10, 11, 12 and 13 all contain in their titles words to the effect of "in opposition to Aristotle." Chapter 14 com-prises the important section on motion on inclined planes; chapter 15 resumes the proofs in opposition to Aristotle. In none of these is the question of persis-tence of motion discussed; not even in chapter 14, where it is proved that motion on the horizontal plane can be induced by a minimal force. In chapter 16 the question is raised whether the rotation of a non-homogeneous sphere situated at the center of the universe would be perpetual or not, but the answer is postponed and never taken up again. At the end of the same chapter it is intimated that a non-homogeneous sphere situated outside the center of the universe cannot rotate perpetually, but again a promise of future explanation is not carried out.

It is only in chapter 17 of *De motu*, when Galileo comes to the matter of pro-jectiles, that persistence of motion becomes a topic of discussion. Galileo adopts the received impressed-force theory as a refutation of Aristotle. Circular motion is not a central question; in a single brief reference to rotary motion, Galileo says that it would last a long time, not that it is inherently perpetual. The rest of the book has nothing to say about persistence of motion as such.

What preoccupied Galileo at Pisa was not the persistence of motion, as Koyré asserted. Even at Padua, that is not a question on which Galileo left any-thing written. The *Mechanics*, composed there, contains a passage that implies persistence of motion on the horizontal plane, but only as an aside, and only by implication. There is evidence that Galileo developed that implication in lec-tures or conversations before 1607; but it was only in 1613, after he had returned to Florence, that he clearly stated an opinion on persistence of motion. What really preoccupied Galileo at Pisa was the refutation of Aristotle's phys-ics. That is evident from the structure of *De motu*, its chapter headings, its tex-tual criticisms of Aristotle, and its general polemic tone.

Returning to Koyré: "Now, it is clear that when he [Galileo, in *De motu*] studies the case of movement (rotation) of a sphere placed at the center of the universe, as of that of a sphere placed outside that center, he has in view the situation created by the Copernican doctine; the marble sphere whose movements he analyses represents, without any doubt, the earth; its movements are those of the earth."

Not only is this far from being evident; it is in a way self-contradictory. If Galileo had been a Copernican at the time he wrote *De motu*, he would not have placed the earth at the center of the universe, even metaphorically; and when he places his sphere outside that center, he assigns to it no revolution about that center. What is clear is that in his anti-Aristotelian *De motu*, he wished to destroy even Aristotle's fundamental division of all motions into natural and violent motions. To do this, he showed that certain bodies might be in movement without their motion being either natural or violent. It is *possible* that Galileo, while still at Pisa, was already interested in the Copernican implications, but it is by no means *clear* that he was. In *De motu* there is no reference, even in passing, to any astronomical topic – except to the belief of some philosophers that the addition of a single star to the heavens would slow down or stop them. In refuting this, Galileo treats the fixed stars as in daily circular rotation. There is not even a parenthetical remark to indicate that others (or he himself) believed that the fixed stars might not be in motion at all.

But the result at which he arrived – in contradiction, moreover, with the essential premises of the physics of impetus – reveals to us in a striking way the difficulties, and the source of the difficulties, in the path of the physics and the new astronomy ...

The result to which Koyré here alludes is specified in his next sentence, discussed below. It is, as will be shown, at least partly not Galileo's result at all. Whether or not it, or Galileo's own result, was in contradiction with the premises of impetus physics, Galileo certainly did not see any contradiction between the two at the time he wrote *De motu*. For in *De motu*, Galileo adopted impetus physics not only for projectiles, but as the basis of his own attempted explanation of the phenomenon of acceleration in free fall. He saw no contradiction between a wasting impressed force and his analysis of rotations in the *De motu*, nor is there any evidence that he saw any connection between that analysis and the new astronomy. For such a connection, we have only Koyré's hypotheses.

In fact, the result to which the Galilean analysis led is the natural persistence, or more precisely the privileged character, of *circular* movement.

This is the thesis put forth by Koyré and adopted by most historians of science after him. Koyré appears to consider it a result that Galileo reached at Pisa, rather than later, though *aboutit* may refer to some later time. The passages of *De motu* cited by Koyré in his footnote to the above assertion do not assert anything about the persistence of circular motion, nor do they deal with real physical spheres; still less do they assert that circular movements have any privileged character. In the same footnote, Koyré invites us to compare a passage in Galileo's *Mechanics* (a work which came after the Pisan period). It is the passage already mentioned above, and it still does not assert, though it does imply, the persistence of motion on horizontal planes.

Koyré next reviews Galileo's analysis of the motion of a hypothetical sphere located at the center of the universe, as a motion neither natural nor violent. *Mais le cas de la sphère située dans le centre du monde est loin d'être unique : à vrai dire tout mouvement circulaire (autour du centre) est un tel mouvement qui n'est ni naturel ni violent*, adds Koyré.[4] This is where trouble truly begins. Galileo made no such assertion, which would include the case of circulation around the center of the universe, as well as the case of the rotation of a sphere located there. No circulations, and not even all rotations, are said by Galileo to be "neither natural nor violent." Yet it is upon circulations (not mere rotations) that the whole question of "circular inertia" has been made to hinge. We shall return to this point shortly. First, however, it is important to note that Koyré's thesis attributes to Galileo a lifelong constancy to this principle: *Le mouvement circulaire occupe dans la réalité physique une position absolument privilégiée ... Telle est la situation Galiléenne. Elle est presque la même à Pisa, à Padoue, à Florence.*[5]

Now in fact, circular motion of any kind occupies only a small part of *De motu*. One chapter is dedicated to it, and that chapter deals hardly at all with physical reality. It is principally concerned with imaginary marble spheres, introduced for the purpose of showing that instances of motion can be adduced that will not fit into Aristotle's classifications of natural and violent. Part of one paragraph, discussing homogenous spheres rotating elsewhere than at the center of the universe, remarks that such spheres must be supported and will therefore be subjected to frictional resistance. That is the only case in which physical reality appears, and such circular motion is most certainly not given a privileged position there. Heterogenous spheres, moving just as circularly, would not continue to move forever, according to Galileo. In another chapter, discussing projectiles, antiperistasis is refuted by a paragraph concerning grindstones. These are the sole discussions of circular motion, which is not given a privileged position in *De motu* at all; it is merely seen to require separate discussion.

That Galileo held any single unifying view, virtually the same at Pisa, at

Padua and at Florence, is an inference for which no factual evidence exists. Growth and change characterized his thought; witness the evolution of his view of acceleration. As to circular motion, it is not a principal topic in any of Galileo's published books before 1632, nor in any of his letters. (The persistence of motion is not such a topic either, though he brought it at one point into the *Letters on Sunspots*.) In the *Assayer*, Galileo denied circular orbits to comets, and he ridiculed the idea that any shape (particularly the circle) had any priority over any other; none had patents of nobility, and none were absolutely perfect: perfection had to do with the intended use, and not the form.[6] So it is gratuitous, if not hazardous, to assume that when Galileo composed the *Dialogue* he had spent a lifetime ruminating over the persistence and absolute privilege of circular motion in the world of physical reality.

The underlying philosophical preoccupation attributed to Galileo by Koyré is equally unsupported by the surviving correspondence. Galileo's letters are as devoid of metaphysical discussions as they are of encomiums on the marvels of the circle. Only after he was blind, and had published his last book, do we find him debating (with Liceti) on any abstract principles. Since in that debate he sides with Aristotle, his final position can hardly be called a lifelong unifying philosophical view.

II

With the philosophical interpretation of passages from the *Dialogue* in the light of Koyré's thesis, we are not concerned here. Quite different interpretations are capable of being made on other assumptions, and Koyré's interpretation does not easily apply to other passages. Let us therefore pass on to the problem with which Koyré was left as he turned to discuss the successors of Galileo.

Si – ainsi que nous croyons l'avoir montré – Galilée n'a pas formulé le principe d'inertie, comment se fait-il que ses successeurs et disciples on pu croire le trouver dans son oeuvre? Et un autre [problème]: si, ainsi que nous croyons l'avoir également démontré, Galilée, non seulement n'a pas conçu, mais encore n'a pas pu concevoir le mouvement inertial en ligne droite, comment se fait-il, ou, mieux, comment s'est-il fait, que cette conception, devant laquelle s'est arrête l'esprit d'un Galilée, a pu paraître facile, évidente, allant de soi, à ses disciples et successeurs?[7]

Let us try to resolve the paradoxes seen by Koyré. And first, let us remove from them the exaggeration inherent in the statement he has made of them; that is, the word "demonstrated." For if it were indeed demonstrated that Galileo did not, and could not have, conceived or formulated the idea of continued and undiminished rectilinear motion, the paradoxes would have to be admitted. As things stand, however, we may for the present take Koyré's position not as demonstrated, but only as one of several plausible hypotheses. In that case, the paradoxes named give way to ordinary questions, questions that may serve as the basis for further fruitful research. These questions would be, for example:

If Galileo could and did conceive of the possibility of continued and undiminished rectilinear motion, why did he not formulate that conception as a physical law in unequivocal terms? And since he did not so formulate it, whatever his reasons, how did it come about that his pupils and successors not only adopted it as a matter of course, but treated it as something derived from Galileo's own work?

An answer to these questions should be sought in Galileo's own works, not in general philosophical and historical principles such as those to which Koyré appealed in support of his attempted demonstrations, demonstrations which led him in the end to paradoxes. Charming as paradoxes are to the philosopher, they are dead ends to the historian, who deals rather in real problems.

That Galileo could and did conceive of the *possibility* of continued uniform rectilinear motion is made evident at various places in the *Dialogue* and the *Two New Sciences*. The treatment of these in Koyré's monograph is unsatisfactory; he neglects to mention certain relevant passages in the *Dialogue*, and he explains away others in the *Two New Sciences* by an argument that proves only Galileo's denial of the *fact*, and not any denial of (or failure to recognize) the *possibility*, of rectilinear inertia. Thus Koyré correctly says, of the celebrated passage in the *Two New Sciences* deriving the parabolic trajectory, that continued uniform motion in the horizontal plane is not, for Galileo, rectilinear motion at all; fundamentally it is circular motion, for the horizontal plane is not a mathematical plane, but the surface of a sphere concentric with the earth. This argument shows, not that Galileo overlooked the possibility of rectilinear inertia, but merely that he avoided any assertion of its existence in fact. Since it does not exist experimentally, that is quite correct. The *possibility* is here neither affirmed nor denied; for its recognition, we must look elsewhere.

A single instance will suffice to show that Galileo recognized the *possibility* of rectilinear inertia. The clearest and best example, as well as the most important, is the very existence of the discussion in the Second Day of the *Dialogue* in which Galileo replies to the argument that rotation of the earth would cast off objects resting on its surface. In preparation for his argument, Galileo established this physical proposition:

The circular motion of the projector impresses an impetus upon the projectile to move, when they separate, along the straight line tangent to the circle of motion at the point of separation ... and the projectile would continue to move along that line if it were not inclined downward by its own weight, from which fact the line of motion derives its curvature.[8]

Thus there would be no curving of the line in the absence of gravity. The ensuing discussion assumes that the rectilinear motion, if it could continue, would be uniform, and this conception is confirmed by Galileo's diagram, in which equal times are laid off along the tangent. The case of a projectile impelled by a cannon had been previously discussed and is summed up in a postil: "Projectiles continue their motion along a straight line which follows the direction of the motion that they had together with the thing projecting them while they were connected with it."[9] Here, as elsewhere, it is not asserted that an actual cannon ever *is* at rest, or that the absolute path of the ball would ever actually *be* straight, but the *possibility* of a straight path is pointed out by the statement.

On the other hand, the real existence of "circular inertia," in the sense of an indelibly impressed impetus to follow the diurnal rotation of the earth, is excluded for Galileo by his argument against extrusion. Even the possibility of such an impetus is in effect denied by the omission of the Copernican argument. Had Galileo considered it a fact that every body resting on the earth is indelibly impressed with a circular motion identical with that rotation, his entire complicated mathematical argument could have been avoided, and in its place we should have had merely a reiteration of the Copernican position that such a circular motion must inhere naturally in all earthly bodies. These are powerful considerations in the search for Galileo's own views on the subject. The highly technical demonstration that Galileo attempted in explanation of the quiescence of detached bodies resting on a rotating earth could serve no purpose for himself, and still less for his readers, if he believed literally in "circular inertia." It is therefore significant that Galileo did not give this demonstration in addition to the Copernican "circular inertia" argument, but in place of it. There is simply no reason that Galileo should ever have worked out this demonstration in the first place, if his own conviction was carried by "circular inertia."

Here, however, an interesting point arises. Elsewhere in the *Dialogue*, particularly in connection with the fall of a weight from a tower or a ship's mast, the body is said by Galileo to follow in its descent the circle swept out by the tower in consequence of its diurnal motion, or of the ship in consequence of its own motion. This is indeed "circular inertia," and it is adopted in several places, for example: "Now as to that stone which is on top of the mast; does it not move, carried by the ship, both of them going along the circumference of a circle about its center? And consequently is there not in it an ineradicable motion, all external impediments being removed? And is this motion not as fast as that of the ship?"[10]

If we have to decide which of the two different conceptions Galileo himself held as an inner conviction – that is, whether the ineradicable motion impressed on a heavy body is the uniform rectilinear motion of which he spoke in connection with cannon-balls and stones flung from whirling slings, or the circular motion about the earth shared with the ship's mast from which a stone is dropped – there is at least one strong hint to be gained from the discussions in which those two different ideas occur in the *Dialogue*. In the first-named case, whether the original motion is rectilinear (cannon shot) or circular (whirled object), the discussion includes a comment to the effect that the object would continue along the straight line if it were not immediately drawn down by its own weight, as in the passage cited above. But in the ship-mast or related examples, in which a circular motion is shared by the object with the whole earth, or is imparted to it solely because of the earth's spherical shape, Galileo never adds an analogous statement that the body would continue along that (circular) line if it were free of weight.

In short, Galileo does not say positively anywhere that a heavy body near the earth would continue in circular motion if deprived of support, in any circumstances, whether dropped from a ship's mast or rolled off the end of a level plane; but he says positively in several places that such a body would continue any rectilinear motion imparted to it, if it were not for the downward action of its weight. Wherever it appears that he asserts the continuance of a circular motion for terrestrial bodies, the motion is of a character that is indistinguishable from rectilinear motion by reason of the huge size of the earth. It has become customary to suppose him to have been thinking "and yet, of course, the motion really *is* circular, though we can't see that it is." But in that case, one would expect him to have made this specific qualification for the cannon-ball and sling-shot; and that is the exact reverse of what he did do. We, on the other hand, might equally well suppose him to have been thinking, "but of course the circular character of the ship-mast motion or of the actual horizontal plane is merely accidental, existing only because the earth happens to be round"; and then all his statements concerning terrestrial inertia would be consistent and rather easy to explain. The prevailing preference among historians of science for believing Galileo inconsistent and paradoxical on this matter is guided not so much by necessity as by a belief that the history of science receives more light thereby.

The sole exception, for terrestrial objects, might seem to be the discussion of semicircular fall in the *Dialogue*. Certainly that seemed to me decisive for many years. As I have previously explained, however, the entire purpose of that discussion is remote from any consideration of inertia in its physical aspect; the continuation of motion there is introduced as a mathematical assumption rather than an observed fact, and is used for the purpose of a geometrical speculation about an apparent path rather than for a physical analysis of any projectile motion.[11] It corresponds, moreover, to the ship-mast type of circular motion at most, and hence to a line indistinguishable by observation from a straight line for the amount of arc under consideration. To justify the treatment of such arcs as straight lines, Galileo appealed all his life – from the unpublished *De motu* of 1590 to the final *Two New Sciences* of 1638 – to the treatment by Archimedes of the pans of balance as hanging by parallel lines, though in fact they converge toward the center of the earth. Hence it is, in my opinion, absurd to suppose that Galileo himself considered the curvature of the earth as introducing an *essential* circular property into all imparted motions near that surface. Rather, his consistent life-long statements show that he took Archimedes as his model in the treatment of physical problems, ignoring inconsequential discrepancies in the choice of physical postulates for his mathematical deductions.

There are, moreover, additional reasons for believing that inconsistencies in the *Dialogue* concerning a choice between ineradicable circular impetus or ineradicable rectilinear inertia in free fall to a rotating earth are only apparent, and not real, inconsistencies. Considering the order in which the references occur, and their relative simplicity and complexity, there is some reason to suppose that Galileo used the less involved (Copernican) assumption earlier in the *Dialogue*, to lead his readers easily along, and then refined it to tangential motion when the argument became more precise. That view is supported by the many instances in which he treated the earth's spherical surface as an approximation to the horizontal plane. It is also supported by considerations of style, since a precise statement of the tangential character of conserved motion would have been necessarily clumsy in his early illustrations as compared with the expressions used. On the whole, it is more plausible that Galileo personally held to the rectilinear motion as essentially true for terrestrial bodies.

But we do not really have to decide for one view or the other; if anything, we should avoid such decisions. What we do have to do is to see that Galileo consistently leaves open the *possibility* of both circular and rectilinear continuations of uniform motion in the case of free fall (a "natural" motion), whereas he specifies unambiguously the *rectilinear* continuation for the "violent" motion of cannon balls and of terrestrial objects released from slings or flung from rapidly rotating wheels.

Now, this point becomes still more significant when we turn from the behavior of heavy bodies on or near the earth's surface (terrestrial physics) to Galileo's cosmological speculations (celestial physics). I paraphrase Galileo's thought thus: "For terrestrial physics (the physics of heavy bodies), the ineradicable *tendency* is for uniform and perpetual motion in the horizontal plane, but the actual *motion* of a supported body cannot be in any true plane, mathematically, since the earth happens to be round. Again, the ineradicable *tendency* of terrestrial projectiles is to follow the line of the cannon, or the tangent to the circle of the sling, but the actual *motion* cannot be so because the body has weight. The *essential* motion of a terrestrial body is one thing; its accidental path is another. Where we cannot distinguish them, we do not have to decide between them, and such is the case of the impressed diurnal motion, where the straight tangent and the circular arc do not differ by an inch in a thousand yards." But where the distinction is clear, the *essential* rectilinearity is easily recognizable: "When the stone escapes from the stick, what is its motion? Does it continue to follow its previously circle, or does it go along some other line? – It certainly does not go on moving around ... It is necessarily along a straight line, so far as the adventitious impulse is concerned."[12]

These and similar passages answer the question whether Galileo could perceive the possibility of continued uniform rectilinear motion for terrestrial (heavy) bodies, and the question why his pupils and successors treated the principle of inertia as a part of his work even though he did not formulate it clearly and unambiguously. For a possible answer to the question why he did not so formulate it, we must look also at his cosmological speculations; for it is evident that his failure to do so cannot have been, as Koyré supposed, an inherent incapability of perceiving the possibility.

The principal (if not the only) source of knowledge concerning Galileo's cosmological speculations is the First Day of the *Dialogue*, and particularly its opening section. Having begun with Aristotle's position, Galileo makes his spokesman, Salviati, agree with Aristotle, supplying additional arguments in favor of the Aristotelian position up to a certain crucial point. But where Aristotle deduces the perfection of the heavens from the perfection of the circle, Salviati introduces instead the postulate that the universe is perfectly orderly. From this he deduces that it is impossible for integral world-bodies (the celestial bodies and the earth itself) to move in straight lines: for in that way the cosmic order would be changed. (Note that this postulate does not commit Galileo with respect to terrestrial objects.) Cosmic order can be preserved only if the heavenly bodies are at rest or in circular motion. Only that motion, he says, is capable of true uniformity; straight motion is either accelerated, decelerated, or infinite, the first two being nonuniform, and the last being inadmissible in a well-ordered universe.

It is here that Galileo introduces his "Platonic concept" of cosmogony: the planets having been created at a certain place, they were moved with straight accelerated motion until each had received its assigned speed, at which point its motion was converted from a straight to a circular path by God, who willed that the planet keep that same velocity perpetually thereafter.[13] On the basis of this cosmogony and its attendant arguments, it is now widely held that Galileo attributed the motions of the planets in their orbits to "circular inertia." That, however, contradicts several clear statements by Galileo himself; moreover, it implies certain beliefs on his part that he cannot have held if he was even a competent, let alone a gifted physicist and astronomer.

First we must note that in this passage, Galileo attributes the continuance of the planets in their orbits to the will God, and not to any physical principle whatever. Much later in the *Dialogue*, he expressly denies that he (or anyone else) knows by what principle the planets are moved.[14] Equally important is his refusal to grant that the universe has a center: "We do not know where that may be, or whether it exists at all. Even if it exists, it is but an imaginary point; a nothing without any quality."[15] Now, Galileo's entire argument for perpetual uniform movement, wherever it is found in his writings, consists always in the body's path being such as not to approach or to recede from some center toward which it has a natural tendency to move. For actual heavy bodies, that is the center of the earth. Only by the most tenuous arguments, and in contradiction with Galileo's own words, can a case be made that he believed the planetary circulations to be analogous to such motions. And having made out such a case, one would be confronted with his denial of a center for their circulations.

Nor is it possible to argue that Galileo believed the actual planetary motions to be literally circular around a common center such as the sun. No competent astronomer since Aristotle had believed in homocentric orbits, with the possible exception of Fracastoro. Certainly no Ptolemaic or Tychonian, let alone Copernican, believed in literally concentric paths for the planetary bodies themselves. Galileo was no great theoretical astronomer, but he certainly was aware of the classic problems of planetary motions, variously solved by eccentrics, epicycles, ovals, and ultimately by ellipses. If anyone wishes to contend that Galileo was actually so ill-informed as to believe that some set of perfect concentric circles – centered, moreover, about an occupied point – would fit the actual observed motions of planets, and that the "Platonic concept" of the *Dialogue* presents us with mature astronomical convictions, then he must also contend that Galileo believed those circular motions to be absolutely uniform. But in the Fourth Day of the *Dialogue*, Galileo argued that the circuits of the sun, the earth and the moon are not uniform in speed.[16] Hence the attribution to Galileo of a belief in planetary motion by reason of "circular inertia," meaning by that absolutely uniform motion in perfect circles, can be maintained only at the price of rejecting his own words and his astronomical competence, or supposing him to have been so devout a Platonist as to hold to a cosmology refuted by his own perceptions.

III

We have now seen some of the factual weaknesses behind Koyré's thesis, and some of the serious difficulties to which it gives rise. If we abandon his interpretation of "circular inertia" as the unifying principle of Galileo's physics and his astronomy – as the unchanging core of his work at Pisa, Padua, and Florence – what interpretation shall we put in its place? To this question I wish to reply only tentatively at this time, putting forth possibilities rather than conclusions.

The contradictions, or seeming contradictions, in the *Dialogue* are best resolved by paying very close attention to Galileo's exact words. Nothing is gained by postulating some unifying conception behind them in Galileo's own mind and then saying in effect: "When he said this, he really meant thus-and-so." Let me give an example. When Galileo had occasion to speak of the terrestrial horizontal plane, he usually went to the trouble of pointing out that no true geometrical plane existed on a spherical surface, and that the so-called horizontal plane of our experience was really a portion of the sphere. Yet when Galileo argued that a projectile released from a whirling sling tended to move along the straight line tangent at the point of separation, he did not add that the line was not really straight, but must share in the circularity of its previous path, or in that of diurnal rotation, or anything of the sort. Therefore it is useless to argue that he did not, and could not, conceive of uniform rectilinear inertial movements. The fact that such movements are made nonuniform by air resistance and are curved by downward action of weight has nothing to do with their essential character, and Galileo was perfectly clear about that.

On the other hand, when he spoke of free fall along a tower or the mast of a ship, he plainly said on several occasions that the true path was a compounding of the straight motion toward the earth's center and the general circular terrestrial motion shared by the object before the commencement of its descent. So there is no use claiming that he really always meant a straight tangential motion. It may be that he had such an idea in mind and did not adduce it in every case, for reasons of style; but there is no more evidence that he always thought of even terrestrial inertial movements as essentially straight than that he always thought of them as essentially circular. So, if we pay strict attention to Galileo's own words, we shall have to say that he had no unifying principle for all cases of inertial movement.

It does not necessarily follow, however, that Galileo was inconsistent in the matter. For example, if there is some element common to all the cases in which he specified an essentially straight inertial movement (essentially, that is, by the very nature of things), and some other element common to all the cases in which the motion is spoken of as essentially circular, then the apparent inconsistency might vanish as thoroughly as did Koyré's supposed unifying principle.

Now it appears to me that Galileo is pretty consistent in applying the idea of "circular inertia" to instances in which the motion is a "natural" one in his sense; that is, a motion induced by an innate tendency of the body to move when it is set free. The idea of rectilinear inertia, on the other hand, he applied most specifically to instances of "violent" motion – cannon balls and projectiles thrown by slings. One may see that even in this he is not entirely consistent, though, for his long discussion of bodies that ought to be flung from the earth by its rotation is an apparent exception. They are not subjected to an external force, yet they are treated as projectiles from slings would be treated; that is, as potentially moving along the tangent in a straight line. Whether the exception is real or apparent may be argued. Galileo's question was how such objects would move if they moved at all, and that question he treated on the analogy of the sling, or wheel.

In short, I doubt on the one hand that Galileo had a unifying principle in this matter, and on the other hand that he was vague and inconsistent about it. On one point he was quite definite and consistent, though we have made it hard for ourselves to see this. Tangible terrestrial objects subject to observation, to which a violent impulse imparted either by a straight push or by release from whirling, conserved the previous impetus in the form of uniform rectilinear motion. Stated several times in the *Dialogue*, usually together with the idea of composition of independent motions, this conception was applied again in the *Discorsi*, and it was understood and adopted by Galileo's pupils and successors. He seems to me to be equally consistent in attributing "circular inertia" to terrestrial objects in "natural" motion only, where it happens that an observer could not actually distinguish the tangential from a very slightly arcal path.

It remains a question whether Galileo had an ulterior purpose in making the distinction suggested above, and whether it had anything to do with a belief on his part about the planetary orbits. It seems not unlikely to me that he did have a reason, but one that had nothing to do with any cosmology. This reason was that for Galileo, heavy bodies on or near the earth strove by an innate tendency to reach its center, or rather, the common center of gravity of all such bodies; and this they would never reach by the parabolic trajectories implied by inertia, though they would reach it along a suitably chosen circular arc.[17] Violent motion could be permitted to disturb natural order, but natural motion could not. If Galileo took such a view, his seeming vacillation between inertia and "circular inertia" in the *Dialogue* would be reasonably explicable without recourse to a unifying principle on the one hand, or disparagement of his acuity on the other.

Galileo's cosmological speculations, which occur almost entirely near the beginning of the *Dialogue*, are usually interpreted as evidence that Galileo believed in an extension of "circular inertia" to the planets. It has been pointed out above that taken literally, this not only implies an extraordinarily poor knowledge on his part of the locations of planetary orbits and the speed of planets in them, but contradicts various statements and denials of his own. Elsewhere, I have set forth my view concerning the motivation for those cosmological speculations in that place, giving attention to their polemic value and remarking that no similar passages are to be found in any of Galileo's voluminous writings and correspondence.[18] The fascination with circles that Koyré makes fundamental to the understanding of Galileo's physics is not evidenced by lifelong discussions, as is, for example, Kepler's fascination with musical harmonies in the universe. Without repeating what I have said before, I shall merely point out here that in the *Dialogue* itself, Galileo ridicules the Peripatetic insistence on a mathematical perfection of sphericity for the heavenly bodies, and cheerfully admits that perhaps no perfectly spherical body can be formed of actual matter. This, however, does not invalidate mathematical reasoning about physics; it merely cautions the calculator to adjust his accounts as necessary.

It is in that light, I think, that we should interpret such cosmological statements as this: "I therefore conclude that only circular motion can naturally suit bodies which are integral parts of the universe as constituted in the best arrangements."[19] Koyré and his follows (indeed, some of his critics as well) want us to believe that this proves Galileo to have believed in his heart that absolutely perfect uniform circular motions carried the planets around the sun. I have remarked above that taken literally, this would require him to have been alone among all the astronomers after Aristotle to believe that observed positions of the planets were compatible with any set of simple rotations about a single center. For Galileo does not say "circular motions," as if to allow for resultant epicyclic paths made up of a plurality of circular motions.

Against the prevailing view, which would make Galileo not only a mystic but a foe of actual observation, I take his phrase "circular motion" as meaning simply "circulation"; that is, recurrent motion over a closed path. That is perfectly consistent with his manner of arriving at the proposition, which is deduced from the orderliness of the cosmos. It is also consistent with his obvious motive in the First Day, which is to prepare a basis for attributing to the earth a revolution about the sun.

The impropriety of Koyré's thesis as a basis for judging Galileo's beliefs about planetary motions is conclusively shown by Galileo's mature rejection of the quest for causes in physics, the very attitude for which Descartes most criticized Galileo. To offer "sympathy," "antipathy," or any other occult quality in explanation of a physical effect was repugnant to Galileo. It was for this kind of "causation" that he criticized Kepler's theory of the tides. In the *Dialogue*, he reproved Simplicio for offering "gravity" as a cause of the fall of bodies, in a passage that overtly rejected all purely verbal attempts to assign causes to planetary motions. And if the phrase, "circular inertia," had existed in Galileo's day, in the vague sense in which today that phrase is offered by most writers as Galileo's own explanation of celestial motions, I cannot doubt that Galileo would have laughingly included it with the "informing spirits" and "guiding intelligences" that he ridiculed as explanations in his own time.

It is quite correct to say that Galileo did not offer us a complete system of the universe. It is not only incorrect, but in violation of Galileo's whole approach to physics, to say that he offered a complete system based on the false principle of circular inertia – or any other single unifying principle, true or false. If the term had been explained to him, I think he would have said:

> The introduction of such a phrase is in no way superior to the "influences" and other terms employed by some philosophers as a cloak for the correct reply, which would be, "I do not know." That reply is as much more tolerable than the other as candid honesty is more beautiful than deceitful duplicity.[20]

NOTES

1 "Galileo and the Law of Inertia," *American Journal of Physics*, 32 (1964), pp. 601–608; "The Concept of Inertia," *Saggi su Galileo Galilei* (Florence, 1967).

2 My call for a re-examination of this question was presented orally at the 12th International Congress of the History of Science at Paris in August, 1968, at which time I learned that Mr. J.A. Coffa had already prepared a paper in which many objections to the prevailing views were set forth. Mr. Coffa kindly sent his paper to me, which deals with aspects of the problem as they relate to post-Galilean physics. It has been published as "Galileo's Concept of Inertia" (pp. 261–281 of this issue of *Physis*). The present paper complements that study with respect to purely Galilean matters and the place of the question in the historiography of science.

3 A. Koyré, *Galilée et la loi d'inertie* (Paris, 1939), pp. 195–196.

4 Ibid., pp. 197–198.

5 Ibid., pp. 198–199.

6 S. Drake and C.D. O'Malley, *The Controversy on the Comets of 1618* (Philadelphia, 1960), pp. 191–197; 279. Galileo, *Opere*, VI, pp. 242–244; 319–320.

7 Koyré, op. cit., p. 282.

8 Galileo, *Opere*, VII, p. 220; *Dialogue* (tr. Drake, Berkeley, 1967), p. 103.

9 *Opere*, VII, p. 201; *Dialogue*, p. 175.

10 *Opere*, VII, p. 174; *Dialogue*, p. 148. Mr. Coffa has raised a neglected point of great cogency in this connection; if the ship is not sailing in the direction of the earth's rotation, what circular motion "as fast as that of the ship" is involved?

11 This was discussed at length in the previous paper of this series. *Physis*, X (1968), pp. 89–100.

12 *Opere*, VII, pp. 217–18; *Dialogue*, p. 191.

13 *Opere*, VII, pp. 44–45, 53–54; *Dialogue*, pp. 20–21, 29.

14 *Opere*, VII, pp. 260–261; *Dialogue*, pp. 234–235.

15 *Opere*, VII, pp. 58, 61; *Dialogue*, pp. 33, 37.

16 *Opere*, VII, p. 478; *Dialogue*, p. 453. For his doubts concerning the accuracy of existing planetary theory, see *Opere*, VII, pp. 480–81; *Dialogue*, pp. 455–56.

17 *Opere*, VII, pp. 190–92; VIII, pp. 283–284; *Dialogue*, pp. 164–166; *Two New Sciences*, pp. 261–262.

18 See the paper cited in note 1, above.

19 *Opere*, VII, p. 56; *Dialogue*, p. 32.

20 Cf. Drake and O'Malley, op. cit., p. 197; *Opere*, VI, p. 244.

6

The Tower Argument in the *Dialogue*

The Second Day of Galileo's *Dialogue* was concerned mainly with the removal of objections raised against daily rotation of the earth. His novel doctrine of motion was preceded by twenty pages of preliminary conversation about cosmological aspects of the Copernican and Ptolemaic astronomies. About forty pages were then devoted, directly or indirectly, to the tower argument, and another fifty pages followed on matters relating to fall.

By the tower argument is meant a certain supposed proof against daily rotation of the earth derived from the fact that a body falling from the top of a tower is seen to graze its side to the ground. If rotation of the earth moved the tower eastward at a thousand miles an hour, it seemed that a weight dropped from its top should hit the earth considerably to the west of its base.

The merits and demerits of Galileo's treatment of the tower argument have been discussed in recent years by Paul Feyerabend[1] from a philosophical base, and by Maurice Finocchiaro[2] from a logical and a rhetorical viewpoint, but the history of the tower argument has never been investigated. Feyerabend, in a note at the end of his discussion, asserted incorrectly that it is found in Aristotle's *De caelo* and Ptolemy's *Almagest*, and that it was discussed by Copernicus.[3] Finocchiaro called the tower argument a classical objection against motion of the earth. Philosophers tend to regard historical origins as irrelevant to the problems that concern them, though in this case both writers ventured to render moral judgments about Galileo's handling of the argument, which is surely an irrelevant and gratuitous question within the domain governed by philosophers of science and logicians.

Had Galileo wished, a single paragraph would have served him to show that the tower argument proved nothing against rotation of the earth as such, if its

Reprinted from *Annals of Science* 45 (1988), 295–302, by permission.

center remained fixed. Without any risk of offending Aristotelian natural philosophers, that would have been easy to show if Galileo had merely written:

> Aristotle implied that fall of our heavy bodies is always along the line directed toward the earth's centre. The side of a tower lies along such a line at all times, whether the earth rotates or not. That is accordingly the line along which fall must take place from a tower on earth, from which path a body can depart only by its falling in some direction other than straight towards the centre of the earth – in contradiction of Aristotle. If the earth were in fact now rotating about its centre, heavy bodies free to move would be seen to fall as we see them to fall, and as Aristotle implied that they must fall.

That succinct reply, an exercise in semantics rather than a physical refutation, contains nothing that Aristotle could not have said *if* rotation of the earth had been suggested up to his time. Aristotle held that the earth's centre coincides with the centre of the universe, not by necessity but in fact, and that a heavy body free to move always moves naturally to that centre. Not until a century after Aristotle is anyone (Aristarchus) known to have proposed a cosmological scheme or astronomical hypothesis that necessarily implied diurnal rotation of the earth, so that was not yet in debate. Aristotle wrote nothing about a possible rotation of the earth around its fixed centre when he reviewed and criticized the various systems that had been put forth up to his time.

Here it should be noted that medieval natural philosophers freely debated the possibility of diurnal rotation of the earth with its centre fixed at the centre of the universe. They had read *De caelo* with great care, and had found nothing in it that excluded the possibility. But even in those debates, no one is known to have advanced the tower argument against rotation of the earth. It is probable that some medieval natural philosophers had also read what Ptolemy had written about vertical fall in the opening section of the *Almagest*, and had not found there anything that suggested the tower argument against rotation of the earth.

In the *Dialogue*, however, Galileo literally attributed the tower argument to Aristotle, and in his syllabus on cosmography written three decades earlier he had ascribed it to Ptolemy. Or rather, he had included it with a number of objections against rotation of the earth following a correct summary of what Ptolemy had written:

> Ptolemy, considering this opinion [of diurnal rotation], argues in this fashion. If we, together with the earth, were moved eastward with such speed, it would follow that all other things disjoined and separated [from earth] would appear to be moved westward with like speed; and so birds and clouds suspended in the air, being unable to follow the earth's motion, would remain behind to the west. –*– Likewise, things allowed to fall

down from high places as for example a stone from the top of a tower would never fall at the base of the tower, because in the time that the rock coming vertically down was in the air, the earth, getting out from under it and moving eastward, would receive it at a part very far from the foot of the tower, in the way that on a rapidly moving ship the rock falling from the top of its mast does not fall at the foot but more towards the poop. ...[4]

As will be seen, Ptolemy was correctly summarized up to the point starred, but he did not continue with fall from a tower, or ships (or arrows, or projection from the earth, as the paragraph continued). Galileo's phrase "argues in this fashion" (*argumenta in questa guisa*) did not mean "writes the following." Later, when the tower argument was put into the *Dialogue*, a similar mixture of correct summary followed by interpolations was used in attributing it to Aristotle.

When Galileo dealt with the tower argument in the *Dialogue*, he departed from his usual style by multiplying refutations when a simple exposure of fallacious reasoning would have sufficed. Both that, and his misattributions of source, need explanation. His reasons may be related to the possibility that the *Dialogue* was the first printed book to set forth the tower argument, with or without a refutation of it. Such reasons might be detected by Galileo's biographers, familiar with his purposes in writing the *Dialogue*, the state of his knowledge in 1624–30, and his bad relations with professors of Aristotelian philosophy, things not always known to philosophers, and not deducible by logicians.

The possibility that Galileo devised the tower argument only to refute it may seem remote, but it cannot be excluded. If he did, that would not have been the first time that a Copernican framed a novel objection against daily rotation of the earth and then ascribed it to an ancient predecessor. Copernicus himself raised the objection that if the earth rotated at great speed, heavy bodies resting on it would be cast off to the skies by that motion. He appears to have believed that the projection argument, as this is called,[5] had been already put forth by Ptolemy in the *Almagest*. In that, Copernicus was mistaken, but it is worth noting that Galileo also seems to have ascribed the projection argument to Ptolemy in his syllabus on cosmography.

Logic sufficed to reveal the fallacy of the tower argument. To see that the projection argument is fallacious requires some knowledge of physics, though only of the most elementary variety. To hurl a resting heavy body into the heavens, a great force must be applied. No force is exerted on a resting body by rotation of the earth, beyond that which keeps the body from sinking into the ground on which it is resting. If it were to lose all its weight by some higher speed of rotation, the body would be in orbit and no greater speed could then be imparted to it by friction with the earth. Dynamically, there is no source of a force to project a body from the earth's surface, no matter how rapid the earth's rotation might be, or might become. Galileo set forth a purely kinematic demonstration against the projection argument in the Second Day of his *Dialogue*.[6]

If Copernicus contributed the projection argument simply to forestall others who might raise it, and ascribed it to Ptolemy by extending to circular motion a Ptolemaic argument against fall of the earth as a heavy body, Galileo had like reasons for using the tower argument and taking liberties with what Aristotle wrote to make him responsible for the fallacy involved. Galileo's goal in the Second Day was to introduce and clarify his new physics of motion whose three main principles were the relativity of motion, conservation of motion, and independent composition of two (or more) tendencies to motion in a heavy body. Each principle was completely contrary to Aristotle's physics. Persons who opposed Galileo's new science of motion (and everything else of his) were Aristotelian professors of natural philosophy. Though the tower argument was not technically Aristotle's, calling it his was not a wholly unjustifiable way of symbolizing their opinions. The tower argument, Galileo knew, would be accepted as authentic by the Aristotelians of his time; for if they were to reject it as non-Aristotelian, they would find themselves compelled to agree with much of Galileo's physical reasoning for a new physics of motion that was very different from Aristotle's.

Aristotle took it as a principle that any heavy body free to move seeks its natural place, at the centre of the *universe*, moving straight toward that. The earth *happened* to be concentric with the universe, in Aristotle's view, implying that bodies must fall straight toward the centre of the earth until obstructed by the earth's surface. Aristotle cited experience as evidence that heavy bodies strike the earth vertically at the end of free fall, wherever the observer might be on its spherical surface. He then made the fact of vertical fall one of his principal arguments for placing the earth's centre at the centre of the universe. The earth had no "local motion" as long as its centre remained fixed, so Aristotle had no reason to say anything about rotation (not a "local motion" because rotation of a sphere does not change its place). If Aristotle implied anything, as Galileo presented the matter in his *Dialogue*, that was incidental to Aristotle's chief purpose of establishing a central earth in the universe.

The tower argument was admirably suited to Galileo's purpose of introducing to readers of the *Dialogue* all three principles of his own new physics of motion. Relativity of motion came in at once, through the fact that if the tower is moved, so is the observer. Conservation of motion came in indirectly, through discussion of the composition of motion downward along the tower and motion with the tower if the earth moved it. That likewise justified further discussion of the independent composition of motions, to which Galileo gave much space in the Second Day.

It may be that the tower argument first appeared in 1632, in Galileo's *Dialogue*, as a convenient way to introduce his physics. If so, its subsequent history is even more curious. After it had been refuted by Galileo, a proposition equivalent to it[7] came to be adopted as serious physics confirming fixity of the earth by a Jesuit astronomer, G.B. Riccioli, who remains in high esteem as a scientist and a historian. Riccioli was in turn refuted by two mathematical physicists, Stefano dell'Angeli and G.A. Borelli.

Evidence that the tower argument made its first appearance after 1543 will be amplified later. It seems not to have been formulated before the seventeenth century, to judge from sixteenth-century printed books written by anti-Copernican authors.[8] As late as the year 1611 Lodovico delle Colombe wrote a manuscript treatise against Copernicans into which he probably put all the physical and cosmological arguments he could muster. Since he began his treatise by expounding Tycho Brahe's arguments drawn from the supposed differences in carriage of artillery shots when fired to the east and to the west, he probably knew all the arguments that were grounded in Aristotelian physics up to his time. The tower argument as such is not in Colombe's treatise, though perhaps it was implicit in one of his others – that a heavy weight and a light one, if released together from a high place, would arrive on the ground at different places.

Another argument closely related to the tower argument was also included. Some, Colombe wrote, asserted that a weight when dropped from the mast of a ship fell to its foot, whether or not the ship was moving, and said they had made the experiment. That result Colombe simply denied as impossible. Quite possibly he intended this as a direct challenge to Galileo, when he sent his treatise to him and expected a reply, for Galileo was almost certainly included in Colombe's list of those who had made the assertion that he regarded as an imposture.

In 1624 Galileo stated in a letter that he had made ship experiments,[9] having philosophized twice as well as people who reasoned that if the ship were moving, the weight would alight sternward from the mast. He remarked also that before his tests he had known by reason what would happen. Galileo's spokesman in the *Dialogue* phrased this differently, telling Simplicio that "Without experiment, I am sure that the [ship] effect will happen as I tell you, because it must happen that way ... and you yourself also know that it cannot happen otherwise, no matter how you may pretend not to know it."[10]

A ship's mast, like the side of a tower, lies along a straight line through the earth's centre; hence the same reasoning applies here as in refuting the tower argument by simple logic and without contradicting Aristotle. The ship argument was related to the tower argument in the *Dialogue*, and when Galileo was composing that book he probably made use of the treatise sent to him by Colombe, for the order of the Aristotelian arguments presented follows pretty closely the order in that treatise, in which Galileo had written a number of marginal comments. It is not at all impossible that Galileo's choice of the tower argument as an organizing theme for the Second Day was made when he re-read the most complete compendium of anti-Copernican arguments at hand. Its author had omitted the tower argument, which he had not read into Aristotle. Galileo supplied it as Aristotle's not because it was, but because it exemplified many fallacies accepted by Aristotelian natural philosophers who opposed his physics. That may well be why Galileo multiplied refutations against the tower argument when a simple exposure would have destroyed it, but not without removing any plausible reason for presenting his own three principles of kinematics.

It is worth calling these neglected possibilities to the attention of students truly interested in Galileo, his *Dialogue*, and the seeming lack of organization of that book to many of its critics, from Descartes, down to the present. It will also be worth their trouble to review carefully what Aristotle said, and likewise Ptolemy, who wrote long after Aristarchus and hence had reason to take up diurnal rotation of the earth, and who clearly took his clue from Aristotle but dealt with the matter as an astronomer, not as a philosopher.

At the beginning of *De caelo* II, xiv, Aristotle summarized his arguments for placing the earth motionless at the centre of the universe, writing:

1. Some make it one of the stars [i.e. distant from the centre of the universe]; others put it centrally [see below] but winding and moving about the pole as its axis. But ... whether [it be] at the centre or at a distance from it, if the earth moves, its movement must be forced ... [continuation below][11]

Strictly speaking, the earth could not move if literally *at* the centre of the universe, for which reason I put "centrally" in place of the phrase "at the centre" where those words first occur in the Loeb translation. That cannot be done in the second occurrence, but Aristotle's meaning is clearly "no matter where the earth is said to be."

There was surely no allusion here to *rotation* of the earth, for neither "winding [about the pole]" nor "moving about the pole" describes rotation around the earth's centre. The "pole" meant is not a point on the earth, but a line in the heavens, an axis about which the earth could wind and move, such as the axis of the zodiac. Aristotle's description here suggests a helical motion, which makes good sense astronomically. That was the way in which some had placed the earth centrally but not immovably. The earth's alleged motion of revolution, Aristotle went on to say,

[1. ... must be forced]: it is not [the natural] motion of the earth itself, for otherwise each of its parts would have that same motion; but as it is, their motion is invariably in a straight line towards the centre.

Motion of the earth being forced motion in either case, both alternatives were now refuted in Aristotle's view; for him forced motion could not be eternal, whereas the order of the universe is eternal. This was his first argument for establishing the earth motionless at the centre of the universe. He said "motionless" because the earth as a whole does not change place. The simple *rotation* of a sphere would not be local motion, for the rotating sphere need not change place as a whole, and local motion was the only issue considered by Aristotle in this argument. Next, he went on:

2. All bodies having circular celestial motion, except the primary sphere, are seen to have a double motion. If the earth had such a double motion, we would see passings and turnings among fixed stars; we do not, so the earth does not move as planets do.

Again, rotation of the earth as a whole has no place in this argument, in which the double motion meant is the general diurnal revolution combined with the contrary motion belonging to each of the planets. Indeed, rotation of the earth was implicitly denied here by assignment of revolution to the fixed stars. Next:

3. The natural motion of the earth as a whole is toward the centre of the universe, like the natural motion of its parts. Heavy bodies move toward the centre of the earth only because that incidentally coincides with the centre of the universe. We see that weights moving towards the earth move not along parallel lines, but along lines normal to its surface, proving all weights to move towards the same centre.

"It is now clear that the earth must be at the centre and immobile," Aristotle wrote. "To our previous reasons we may add that

[4.] Heavy objects, if thrown forcibly upwards in a straight line, come back to their starting-place, even if the force hurls them to an unlimited distance."

Aristotle's added argument for immobility of the earth, by its very nature, could have no bearing on a possible rotation in which the centre of the earth remains fixed. Aristotle went on to prove the earth spherical, and other matters. If any argument against rotation of the earth was intended by Aristotle, it was at most only weakly implied by [4.].

Galileo ascribed the tower argument to Aristotle not as a textual statement, but only as an *implication* of Aristotle's afterthought [4.]. Literally translated, what he wrote was:

... So Aristotle says that a most certain argument of the immobility of the earth is our seeing things projected vertically upward to return by the same line to the same place from which they were thrown, and this, even if the movement is very high; the which would not be able to happen if the earth were to be moved, since in the time during which the projectile is moving upward and then downward, separated from the earth, the place from which the motion of the projectile had its origin would go, –*– thanks to the revolving of the earth, a long way toward the east, and by that much space, the projectile, in falling, would strike the earth distant from the said place ... [12]

The interpolated "thanks to the revolving of the earth" here seems[13] gratuitous on Galileo's part. Had Aristotle been thinking of the tower argument, it would have been much easier for him to have simply stated it than to write [4.] as he did in *De caelo*, and still simpler to dismiss it as incapable of proving anything about rotation or non-rotation.

In [4.], motion is started by throwing a heavy object straight upwards. The way to judge that this act had been carried out successfully would be to stand close to a high tower whose side had been erected with the aid of a plumb-line. If the body nearly grazed the side of the tower to its top, one could be certain that it had been thrown upwards in a straight line. There would accordingly be no reason for surprise when a body *carried* to the top of the tower, instead of being thrown to that height, fell grazing the side of the tower to its base. Quite the contrary; great surprise would be felt if anyone had asserted that without the earth's centre moving, a body could fall from the top of a tower along a line not passing through its base, and had offered in support of this no more than a hypothetical earth rotating on its centre.

What was argued in [4.] was that the earth as a whole does not change place, carrying its centre elsewhere while a body detached from it moves up and down in a line passing through its centre, first in forced and then in natural motion. If that centre changed place, as Aristotle saw things, the body would at all times be moving straight toward the centre of the *universe*, without regard to the earth's centre, which did not remain at *its* starting-place. So long as the centre of the earth remained at the centre of the universe, the body whose motion he described would return to its starting-place – defined in respect to the universe, of course, by reference to its centre.

Ptolemy, who later considered possible daily rotation of the earth, expressly rejected that without appealing to vertical fall from a tower in his refutation. What Ptolemy wrote was this:

It seems to some people that there is nothing against their supposing the heavens immovable and the earth as turning on the same axis from west to east very nearly one revolution a day ... They would have to admit that the earth's turning is the swiftest of all movements ... because of its making so great a revolution in a short time. So that all things not resting on the earth would seem to move contrary to it, and never would a cloud be seen moving eastward, nor anything else that flew or was thrown into the air. For the earth would always outstrip them by its eastward motion, so that all other bodies [in the air] would seem to fall behind and move westward.[14]

As an astronomer, Ptolemy argued from relative speeds of motion, rather than absolute directions of natural (and forced) motions as did the Philosopher. He said that some people had proposed rotation of the earth on the axis of the heavens. If any of them had introduced the tower argument, Ptolemy would surely have discussed that too; he did not, so it appears that at his time the tower argument had not been used. He himself, of course, had no reason to invent it, because his treatment of objects thrown into the air applied to any direction, including the vertical.

Nor did Copernicus reply against the tower argument, though one would expect that, if the idea had been proposed before 1543. He did reply against Ptolemy's argument, adducing a reason which could be called semi-Aristotelian – that any motion natural to the earth would be natural also to heavy bodies containing earthy matter. Aristotle could have granted that; he simply denied that any motion except straight is natural to the earth. Copernicus held circular motion natural to a sphere, whence rotation was natural to the earth as a sphere; but that has no bearing on the tower argument as such.

David K. Hill drew from Galileo's handling of the projection argument some damaging conclusions about his intellectual honesty. Feyerabend concluded from Galileo's handling of the tower argument that he had used (and was obliged by the nature of science to use) deceit, trickery, and persiflage to persuade his readers of the truth. Galileo's own view was that truth is such as to make itself known in a variety of consistent ways, while the defender of false opinions is obliged to resort to quibbles and unsound reasoning. No reason appears why anyone who said that should deliberately jeopardize his truths by deceiving his readers.

Finocchiaro says that the purpose of composing the *Dialogue* was to prove motion of the earth. Proof, however, is the logical equivalent of assertion, and assertion had been forbidden by the Church edict of 1616. One might expect a modern logician to seek a different purpose. Discussion without positive assertion was not forbidden, and that supplied adequate purpose for composing the *Dialogue*.

Galileo's rhetoric, Finocchiaro further says, did not fool anyone except two or three censors who allowed the *Dialogue* to be printed. Those two or three were not ordinary censors, but the Master of the Holy Palace at Rome and the Chief Inquisitor at Florence, authorized by the Roman authority to act according to his own conscience. The question was one of Church law, on which both were fully informed and arrived at the same conclusion. It may be good logic, but it is not good history, to postulate for Galileo a forbidden purpose and then draw historical conclusions.

NOTES

1 *Against Method* (London, 1975), pp. 70–92 passim.

2 *Galileo and the Art of Reasoning* (Dordrecht and Boston, 1980), pp. 192–9, 250–1, 282–8.

3 Passages that Feyerabend misconstrued as implying the tower argument are cited in full and discussed toward the end of the present paper.

4 *Opere di Galileo, Edizione Nazionale*, edited by A. Favaro (Florence, 1932), II, 223–4.

5 See D.K. Hill, "The Projection Argument in Galileo's *Dialogue*," *Annals of Science*, 41 (1984), 109–33; S. Drake, "Galileo and the Projection Argument," ibid., 43 (1986), 77–9.

6 It was possible to do this by using the times-squared law of distances from rest in free fall. Again, the situation is made clear by Newton's statement that the motions of bodies due to gravitation do not change their common centre of gravity.

7 Cf. A. Koyré, "A Documentary History of the Problem of Fall ... " *Transactions of the American Philosophical Society*, new series, 45 (1955), 329–95. Equivalence of Riccioli's argument to the tower argument is not complete, but see p. 352, n. 107, concerning his inability to grasp Galileo's principle of relativity of motion.

8 I have not seen every such book, to be sure, but it seems likely that so plausible an argument would be widely taken up, and that in particular it would not have escaped Colombe, discussed next. For his treatise see *Opere*, III, pt. 1, 253–90.

9 His experiments had been made while he was living at Padua, where a Jesuit had heard him make the same assertion. They had probably been carried out by 1604; see my *Galileo at Work* (Chicago, 1978), p. 84.

10 Galileo, *Dialogue* ... translated by Drake (Berkeley, 1953), p. 145.

11 In the Loeb edition (London, 1960), see pp. 241–5.

12 See *Dialogue*, p. 139 for a freer translation. The star marks the place at which Galileo, as previously in his syllabus, began to extrapolate from the text of an ancient writer.

13 I say "seems" even though I believe it was, because Galileo's word *rivolgimento* literally means "revolution" rather than "rotation," and of course Aristotle's reasoning depended on a local motion of the centre of the earth.

14 *Almagest* I: 7; *Great Books of the Western World*, edited by M.J. Adler (Chicago, 1952), xvi, 12.

Galileo Gleanings – X
Origin and Fate of Galileo's Theory of the Tides

I

Galileo's correspondence with Kepler opened with a letter from Galileo dated 4 August 1597, in which he declared that he had been an adherent of the Copernican theory for several years, and stated further that by means of this theory he had been able to discover the causes of several physical effects which could perhaps not be satisfactorily explained under the geostatic theory.[1] Had he written soon again to comment on Kepler's *Prodromus*, as he said he would, he might also have disclosed precisely what effects he thus referred to.[2] Scholars who have pondered over this question have generally concluded that Galileo must have had in mind the tidal theory he developed at length in the *Dialogo* nearly half a century later. As Emil Strauss pointed out, no terrestrial phenomenon except that of the tides was ever maintained by Galileo to be incapable of explanation (without recourse to the miraculous) under either cosmological theory.[3] Nevertheless, the earliest surviving mention by Galileo of his tidal theory came in 1616, when it was embodied in a discourse addressed to Cardinal Orsini.

Until quite recently no concrete evidence had been published, so far as I know, that anyone had formulated such a tidal theory as early as 1597. This theory, which was highly esteemed by Galileo and was vigorously claimed by him as his own invention, remains virtually the only thing he ever published for which no rival claim of priority was ever entered, either in his own time or by recent historians, on behalf of someone else. Probably it has gone unchallenged because in his day it was dangerous to champion the assumptions of the earth's motions on which the theory rested, and since his death the theory itself has

Reprinted from *Physis* 3 (1961), 185–94, by permission.

generally been considered worthless. A review of these matters is the purpose of the present paper.

Kepler himself, as early as 1598, conjectured that the physical effects hinted at in Galileo's first letter were the tidal phenomena.[4] Kepler, however, assumed that Galileo had in mind only the mistaken idea, broached by Kepler's friend Herwart, that any motion of the earth would cause a commotion in the waters and in the air, producing tides and winds.[5] Accordingly he gave no serious thought to the matter, remarking only that the evident relation of the moon's motion to the tidal periods made it necessary to base any rational explanation of the tides on the action of the moon. Now Galileo's theory depended not on a single motion of the earth, but on a composition of the diurnal and annual motions, and Galileo thought he could account for an apparent influence of the moon on mechanical grounds without need to invoke any special dominion of the moon over the waters as such. It was for Kepler's apparent support of the latter idea that he was respectfully reproached by Galileo later in the *Dialogue*; for to Galileo, such occult qualities were the bane of natural philosophy.[6]

The conjecture that the tidal theory later put forth by Galileo had already been formulated as early as 1597 was finally verified by publication in 1951 of certain entries from the notebooks of Paolo Sarpi concerning philosophical, physical, and mathematical problems. Sarpi's original notebooks have perished, but reliable copies are extant which preserve even the occasional marginal notations of dates that serve to establish a rough chronology of the entries. The notebooks contain 655 numbered paragraphs written during a period of about thirty-five years, but chiefly from 1578 to 1597. Among these there are three consecutive paragraphs dealing with the tides, written in all probability in 1595. The content of these paragraphs may be translated as follows:

569. – Any water carried in a basin, at the beginning of its transport remains behind and rises at the rear, because the motion has not been thoroughly received; and when stopped, the water continues to be moved by the received motion and rises at the front. The seas are waters in basins and by the annual motion of the earth make that effect, being now swift, now slow, and again average through the diurnal [motion], which is seen in the moving of the basin diversely. And if the seas are so large that they have a [quarter-sphere (?)] so that part is in the swift and part in the average [motion], there will be a greater diversity, and still greater if they have a half-sphere, so that part is in the swift and part in the slow [motion].

570. – Hence it is manifest that lakes and small seas do not make this effect, by being insensible basins. It is also manifest how the variation of the eccentricity [i.e., differences in latitude (?)] varying the ratios of the annual and diurnal motion equalizes [*sic*] the augmentations and decrements. Also it is manifest how the various positions of the

shores may cause variation, if their length be along or across the motion. Finally it is manifest how the motion of the seasons, carrying the shores now to one site and now to another, makes an annual variation of the augmentations and decrements.

571. – The motion of the sea will therefore be a composite of two things: first, of the resting behind and rising of the water when the motion passes from slow to swift [in opposition] to its natural tendency to return to equilibrium; the other, of following the motion when it passes from slow to swift [*sic*], and hence of rising by its natural [tendency] to return to equilibrium.[7]

The notebooks contain nothing else to suggest that Sarpi was a Copernican. There are but two other references to any motion of the earth, and one of these is in a Cusanian rather than a Copernican sense, while the other is dubitative; there is only a single mention of the name of Copernicus, and that occurs in connection not with the motion of the earth but with a theorem of Apollonius and its relation to a dictum of Aristotle's.[8] The absence of any previous or subsequent notes relating to the tides or to arguments for or against Copernicus suggests that these paragraphs may record an interesting theory heard from Galileo, with whom Sarpi often talked at Venice or at Padua in these years, rather than a theory which Sarpi himself had thought out. The compactness and relative completeness of the summary tend to support such a view. In no place is the name of any of Sarpi's contemporaries mentioned, and though in general the thoughts recorded may be presumed to have been his own, there is nothing to obviate the assumption that on occasion he noted down the ideas of others.

An examination of the text of the three paragraphs cited also lends support to this conjecture. In particular, the confusions or contradictions marked by the word *sic* occur at places where it would seem that the author of the ideas could scarcely fail to be precise. In the first of these, whatever may have been intended by the word "eccentricity," the context seems to demand the phrase "renders unequal" rather than the word "equalizes." In the second, it is apparent that the motion of the container and not of the water is the subject of the verb "passes," and hence that the order of the words "slow" and "swift" should be reversed. The examples of the quarter-sphere and half-sphere seas are trademarks of Galileo's two later expositions, one of which was widely circulated during Sarpi's lifetime. And the reference to the seasonal motion (Copernicus's "third motion") is suggestive. If the originator of this tidal theory was a literal adherent of Copernicus, he ought to have expected a special tidal effect from that motion. Galileo, however, recognized it as an inertial phenomenon and not a true motion, so he limited its effect to a gradual annual alteration of the components in his primary cause of the tides, precisely as the writer of the summary has done. It is highly improbable that Sarpi should have won through to this conception in a single stroke, without having made any critique of or other note on the "third motion" of Copernicus.

Yet if Sarpi made these entries to record a theory heard in conversation, it is undeniably odd that three separate paragraphs rather than a single one should be devoted to it. Perhaps the last two entries represent additions made after Sarpi had raised further questions which were clarified in subsequent conversatons. In support of this idea, attention is called particularly to the third paragraph, for what it says is not quite what it appears to be at first glance. This paragraph is not an attempt to summarize the general cause of tides, which was said in the first paragraph to be the unequal motions of parts of the earth. In the final paragraph the word "sea" (*mare*) is used in the singular, and refers not to seas in general (as does the plural in the first paragraph), but probably to the Adriatic (or perhaps the Mediterranean) in particular. The period of the tides there was known to Sarpi, and it would be quite natural for him to object that this did not conform to the implications of the proposed primary causes, after he had had time to consider them. And in fact the principal explanation Galileo offered for this was the reciprocal coursing of water seeking equilibrium, which varied with the length of the basin. A person recording the answer to a specific query would be far more likely to abbreviate this explanation, as Sarpi has done here, than would the originator of the theory when working it out.

Considerable indirect evidence that Sarpi was not the author of this tidal theory is afforded by writings of his intimate friend and first biographer, Fulgenzio Micanzio, who was also a fast friend of Galileo's. Micanzio had seen a treatise of Sarpi's on the tides (now lost), addressed to one Marioti.[9] He was also thoroughly familiar with Galileo's theory as set forth in the *Dialogue* and with his uncompromising claim to its authorship in the preface to that book. Micanzio not only praised the *Dialogue* for its clarity and originality, but undertook at Galileo's request to carry out further observations for him of the tides at Venice. Yet nowhere in his biography of Sarpi, published after Galileo's death, or in his correspondence with Galileo, is there any suggestion that Galileo owed to Sarpi any part of his tidal theory. In contrast, Micanzio makes it clear in the biography that Sarpi had more than a little to do with Galileo's success in constructing his first telescope.[10] Furthermore, he is very severe with Fabricius of Aquapendente for not giving Sarpi proper credit for his discovery and explanation of the valves in the veins.[11]

II

The fate of Galileo's tidal theory is a curious one, and deserves even more attention than can be given to it here. It was warmly accepted by friends of

Galileo's like Sagredo, who had the critical ability and honesty to challenge certain other ideas of his esteemed teacher. It continued to find some support from other able men up to the promulgation of the theory of universal gravitation by Newton in 1687. Foremost among these was John Wallis, who admired the device by which Galileo attempted to account mechanically for the well-known relation of the motion of the moon to the tides, and he attempted to improve on this.[12] After Newton, of course, Galileo's theory fell into oblivion except for occasional comments by biographers and historians.

The opinion of most of these writers has resembled that of Drinkwater, who recognized the ingenuity of Galileo's theory but assured his readers that despite its plausibility, it did not in the least account for the tides, for reasons too technical to set forth.[13] Emil Strauss, however, opined that the causes alleged by Galileo might have some tidal influence which was too small to be detected under the much greater effects of gravitation, and even of winds and currents.[14] Most subsequent historians have preferred the view of Ernst Mach, who maintained that the Galileo effect was entirely illusory, and attempted to illustrate this by means of an ingenious thought-experiment utilizing the Galilean principle of relativity of translatory motion.[15] Emil Wohlwill, writing in 1909, remarked on Strauss's credulity but stated that he himself had encountered technically trained persons who leaned toward a similar view.[16] The most recent writers have agreed that no such effect is possible, and have suggested that Galileo either blundered into an absurd contradiction of his own principles of mechanics out of fondness for his own invention, or saw the contradiction and deliberately played upon the ignorance of his readers.[17] Some say that Galileo's refusal to accept the theory of Kepler, with its remarkable anticipation of the theory of gravitation, is a sad commentary on Galileo's attitude toward his great contemporary. It does not appear to me that any of these widely varied commentaries hit the nail on the head, while some of them merely illustrate the perils that assail the historian of science when he fails to assess the views of earlier periods in terms of the knowledge available to the men who first expressed them, or attempts to construe too literally some of the generalizations of modern mathematical physics.

Let us consider first the notion that Galileo's theory was absurd or is demonstrably wrong in the light of modern physics. This may be so (though I am inclined to doubt it), but whether true or not, it is certainly not settled by the thought-experiment of Mach. In this, he imparts to an isolated rotating globe of water a translatory motion, and correctly asserts that this will have no further effect. One may feel that he is justified by the fact that the motion of the earth round the sun for a single day is pretty much a straight line. But this empirical fact is irrelevant to an evaluation of Galileo's theory, which did not attempt to discuss quantitative effects, partly because Galileo believed them to be the product of the widest diversity of causes independent of his primary thesis.[18] Galileo's theory depends specifically on the compounding of two circular motions, and not of a rotation and a translation. Hence in order to attack his theory, one is obliged to consider both the annual and the diurnal motions as circular, and Mach did not do this. With that in mind, let us attempt our own thought-experiment, confining ourselves to a framework which Galileo himself might have used.[19]

Imagine a grindstone pierced by two fairly large holes placed near the rim and about sixty degrees apart, connected by a cylindrical passage also near the rim and of fairly large diameter. One of these holes is to be filled with colored water, and both are then to be covered with glass to permit observation. Let the grindstone be rotated horizontally on an axle held in a U-shaped fork to which a long cable is attached, provided with a handle at one end and a saddle near the fork. As we crank the stone to a moderate speed, we shall see the water recede into the hole which is further back in the direction of rotation, but once we have got it to the desired speed and let it rotate freely, the water will distribute itself along the passage and between the two cavities in such a way as to get as close to the rim as possible. Next we mount the saddle, and at our command an Olympic hammer-slinger begins to revolve the whole framework while we watch the grindstone from our seat in the saddle. With each rotation of the stone on its axle, the distribution of the water between the two holes and their connecting passage will be altered cyclically. The tidal effect thus produced would vary considerably for different relative radii and speeds, bur reciprocating flows would undoubtedly take place.

It is true that our thought-experiment appeals to a physical intuition of tangential velocities (or centrifugal forces), whereas Galileo appealed to differential velocities at the ends of the cavities. Also ours does not produce its effect on external seas, though it may easily be modified to do this.[20] And the reciprocal inertial motions of the waters, which play a large part in Galileo's theory, will scarcely be perceptible in our model at most speeds. These and similar objections, however, seem to me inconsequential for our present purpose, which is to induce some spirit of scepticism in historians of science with regard to the application of physical principles of great generality when attempting to criticize the thought of early physicists. Whether the "Galileo effect" could be observed in some hypothetical universe in which the attracting bodies were given such masses, distances, and periodic positions as virtually to cancel out their tidal effects, I do not know; but I should not wonder if mathematical physicists could produce such a universe to order. It is less to the discredit of Galileo that he happened not to live in such a universe, than it is to the discredit of his modern critics that they believe such a universe to be impossible, or submit his ideas to critiques inconsistent with his stated postulates.

A typical criticism of Galileo's tidal theory, for instance, asserts that it would imply a high tide precisely at noon every day. One critic has gone so far as to say that Galileo postulated such a condition.[21] As a matter of fact, Galileo postulated only that the tides must be accounted for by motions of the earth; as to periodicity, he asserted that there was no general period for tides, but that each basin had its characteristic period. In any case, a contemporary of his had pointed out to Galileo himself this supposed implication of his theory of the tides before he published it.[22] Hence it is evident that he believed he had freed the theory from this objection. Galileo believed that virtually no rising or falling of water was to be found except at the extremities of large seas; in the central parts, he thought, the water wold merely course more rapidly in one direction or the other. Hence it would make a good deal of difference in the time of high tide whether the observer happened to be at the end of a sea-basin or on a tiny island in a large ocean. Perhaps it would also make a difference whether he was at the eastern or western end of a given basin, for it is not at all obvious that just because the maximum high tide would occur at noon (or shortly thereafter) at the eastern end, when the western end (on Galileo's theory) was still retarding its motion as it approached noon, it would also follow that the highest tide of the day would occur at noon at the western end, when the eastern end was accelerating away from it.

But let us suppose that Galileo's theory did entail only a single daily cycle of tides, and that he failed to note this or to offer any explanation of the discrepancy between this consequence and the most elementary observations of the phenomena at Venice. Suppose further that Galileo read and pondered over Kepler's theory as set forth in the preface to the *Astronomia Nova* (as indeed he had done). Now on what grounds might one ask him to abandon his own theory and accept that of Kepler? How many daily cycles of tides would Kepler's moon-attraction theory produce? Not even one. Kepler's theory, instead of accounting for two tides a day in the Adriatic, would account for but one tide anywhere in nearly twenty-five hours, so he would be even farther from the truth than adverse critics assert that Galileo was (neglecting all but the primary causes in his theory). In short, Kepler made no effort whatever to square his theory with the facts, which is perhaps fortunate, when we consider that one of his forerunners had accounted for a double tide under the moon's attraction by endowing the moon with power to confer upon the opposite sign of the zodiac its attractive influence – a theory which Galileo also duly noted and rejected, in terms much less complimentary than those which he applied to Kepler when he reproached him for lending his ear to the doctrine of occult properties.[23]

An amusing aspect of the whole affair is that, in a Pickwickian sense at least, Galileo was correct in his fundamental postulate – that without invoking the miraculous, one cannot explain the tides on the assumption of a truly immovable earth. For the double period of the tides is now attributed to a literal motion of the earth away from the waters on the side opposite the attracting body. In place of criticizing either Kepler or Galileo, one might well say that both were correct in their physical intuitions, the one seeking an explanatory motion of the earth and the other seeking an explanatory attraction of the moon. Neither found precisely what was required; yet to anyone who might have accepted Galileo's postulate and Kepler's attractive force, allowing the earth then a motion in response to that force, the essential cause of the basic tidal movement should have become apparent. Only prejudice leads men to see profound insight in one approach and blind recalcitrance in the other.

NOTES

1   *Opere* X, 58. Kepler is usually said to have opened the relationship by sending to Galileo a copy of his *Prodromus*. There is no evidence that either Kepler or his teacher Maestlin had yet heard of Galileo, who had published nothing at this time. Kepler had asked a friend, Paul Hamberger, to take two copies of the newly published book to Italy, where he left them with Galileo. It is quite clear that there was no letter of transmittal, and when Kepler mentioned Galileo's letter of thanks he spoke of him to Maestlin as "*Paduanus Mathematicus, nomine Galilaeus Galilaeus, uti se subscripsit.*"

2   Because Galileo's failure to do this has recently been construed as churlish, it may be remarked here that on the contrary it was an act of charity. Having read beyond the preface, Galileo could only have found himself in a sea of Pythagorean speculations of the sort most distasteful to him; the Kepler of 1596 and the *Prodromus* was by no means the Kepler of 1609 and the *Astronomia Nova*. A contemporary troublemaker accused Galileo to Kepler of appropriating his discoveries as his own. But what discoveries are there in the *Prodromus*? Surely Galileo was not the man to pretend he had discovered that God arranged the planets according to the number and dimensions of successive regular solids. It is painful to jeer at Kepler even in his wilder unscientific flights; I should not do so in the absence of unprovoked attacks on Galileo's character, and I hope others will note that Galileo himself did not do so at all, for which his motives have recently been impugned.

3   *Dialog über die beiden hauptsächlichen Weltsysteme* (Leipzig, 1891) XVII.

4   *Opere* X, 72. E.J.Aiton, "Galileo's Theory of the Tides," *Annals of Science*, vol. 10, n. 1, p. 48, states that Kepler's rejection of a tidal theory based on terrestrial motion was published in his first book, but in fact it occurred only in a letter to the Jesuit

J.G. Herwart and in response to a specific suggestion; see Kepler's works, ed. Frisch, II, 61; Caspar XIII, 178.

5 Theories attributing the tides in general terms to some sort of terrestrial motion had been put forth even in antiquity. Galileo mentions Seleucus in this connection; E.J. Aiton (loc. cit.) mentions Caesalpinus and states that Galileo was familiar with his work.

6 *Opere* VII, 486.

7 Paolo Sarpi, *Scritti Filosofici e Teologici*, ed. Romano Amerio (Bari, 1951), 115. Many of the *Pensieri* contained in this work are of great interest with respect to Galileo, and will be dealt with on another occasion.

8 Ibid., 29, 104, 85.

9 *Vita del Padre Paolo* ... (Leyden, 1646), 105.

10 Ibid., 211.

11 Ibid., 43–44.

12 See E.J. Aiton, op. cit., 51–54.

13 *The Life of Galileo* ... (London, 1829), 71 c. 2 and 73 c. 2. In his discussion of the tidal theories, Drinkwater cites an interesting anticipation of Kepler's view by the Jesuits of the College of Coimbra.

14 Strauss, op. cit., 566.

15 *The Science of Mechanics* ... [tr. McCormack] (La Salle, 1942), 263.

16 *Galileo und sein Kampf* ... (Hamburg, 1909), 1, 600, n. 2. In the copy of this work presented by Wohlwill to Mach, these passages have been underlined in red by the eminent Viennese physicist.

17 E.J. Aiton, op. cit. 46; Arthur Koestler, *The Sleepwalkers* (London, 1959), 453–454 and 464–466.

18 A few minutes' study of the article "Tides" in the Britannica, 11th ed., has a chastening effect. Speaking of the Laplace-Hough attack, the writer remarks: "This theory, although very wide, is far from representing the tides of our ports. Observation shows, in fact, that the irregular distribution of land and water and the various depths of the ocean in various places produce irregularities in the oscillations of the sea of such complexity that the rigorous solution of the problem is altogether beyond the power of analysis." The reader will find several similar passages in Galileo's *Dialogo*.

19 Galileo remarked that it is not possible to impart different speeds to the ends of a vessel of practicable experimental size (*Opere* VII, 455), and yet on the following page he says he can give the construction of a mechanical model which will exhibit the effects of such compounding of motions as he has described. The contradiction is more apparent than real, and for exhibiting the effects he may have had in mind something similar to the proposal which follows.

20 For instance, one might cut a single long crescent from the edge of the grindstone and fill this with water, held in place by a tough but flexible membrane which when

at rest would conform to the shape of the grindstone. Rotation would then produce a bulge, and when revolution was added, it would cause the bulge to change position cyclically. Whether the tide was high at "noon" or "midnight" in this model would depend on whether the observer (say an ant) stood at the edge of the "sea" or was perched on an island provided at its center.

21 Koestler, op. cit., 454.
22 *Opere* XII, 450.
23 *Opere* VII, 446. The theory was propounded by Marcantonio de Dominis in *Euripus sive sententia de fluxu et refluxu maris* (Romae, 1624).

8

History of Science and Tide Theories

Modern tide theory is inadequate for predictions useful to navigators; such predictions are made, port by port, through the use of ingenious machinery. Planetary theory, on the other hand, is perfectly adequate for useful predictions. Tide theorists are aware of the facts that must be taken into consideration for useful predictions, but they do not possess all the information necessary for making them, nor is it likely that they ever will, by reason of the inordinate complexity of interrelations among those factors. In view of this, the word "inadequate" is a rather ambiguous one when applied to previous tide theories, in a sense that does not apply, say, to the case of planetary theories before Kepler, or Newton, or Laplace.

Two main types of tide theory are recognized as scientific in character: equilibrium theories, which began with Newton and became modified and improved by Daniel Bernoulli and others, and dynamic tide theories, which began with Laplace and have been developed by Airy, George Darwin, and later specialists. Newton's equilibrium theory based on universal gravitation had a precursor in the suggestions made by Kepler in the introduction to his *Astronomia Nova* of 1609, though some may say that without the general gravitational law no pre-Newtonian theory truly qualifies as scientifically akin to Newton's tide analysis. But putting aside questions of the proper use of terms like "adequate" and "scientific" in history of science as at least partly a matter of taste, it may be said that those who regard Kepler's theory as an adumbration of equilibrium theories should also, to be consistent, regard Galileo's theory, set forth in the Fourth Day of his 1632 *Dialogue*, as an adumbration of dynamic theories.

Reprinted from *Physis* 21 (1979), 61–9, by permission.

Absence of the gravitational law in Kepler's theory has its counterpart in absence of the concept of force in Galileo's theory, which may be called a purely kinematic theory – not dynamic, but strikingly similar in its emphasis on accelerations and decelerations to later dynamic theories.

If what has just been said is correct, the question will naturally arise in the minds of readers why it has entirely escaped attention in the past. That it is correct, and will in time come to be recognized as so by historians willing to devote some time to the study of modern tide theory and its history, I am confident from my own intermittent studies over a period of many years. Many reasons, themselves almost as complex and interrelated as the factors that determine tides at a given port, account for lack of recognition of the situation both by scientists and by historians. It is appropriate, and perhaps is enough, to touch on only a few of these and in no great detail, since others, and other aspects of these, will inevitably suggest themselves to those who will study and reflect on the matter.

First, as to scientists, rather few are interested in history of science for its own sake, and fewer still concern themselves with it for the sake of science. What is known of older scientific theories to working scientists is generally the result of desultory reading, not of direct consultation of original sources. Hence it is hardly to be expected that even the most eminent tide theorists after Newton would trouble to examine Kepler's *Astronomia Nova* or Galileo's *Dialogue*, as nothing useful to their advanced work could possibly be found there. It is evident that Laplace and G.H. Darwin, for example, had only vague and incorrect notions of Galileo's theory, derived probably from casual reading of summary accounts of it by historians. Thus J.F. Montucla, writing in French at the time of Laplace, had dealt very briefly with that and had dismissed it as inadequate, while J.E. Drinkwater-Bethune, writing in English before Darwin's time, remarked that Galileo's theory was unsound for reasons too technical to explain in a popular biography. Such accounts would scarcely attract further interest among scientists, though Montucla and Drinkwater provided better summaries of Galileo's tide theory than do most modern historians of science.

Turning, then, to modern accounts by professional historians and philosophers of science, it may be said that from the time of Ernst Mach to the present what has usually been presented as Galileo's tide theory is not that at all, but a greatly oversimplified parody or caricature of it. This has been particularly true during the past quarter-century, for reasons which seem to me detrimental to sound historiography. I shall return to this presently, after first mentioning a few points essential to an understanding of the topic.

Galileo's explanation of the tides was divided into two principal considerations, the first dealing with the existence of tides and the second dealing with their periodicities. To account for the existence of tides, he had recourse to a concept of "absolute speed" of any point on the earth's surface, whether occupied by a solid or by a fluid. This concept was based on the distance traversed, regardless of direction, by a point rotating around the earth's axis while that axis was carried around the sun. It is evident that such distances, for equal units of time, vary continuously during each day, and that the absolute speeds so determined will be appreciably different for points near the two east-west extremities of any very large sea-basin in the east-west direction, though they will not differ appreciably between points having little east-west separation. Hence in a large sea, water at one extremity east or west will move toward the other extremity, giving rise to tidal displacements. The periods of observed tides, however, Galileo considered to be entirely independent of the period of this source of displacement, and in each sea having tides he believed the tidal period to be determined principally by the length and depth of its basin. Their reciprocations were attributed to the weight of water accumulating above the level at which the sea would rest if left undisturbed, which weight overpowered the primary disturbing cause (difference between absolute speeds at the east-west extremities) and introduced observed oscillations whose periods had no necessary theoretical connection with the cyclical period of the original disturbance.

Galileo was wrong in supposing this mutual independence of observed tidal periods from the period of regular disturbance, but that was first proved by Laplace. Not that Laplace paid any attention to Galileo's tide theory, for he did not; but Laplace proved in general that, given sufficient time for the stabilization of frictional effects, the period into which a fluid settles for its oscillations is governed by the period of regular cyclic disturbances to which it is subjected. That interesting fact was known not even to Newton, let alone to Galileo, and inattention to it was no more and no less an "inadequacy" in one theory than in the other. This is of no interest to scientists, who are entitled to condemn past theories as inadequate indiscriminately; historians, however, are obliged to assess the scientific character of old theories in terms of the state of knowledge at their times.

Now, Ernst Mach was an avowed and enthusiastic relativist, whence he repudiated Galileo's conception of "absolute speed" without attempting to analyze it. "In the most naive way," Mach wrote, "he considers the fixed stars as the new frame of reference," using a phrase later made famous by Einstein's relativity theory, but one which is out of place in Galileo's tide theory. For that, any point not sharing in the earth's particular rotation and revolution sufficed; an observer on Venus or Saturn would not ascribe the same "absolute motions" in Galileo's sense to points on the earth's surface as would an observer on the sun, but all such observers would ascribe to them different absolute motions and would recognize the accelerations and retardations essential to Galileo's account of the tides. Mach also remarked that "Galileo's theory" could account for but one high tide each day. I have placed the phrase in quotation marks because without them the previous sentence would be false; Mach never considered Galileo's tide theory as a whole. What Mach meant was that Galileo's primary cause of disturbance cold not account for our observed tides, of which two generally occur between two successive identical positions of the moon with respect to a meridian. That was precisely why Galileo did not stop with his primary cause, but went on to assert the independence of observed tidal periods from his primary cause and introduced a second phenomenon capable of overpowering the primary cause and inducing periodic reciprocations which, in the Mediterranean, happened to be about six hours from high to low tide. Mach did not ignore this, but he dealt with it separately as a lame and inadequate attempt to fit facts to a wrong (because absolutist) theory. This view has been accepted by most who have written on Galileo's tide theory during the past quarter-century, with embellishments which perhaps Mach himself might deplore. I find it not without interest that for historiographical reasons, those same writers are generally scornful of Mach as a historian of science.

Now, it is debatable whether Galileo was obsessed with a theory and would not give it up no matter what his own eyes showed him, suggesting lame apologies for mere facts, or whether he thought hard and well as a physical scientist to achieve an integrated theory that would account for all the facts known to him. The overwhelming majority of professional historians and philosophers of science are presently on Mach's side. But though agreement among specialists is by and large the only known criterion of truth, it has sometimes happened to support error. It was the habit of Galileo to ignore small deviations between theory and observation, but to protest vehemently against the acceptance of theories widely at variance with direct experience. One well-known instance is his defence of Archimedes in the treatment of plumb-lines as parallel; another is his ridicule of those who would support Aristotle's doctrine of falling bodies against equal speeds of fall through air, when actual test showed that four inches separated weights which under the Peripatetic theory should be separated by many yards. There is accordingly reason to believe that Galileo considered his tide theory to be a viable whole, and not a single basic cause with patchwork additions. Mach knew the Laplacian proof mentioned, and perhaps it seemed to him that any physicist should always have known that much about fluid behavior; historically, however, not even Newton or Bernoulli did.

Mach's misrepresentation of Galileo's tide theory was so complete that he offered a refutation of it in which the rotation and revolution essential to Galileo's analysis were transformed into a rotation and a rectilinear translation. He further offered, as a working model of tides induced by an attracting body on a stationary earth, a device which could produce a tidal bulge on only one side, not on both. Perhaps because Mach was a physicist, historians of science since his time appear to have accepted such pronouncements uncritically. Mach, however, abstained from saying that Galileo's tide theory was self-contradictory, or that it contradicted other parts of Galileo's physics; such reproaches have been added since 1953, when Giorgio de Santillana first introduced them, so far as I know. Previously it had been usual for historians to say only that the theory was inadequate; since 1953 it has become acceptable to say that it was falsified by the most ordinary observations and that Galileo probably adopted it only out of overzealous Copernicanism. Arthur Koestler embellished the supposed implication that only one high tide a day could be accounted for by declaring that that tide must occur at noon, and the most recent writer I have read adds that Galileo himself explicity insisted on that, though no time of day was mentioned in the *Dialogue*. Thus the departure of presentations of Galileo's theory from what he wrote goes ever widening, even into areas in which (as in the case just mentioned) it could make no difference to the author's purpose whether Galileo's theory was or was not correct.

The origin of this indifference on the part of historians of science to original texts appears to me to lie in an increasing professional fascination with interpretations of motives underlying the proposal and the acceptance of scientific theories. While I have no reason to oppose that activity, its value depends on accurate knowledge of the scientific theories proposed or accepted, concerning which original texts are of paramount importance.

II

It is of interest that essentially new tide theories were offered early in the 17th century by Kepler and Galileo, both of whom were Copernicans. Kepler's theory was a lunar-bulge explanation, and one might say that it was not truly novel, since the Jesuits of Coimbra College had previously suggested an attraction of terrestrial waters by the moon, and prior to that, from antiquity, the complaisance of the god of waters to blandishments of the moon goddess had been frequently recognized. Yet Kepler's account of the tides was accompanied by a remarkable passage adumbrating a kind of universal gravitation not present in earlier accounts, whence there is reason to regard it as a precursor of the Newtonian equilibrium theory in a sense that others were not. The gravitational rule suggested by Kepler was that moon and earth, if not otherwise held apart, would come together at a point determined by the inverse ratio of their bulks. Unable to move toward the earth or to move the earth toward it, the moon was able only to draw up the seas to some extent, creating a bulge that began to take shape when the moon was well above the horizon; this followed the moon westward until it reached the shore.

In Galileo's theory there was no bulge of water; water rose only against a coast and only because it was moving in that direction relative to the land beneath it, whether because of "absolute speeds" resulting from the earth's doubly circular motion or because of reciprocation from an earlier high tide at the other coast. In Kepler's, or any other lunar-bulge theory, tides should occur on mid-sea islands as well as along mainland coasts, whereas in Galileo's theory mid-sea islands should have no tides, as indeed they do not have. Newton's theory added attraction by the sun to that by the moon, so it was not just a lunar-bulge theory, but any theory of lifted and moving water, taken literally, would seem to imply island tides that do not occur in fact.

I believe that most people are taught in school that the moon raises a bulge in the water beneath it, while the sun does the same to a lesser degree. In pictures illustrating the effect, this bulge is greatly exaggerated, and there may be a tendency to associate its height with that of more or less normal tides of a person's own experience. In equilibrium theories it was assumed that such a distortion of sea level could and would be maintained by gravitational forces, though that would be true only if our seas were much more viscous than salt water. In dynamic tide theories, the water is considered rather to lose some weight under an attracting body than to rise up toward it, because water levels itself very rapidly under the earth's pull. The maximum theoretical rise under optimum conditions in a large sea near the equator is, I believe, about one meter, and the normal rise considerably less. What counts in dynamic theories (and gives them their name) is the action of tide-producing forces, which produce speeds, accelerations, and decelerations more horizontally, so to speak, than vertically. In that regard Galileo, though mistaken about the principal source of such speeds and changes in speed, and silent about forces (which in traditional physics had no place in nature), accounted for tidal rises and falls along long coasts in a manner resembling that of modern theory.

To return to the school children, they are taught (or were in my day) that when moon and sun pull seawater toward them, they also pull the earth, and the earth's center of gravity being some 4,000 miles farther away, they pull it less, by the inverse square law of distances, whence the water on the side near moon or sun makes a net rise. But the water diametrically opposite, being pulled even less hard than the earth (by the same rule), is left behind, so to speak, and bulges out. In this way two tides per day, or lunar day, are accounted for, in a simplified form of Newton's equilibrium theory, though that theory was long ago abandoned in science. Some children, more curious than others, want to know why, if the whole earth is moved a few feet every day closer to the moon, it has not long ago bumped into it. Those children are told not to think of what happens as a literal motion, but as gravitational "force" toward the moon, and to think of this force not really as a force but rather as an acceleration, which Newton had discovered to be the same as a force, or at least always associated with a force. Such attempts to direct children to a dynamic theory starting from an equilibrium theory are usually successful only to the extent of discouraging them from trying to understand the tides, or of convincing them secretly that the teacher does not know the answer himself.

It is possible that people generally, and historians of science in particular, are now much better informed about the ocean tides and their theoretical explanation than when I went to school, but I have seen little evidence of that in recent papers dealing with Galileo's theory. The kinship of Kepler's theory, however primitive, with Newton's theory is widely recognized, because the writers know something about the latter. The kinship of Galileo's kinematic theory with dynamic theory, on the other hand, has never been mentioned by historians to my knowledge, perhaps because they remain unacquainted with the latter. Instead of noting that Galileo appeals repeatedly to accelerations, retardations, and differential speeds of water, they feel the absence of any appeal by him to forces, least of all to "the force of gravitation"; and finding something like that in Kepler, they deplore Galileo's rejection of Kepler's theory as a return to occult qualities. And instead of noting that Galileo lists among the "accidents" (or happenings; events) affecting tides and their periods most of the things included in modern tide theory apart from attractive forces, they treat all these as lame excuses for the failure of the very theory of which they constituted integral parts, in Galileo's view.

The net result of these oversights (or deliberate omissions) in discussions recently published in reputable journals and in books with respected imprints, in my opinion, has been a proliferation of beliefs attributed to Galileo as a physicist which would disgrace a nature mystic, let alone the outspoken opponent of vague speculation as a substitute for careful observation and reasoned explanation.

In conclusion it is appropriate to mention some opinions expressed by others at the time of Mach. In 1892 a German translation of Galileo's *Dialogue* by Emil Strauss was published, with very thoughtful notes by the translator. It was his opinion that it would be hard to fault Galileo's reasoning about the tides, and that a disturbance of the kind described by him should in fact occur in large seas, though of magnitude far too small to account for actually observed tides. It is easy to guess Strauss's analysis; for in modern physics every point on a rotating sphere (except the two poles) undergoes constant acceleration, and when such a sphere is also set in rotation around some other center, every point on it undergoes a cyclical changing acceleration. If parts of the sphere are fluid, free to move but restrained from flying off, differential motions will occur among their particles.

Emil Wohlwill, a professor of chemistry at the University of Hamburg, published the first volume of a monumental biography of Galileo in 1909. There he cited the opinion of Strauss, and though he appears not to have agreed with it, he reported in a footnote that technically and scientifically trained persons, when first told of Galileo's tide theory, usually agreed that disturbances of the seas must take place in accord with it, though not those which give rise to our observed tides. Wohlwill sent a copy of his book to Mach, who underlined these passages in red pencil; yet Mach did not subsequently alter his own remarks about Galileo's theory in latter editions of his historical treatment of the science of mechanics.

Now it is true that much has happened in physics since 1909, but nothing of such a nature as to justify historians of science in ignoring the opinions of scientists around the University of Hamburg in 1909 on a matter of this kind; nor should it be assumed that all historians of science in our time are automatically capable of understanding the kinematic implications of Galileo's tide theory, however assured they may be of his fanatic Copernican zeal, his cynicism toward the public he addressed in his writings, his mystic devotion to perfect circles, and his Platonic indifference to the world of sensible observation.

PART VI

GALILEO: MOTION AND MECHANICS, INCLUDING THE *DISCOURSES ON TWO NEW SCIENCES*

The part of Galileo's work to which Drake devoted the greatest amount of attention was mechanics, principally on motion itself, by general consent the field in which Galileo made his most important contributions. His largest publications were the translations with I.E. Drabkin of Galileo's early treatises in *Galileo on Motion and on Mechanics* (1960), the translation of the *Two New Sciences* (1974, 1989), the edition of the manuscripts of *Galileo's Notes on Motion* in facsimile (1979), an extraordinary work of identification and classification of disparate and obscure sources, and *Cause, Experiment, and Science* (1981), containing a translation of Galileo's *Bodies that stay atop water, or move in it*. However, he also wrote more papers on Galileo's mechanics than on any other subject, and mechanics is the largest section of this collection, in which the papers, many based directly upon the manuscript notes on motion, are arranged in order of publication, which is also more or less by subject. A number of the papers are critical of other interpretations of Galileo's mechanics, in particular the views of Alexandre Koyré that experiment played little or no role in Galileo's work, and of Pierre Duhem that Galileo's mechanics can be understood as an extension of medieval theoretical analyses of motion. But it will also be seen that in several of these papers Drake refined and corrected his own interpretations, in particular, the papers on the law of free fall form an interesting series in showing his own, at times circuitous, path of discovery. A theme that runs through many of these papers is that Galileo did indeed derive or confirm his fundamental principles of the motion of bodies experimentally, that is, by measurement, for which Drake found evidence in Galileo's manuscripts and in repeating some of the experiments himself, as Thomas Settle and James MacLachlan have also done in important studies.

The first two papers, "Galileo and the Law of Inertia" (1) and "The Concept of Inertia" (2), are theoretical and concern Galileo's early formulations of principles close to inertia, that is, "neutral" motions that are neither natural nor violent, as a body moving on a horizontal plane, and the distinction of such motions, both logically and historically, from earlier theories of bodies moving by acquired impetus. "Galileo's Experimental Confirmation of Horizontal Inertia: Unpublished Manuscripts" (3) is an examination of manuscript notes recording Galileo's test of the relation that horizontal distance traversed by a body accelerated through descent on an inclined plane is proportional to the acquired speed in descent. The paper written with James MacLachlan, "Galileo's Discovery of the Parabolic Trajectory" (4), uses the same notes to show how Galileo found the parabolic path of a projectile experimentally and, on the basis of a repetition of the inclined plane experiment, modifies the earlier assumption about the inclination of the plane. In "Galileo's Accuracy in Measuring Horizontal Projections" (15), published twelve years later, Drake, using the modified conclusions concerning the height of the projection and inclination of the plane, showed that Galileo's measurements were more accurate than he had originally believed.

The original study of the notes on horizontal projection along with investigation of notes on the acceleration of falling bodies, collected here in the papers that follow, are the basis of "Galileo's New Science of Motion" (5), which is Drake's most general statement of his conclusion that Galileo's discoveries in the motion of bodies were the result of his own, in part experimental, investigations rather than a continuation of scholastic treatments of uniform acceleration, a subject he also considered in other papers. It was in fact to natural acceleration in fall, the subject of most of the remaining papers in this section, that Drake devoted his most extensive investigations of Galileo's mechanics. The first, "Galileo's 1604 Fragment on Falling Bodies" (6), concerns Galileo's reasoning from an initial assumption that the speed acquired by a naturally falling body is proportional to the distance it falls to the conclusion that the distances passed in equal times are as the odd number series and consequently as the square of the times, as set out both in a page of manuscript notes and a letter to Paolo Sarpi of October 1604. Drake considerably revised his interpretation of Galileo's argument in (9) and (11).

"Uniform Acceleration, Space, and Time" (7) considers two subjects, first the manuscript evidence for Galileo's abandonment of the assumption that speed in fall is proportional to distance for the correct rule that it is proportional to time, and then, at greater length, an examination of the well-known and very much disputed passage in the *Two New Sciences* in which he presents a refutation of the earlier distance rule. In "Velocity and Eudoxian Proportion Theory" (10), the subject is taken up again in light of a criticism by M.S. Finocchiaro that Galileo's argument is defective in using a theorem applicable only to uniform motion. "The Uniform Motion Equivalent to a Uniformly Accelerated Motion from Rest" (8) is an examination of the demonstration in the *Two New Sciences* that the time in which a uniformly accelerated body beginning from rest traverses a given distance is equal to the time in which the same distance is traversed by a body moving uniformly with one half the final speed of the accelerated motion. Here Drake is concerned with a proper understanding of Galileo's proof, including the translation of the text, and its distinction from the medieval demonstrations, in particular that of William Heytesbury.

"Galileo's Discovery of the Law of Free Fall" (9) is an analysis of a page of notes in which Galileo derives the relation that in uniformly accelerated motion the square of the speed is proportional to the distance, and thus if points representing distance are placed on a line, points representing speed will lie on a parabola. Although not explicit in the notes, the proportionality of speed to time is directly implied by this relation, and Drake also revises his interpretation of the letter to Sarpi such that the quantity that Galileo there calls *velocità*, which is proportional to distance, is actually the square of the speed. The subject is taken up again in "Galileo's Work on Free Fall in 1604" (11), in which a page of notes recording measurements of distances on an inclined plane is for the first time considered and Galileo's early use of the square of velocity is again analysed. "Mathematics and Discovery in Galileo's Physics" (12) takes the analysis of the manuscript notes yet further, showing the relation of the investigation of fall to Galileo's early proof of his theorem that the times of descent on any chord in a vertical circle are equal. "The Role of Music in Galileo's Experiments" (13) describes a repetition of the inclined plane experiment using singing as the means of marking equal intervals of time: the results are very close to Galileo's.

Galileo's statement in the *Two New Sciences*, generally understood to say that the number of oscillations of two pendulums swinging in very different arcs remain absolutely equal, has been subjected to strong criticism. In "New Light on a Galilean Claim about Pendulums" (14), it is shown that the passage is generally mistranslated and that the last phrase indicates that a difference of one oscillation will be found, which tests conducted by James MacLachlan have shown to occur in 30 oscillations, after which the pendulums remain in synchronization.

Ten years after the publication of (13), Drake returned to further analysis in four additional papers on the manuscript notes reporting the inclined plane measurements to determine the acceleration of a falling body and notes relating measurements of the times of vertical fall to oscillations of a pendulum. "Galileo and Mathematical Physics" (16) considers notes concerning measurements of the times of vertical fall using a pendulum to measure the times, showing that these provided the essential step from measurement of speeds to the times-square law for distance. The notes are taken up in more detail in "Galileo's physical measurements" (17), in which it is shown that Galileo's measurements came within ½ per cent of the correct ratio of the distance of fall to the length of the pendulum swinging to the vertical, that is, through one-quarter of an oscillation. Drake also shows the remarkable result that in Galileo's units the square of the time of fall in *tempi* is equal to the length of a pendulum in *punti* timing twice the fall by its swing to the vertical. "Galileo's Constant" (18) is a yet more detailed examination of the notes on the timings using the pendulum, with a reconstruction of the order of Galileo's measurements and analysis as far as the notes permit. The reconstruction of Galileo's procedures is summarized in "Galileo's Gravitational Units" (19), which also considers the relation, pointed out in (18), that in Galileo's units the ratio of the square of the time of fall to the length of the pendulum timing twice the fall in one-quarter oscillation multiplied by $\pi^2/8$ is a universal constant for acceleration. The subject of these last four papers, the relation of acceleration to timing by a pendulum, is also taken up in the preface to the second edition of the translation of the *Two New Sciences* (1989), the monograph *History of Free Fall, Aristotle to Galileo* (1989), and *Galileo: Pioneer Scientist* (1990).

1

Galileo and the Law of Inertia

This year marks the 400th anniversary of Galileo's birth. I wish first to thank you for the opportunity of addressing you on this occasion, and to salute you for pausing today to do homage to the man who deserves, if anyone does, to be called the first to take up your distinguished profession in modern times.

Physics as taught in the universities when Galileo was a student was not a separate science, if indeed it was a science at all. It was then still a part of philosophy, a prelude to the much more important study of metaphysics, a discipline which owed its name to the fact that it lay beyond physics – above and beyond, his contemporaries would have said. Not the least effective of Galileo's tactics in securing for physics recognition as a separate science was his habit of undermining the prevailing respect for metaphysical reasoning. This speaks well for his intelligence, his integrity, and his courage, for when the whole society in which one lives is wasting time in idle speculation, it takes brains to perceive the fact, integrity to abstain from employing them in the same way, and courage to oppose so popular and comfortable a pursuit. Equipped with all these, Galileo declared openly to his contemporaries that true philosophy was not to be found printed in books, but was written in the great book of nature that stands open to our gaze;[1] he asserted that just as nature gave men eyes to see her works, so she gave them brains to understand these even without a knowledge of Latin;[2] and he contended that anyone who managed by himself to discover one single truth directly from nature could outmatch a thousand Aristotles and all the skillful orators who came armed with tradition and authorities to the contrary.[3]

In saying that Galileo may be called the first teacher of physics in modern times, I do not mean to imply that he succeeded in getting his new science rec-

Reprinted from the *American Journal of Physics* 32 (1964), 601–8, by permission.

ognized as worthy of becoming the subject of a separate course of instruction. He taught at the universities of Pisa and Padua for twenty years, from 1590 to 1610, but the courses assigned to him there never went beyond the traditional lectures on mathematics, astronomy, and military engineering. Physics remained the property of philosophy departments, the professors of which became Galileo's principal adversaries as time went on. But if Galileo did not bring the science of physics into the university program of his day, he did bring it into his private teaching, both at Padua when he was professor there and at Florence in later years, when he was employed as court mathematician by the Grand Duke of Tuscany. In private lessons, Galileo imparted to his pupils many new discoveries in mechanics and strength of materials; from the later correspondence of these men, we catch glimpses of his effective teaching of physics. So Dr. Raymond Seeger is right in saying, in several recent publications, that it is time for us to stop thinking of Galileo primarily as a pioneer of modern astronomy because of his spectacular telescopic achievements, and recognize him also as the pioneer of modern physics.

The creation of classical physics by Sir Isaac Newton, who was born the year Galileo died, has perhaps tended in English-speaking countries, through justifiable national pride, to obscure the achievements of Galileo. At least this was so not many years ago, for I recall my own puzzlement at the fact that Einstein had christened inertial frames of reference "Galilean," when inertia to me was Newton's first laws of motion. Later, as I became interested in the history of science and in the work of Galileo in particular, I came to realize that the origin of the law of inertia, and Galileo's role in it, involved questions more intricate than is generally supposed, being still subjects of study and debate today. And there are few questions more fascinating in the whole history of physics. One need only try to conceive of a science of physics without the concept of inertia in order to perceive that the introduction of his fundamental notion must have produced a revolution in physical thought as profound as any that have occurred since. One can then hardly fail to wonder who it was that introduced this idea, and how it came to occur to him. I shall try to indicate roughly the part that Galileo played in this first great revolution in physical thought.

Newton remarked in his *Principia* that Galileo, being in possession of the first two laws of motion, thereby discovered that the descent of falling bodies varied as the square of the elapsed time.[4] Historically, and biographically with regard to Galileo, this remark leaves much to be desired, though it is interesting autobiographically, that is, as a clue to Newton's own source of the concept of inertia and to his method of thought. Its historical inaccuracy lies chiefly in the word, "thereby." Because Newton easily perceived that the law of free fall followed directly from a correct understanding of his first two laws together with the assumption that gravity exerted a constant force, it was natural for him to assume that his great Italian predecessor had actually made these discoveries in that orderly fashion. In point of fact, however, Galileo arrived at the law of free fall long before he gave any explicit statement of his restricted law of inertia, and though he was first to recognize the true physical significance of acceleration, he never did formulate the force law. As with Kepler's laws of planetary motion, so with Galileo's law of free fall, Newton was able to produce mathematical derivations and demonstrations of results derived by his predecessors only from long study of sometimes chaotic observational data, assisted by flashes of insight rather than by mathematical or even logical deduction. It is interesting that Galileo had already perceived this to be the normal order of events in science; in reference to one of Aristotle's ideas and its proof, he wrote:

> I think it certain that he first obtained this by means of his senses, by experiments and observations ... and afterwards sought means of proving it. This is what is usually done in the demonstrative sciences. ... You may be sure that Pythagoras, long before he discovered the proof for which he sacrificed a hecatomb, was sure that in a right triangle the square on the hypotenuse was equal to the squares on the other two sides. The certainty of a conclusion assists not a little in the discovery of its proof.[5]

As I have said, there is a question whether and to what extent Galileo is entitled to credit for the anticipation of Newton's first law of motion. Among historians of science, technical priority for the first complete statement of the law of inertia is generally given to Descartes, who published it two years after Galileo's death. In the same year, 1644, it was also published by Pierre Gassendi, who adduced in its support observational data instead of the theological and philosophical argument used by Descartes. But if Galileo never stated the law in its general form, it was implicit in his derivation of the parabolic trajectory of a projectile, and it was clearly stated in a restricted form for motion in the horizontal plane many times in his works. A modern physicist reading Galileo's writings would share the puzzlement – I might say the frustration – experienced by Ernst Mach a century ago, when he searched those works in vain for the general statement that (he felt) ought to be found there. It would become evident to you, as it was to Newton and Mach, that Galileo was in possession of the law of inertia, but you would not then be able to satisfy those historians who demand a clear and complete statement, preferably in print, as a condition of priority. But to physicists, if not to historians, it is ironical that this particular law should be credited to Descartes, whose physics on the whole operated to impede the scientific progress begun by Galileo and continued by Newton. For Galileo possessed in a high degree one special faculty that Descartes lacked; one that teachers of physics quickly perceive, though it might be hard to convince psychologists or philosophers of its existence. That is the faculty of thinking correctly about physical problems as such, and not confusing them with either mathematical or philosophical problems. It is a faculty rare enough still, but much more frequently encountered today than it was in Galileo's time, if only because nowadays we all cope with mechanical devices from childhood on. In Galileo's day, thinking was not continually brought to bear in this (or any other) way on mere physical processes. There were of course skilled technicians, craftsmen, and engineers, but their impressive achievements had been derived rather from the accumulation of practice and tradition than by the deductive solution of physical problems. Thinkers as a class, which is to say roughly the university population, were concerned mainly with medicine, law, theology, and philosophy. Formal instruction in physics consisted, as I remarked before, of the exposition of Aristotle.

Now a cardinal tenet of Aristotle's physics was that any moving body must have a mover other than itself, and since this notion also appeals to experience and common sense, it stood as a formidable obstacle to the discovery of the principle of inertia. It was difficult, of course, to explain under Aristotle's rule the continuance of motion in objects pushed or thrown. The first man to override that rule and to suggest that a force might be impressed on a body, and endure in it for some time without an outside mover, was probably Hipparchus. This idea was developed and advocated by Johannes Philoponus, a brilliant sixth-century commentator on Aristotle. During the middle ages, a few daring philosophers developed this thesis further into the concept of *impetus*, largely in opposition to *antiperistasis*, an idea mentioned (though not clearly accepted) by Aristotle. This was the view that the separate mover for a projectile is the medium through which it travels. In case you wonder how the medium could do anything but resist the projectile, I may explain that the usual form of argument was this: As a body moves, it creates a vacuum in its wake; since nature abhors a vacuum, the surrounding medium rushes in to fill it, striking the object from behind and thus impelling it further. It is easy to see why the theory of impetus gained ground against that of antiperistasis, which was ridiculed by Galileo, incidentally, in virtually the same manner as that which we should employ today. After his time little more was heard of it, or of the medieval impetus theory that had never entirely succeeded in displacing it.

From the end of the nineteenth century until recent years, historians of science tended to regard Galileo's inertial concept as a natural and logical outgrowth of medieval impetus theory. That view is now undergoing review and modification, thanks to a more careful analysis of the actual writings of medieval philosophers, as well as to the complete accessibility of Galileo's own papers, including his long unpublished early studies on motion and mechanics. And indeed, if inertia were nothing more than a simple and logical outgrowth of impetus theory, which theory had been developed and debated over a period of several centuries by astute philosophers, then this outgrowth might be expected to have developed much earlier in the game, as a way out of various difficulties inherent in impetus theory which conservative Aristotelians had always been quick to point out. The belief that inertia grew naturally out of an earlier theory was plausible, as we shall see, but turns out to be historically unsound. It is at best a half-truth, and as Mark Twain said, a half-truth is like a half-brick; it is more effective than the whole thing, because it carries further. I shall try to put matters in a new perspective by explaining the sense in which impetus theory opened a road for acceptance of the law of inertia, though it did not thereby suggest that law, and then indicating the steps by which I believe Galileo actually arrived at his concept of inertia, which went along a quite different road.

Aristotle's idea that every motion requires a moving force, and ceases when that force stops acting, appeals to common sense because it is roughly borne out by experience. In most cases the relatively short persistence of motion in an object after the propelling action has ceased is not nearly so impressive as the effort required to set or even to keep the object in motion. Perhaps that is why many philosophers did not feel the need of any stronger force to account for it than some fanciful action of the medium. But there were other continued motions which could not be explained in that way at all; for example, that of a grindstone. This not only persisted in free rotation for a long time, but strongly resisted efforts to stop it, and in this case no explanation in terms of a push from air rushing into any vacated space could apply. Such instances suggested that circular motions might be exceptions to Aristotle's dictum, and this in turn fitted in rather well with his general scheme of things in which perfection and a variety of special physical properties were attributed to circles and motions of various kinds. This scheme had helped Aristotle to explain the motion of the heavenly bodies. For him, the earth was at absolute rest in the center of the universe, and the motions appropriate to earthly bodies were straight motions up and down. Circular motion, being perfect, belonged naturally to the perfect heavenly bodies. To the fixed stars, Aristotle assigned a special source of motion called the Prime Mover. In later times some attempt was made to explain their regular rotation by analogy to the grindstone, since the stars had then come to be considered as embedded in a solid transparent crystalline sphere. Thus, without great violence to Aristotelian orthodoxy, the conservation of angular momentum could be conceived as a natural phenomenon in which Aristotle's outside mover might be replaced by an impressed force which normally diminished as a function of time. In the special case of the stars, this loss of motion could still be offset by action of the Prime Mover.

While the strictest Aristotelians continued to oppose any postulation of impressed forces, more enterprising philosophers went on to extend this concept to the case of projectile motion under the general name of *impetus*. Loss of impetus by projectiles was likened to other familiar phenomena requiring no special explanation, such as the diminution of sound in a bell after it is struck, or of heat in a kettle after it is removed from the fire, and this accorded with a further important rule of Aristotle's that nothing violent can be perpetual. Hence to the extent that impetus theory paved the way for eventual acceptance of the idea that motion might be perpetually conserved, it was done on a philosophical basis that inhibited the taking of that ultimate step. Not only was an indefinitely enduring impressed force unnecessary to experience, but it was ruled out in theory by Aristotle; and to reconcile a theory with Aristotle's opinions was at that time just as important as to reconcile it with experience. Thus impetus theory, given its philosophical context, did not lead to inertial physics; rather, it precluded the need for that so long as physical judgments remained qualitative and were not replaced by quantitative measurements. And that brings us to the time of Galileo.

I should like to remark in passing that no fundamental revolution in science takes place until the way has been paved, usually by vague or incorrect notions, for the acceptance of a radically new idea. Hence the new idea, when it comes, is very likely to have the appearance of a natural extension of the old ones. That is why it was perfectly plausible for historians of science to suppose, when medieval impetus theory was first brought to their attention, that the inertial concept had originally arisen as a natural outgrowth of that theory. And indeed, it might have arisen so, despite what I have said thus far. But I hope next to show you that the actual road to the first announcement of the inertial principle was not the same road at all as the one which had led people to a point at which they would be able to accept the new idea of indelibly impressed motion as an extreme case of fleeting impressed motion.

During his first years as a teacher at the University of Pisa, Galileo wrote a treatise on motion which he intended to publish. In this treatise, which survives in manuscript, he attacked Aristotle boldly, often in favor of ideas of the medieval philosophers. He opposed various Aristotelian notions about the role of the medium, including that of antiperistasis, but at the outset he adopted the Aristotelian division of all motions into "natural" and "violent" motions. This concept may be clarified by quoting the definition adopted by Galileo in his treatise: "There is natural motion when bodies, as they move, approach their natural places, and forced or violent motion when they recede from their natural places.'[6] Without going into detail, I might say that Aristotle's physics was a theory of natural places, high for light bodies (fire and air), low for heavy bodies (earth and water), and that bodies were supposed to seek these natural places by an occult property inherent in them. Galileo dissented from this view; for him, all bodies were heavy bodies and differed only in density, so all tended to approach the center of the earth. But as he pursued this idea in his treatise on motion, he came to question whether all motions were either natural or violent. He perceived that a body might be moving, and yet be neither approaching nor receding from the center of the earth; and he reasoned that any body rotating on that center itself would be moving neither naturally nor violently – in contradiction of Aristotle, who allowed no third possibility. Galileo went on to show that any rotating homogeneous sphere, wherever situated, would also have this un-Aristotelian kind of motion (assuming no friction of its axis with its support), since for every part of that sphere which was approaching the center of the earth at a given moment, an equal part would be receding from it; thus the sphere as a whole would be moving, but neither naturally nor violently. Others before Galileo had pursued similar reasoning. But Galileo arrived in this way at the idea that there was a third kind of motion which was not a mere mixture of the other two kinds, as his predecessors had called it. And this recognition of a special kind of motion was his first essential step toward the concept of inertia. For in the same treatise, analyzing the force required to maintain a body in equilibrium on an inclined plane, Galileo concluded that horizontal motion of a body on the earth's surface would similarly be neither natural nor violent, in the sense of Aristotle, and in a note added to this section he said that this should be called a *neutral* motion. He then went on to prove that, in theory at least, any heavy body could be moved on a horizontal plane by a force smaller than that required to move any body upward on any other plane, however gently inclined, but he was careful to add that:

Our proofs, as we said before, must be understood of bodies freed from all external resistance. But since it is perhaps impossible to find such bodies in the realm of matter, one who performs an experiment on the subject should not be surprised if it fails; that is, if a massive sphere, even on a horizontal plane, cannot be moved by a minimum force. For in addition to the causes already mentioned, there is the fact that no plane can be actually parallel to the horizon, since the surface of the earth is spherical ... And since a plane touches a sphere in only one point, if we move away from that point, we shall have to be moving upward. So there is good reason why it will not be possible to move a (massive) sphere from that point with an arbitrarily small force.[7]

The manuscript treatise from which this is quoted was written about 1590, but Galileo did not publish it. He was dissatisfied with the attempt he had made to reconcile various observed phenomena with the old impetus theory. Also, he had tried in this treatise to account for the observed speeds of bodies along inclined planes by a formula for their equilibrium conditions, and in this he had not succeeded (by reason of a misconception of accelerated motion which is irrelevant to the subject of the present paper). But in the process he had found a new approach to physics in his concept of a "neutral" motion. It is important to note that this concept was in no way a part of impetus theory, and that it was first suggested to him by quite a different kind of question than that of projectile motion.

In 1593, Galileo moved to the University of Padua, where he continued his experiments with inclined planes and with the pendulum. Here he wrote a little treatise on mechanics for his private pupils, which he also left unpublished. Dealing with the inclined plane in this new work, he wrote:

On a perfectly horizontal surface a ball would remain indifferent and questioning between motion and rest, so that any the least force would be sufficient to move it, just as any little resistance, even that of the surrounding air, would be capable of holding it still. From this we may take the following conclusion as an indubitable axiom: That heavy bodies, all external and accidental impediments being removed, can be moved in the horizontal plane by any minimal force.[8]

From these two propositions, written not later than 1600, Galileo can hardly have failed to deduce the corollary that horizontal motion would continue perpetually if unimpeded. This second essential step in his progress toward the principle of inertia was not explicitly stated until several years later. Yet Galileo must have been teaching it in his private classes, for one of his pupils, Benedetto Castelli, who had left Padua some time before, wrote to Galileo in 1607 mentioning "your doctrine that although to start motion a mover is necessary, yet to continue it the absence of opposition is sufficient."[9]

Galileo's ideas on these matters were probably not widely known until 1613. In that year he published a book on sunspots, in which he argued (among other things) for the sun's axial rotation, and as a preliminary to that argument he wrote:

I have observed that physical bodies have an inclination toward some motion, as heavy bodies downward, which motion is exercised by them through an intrinsic property and without need of a special external mover, whenever they are not impeded by some obstacle. And to some other motion they have a repugnance, as the same heavy bodies to motion upward, wherefore they never move in that manner unless thrown violently upward by an external mover. Finally, to some movements they are indifferent, as are heavy bodies to horizontal motion, to which they have neither inclination ... nor repugnance. And therefore, all external impediments being removed, a heavy body on a spherical surface concentric with the earth will be indifferent to rest or to movement toward any part of the horizon. And it will remain in that state in which it has once been placed; that is, if placed in a state of rest, it will conserve that; and if placed in movement toward the west, for example, it will maintain itself in that movement. Thus a ship ... having once received some impetus through the tranquil sea, would move continually around our globe without ever stopping ... if ... all extrinsic impediments could be removed.[10]

In my opinion the essential core of the inertial concept lies in the ideas, explicitly stated above, of a body's indifference to motion or to rest and its continuance in the state it is once given. This idea is, to the best of my knowledge, original with Galileo. It is not derived from, or even compatible with, impetus theory, which assumed a natural tendency of every body to come to rest.[11] It is noteworthy that this first published statement of a true inertial principle was used by Galileo to support an argument for the conservation of angular momentum, in this case by the sun. He seems always to have continued to associate the phenomena of inertia and of conservation of rotatory motion, which has led most historians to believe that for Galileo, inertia had a circular quality. I return to this point later, but I wish to say here that in my opinion the association had quite another basis; namely, the linkage in Galileo's mind of these two phenomena by the unifying concept of a "neutral" motion which had first led him to an inertial principle.

Statements relating to inertia in Galileo's later books, the *Dialogue* of 1632 and the *Two New Sciences* of 1638, are more elaborate than the above but are essentially repetitions of it. His argument always proceeds from a consideration of motion on inclined planes to the limiting case of the horizontal plane, where motion once imparted would be perpetual, barring external obstacles or forces. The *Dialogue* is of particular interest for a long section dealing with the motion of a projectile, in which it is made clear that this motion would be rectilinear if it were not for the immediate commencement of the action of the body's weight, drawing it downward as soon as it is left without support. But Galileo never explicitly abstracts weight from the body, and consequently does not give a statement of the law of inertia in the form and generality which we accept today. This was added by Descartes shortly after Galileo's death.

Because of the fundamental significance of the inertial concept to the later development of celestial mechanics, I should like to stress its importance in Galileo's arguments in favor of the Copernican theory and to mention some implications that have been drawn from that use of it. A strong objection to Copernicus in those days was that if the earth rotated at the rate of a thousand miles an hour, then any body separated from the earth, such as a bird or a cannonball or an object falling freely from a high place, would be rapidly displaced westward from an observer stationed on the earth. Copernicus had offered as a possible explanation for the absence of such effects some natural tendency of terrestrial bodies to share in the earth's motion wherever they were, but this was not widely accepted, and to Galileo it was no better than those "occult properties" invoked as explanations by the very philosophers against whom he contended. In the *Dialogue*, he replaced this explanation by giving numerous examples of inertial motion, such as that of a ball dropped by a rider on horseback, and he refuted the idea that an object dropped from the mast of a moving ship would strike the deck farther astern than on a ship at rest. These arguments, based on observations that anyone could duplicate and supported by correct physical reasoning, did much to gain a fair hearing for the Copernican system from his contemporaries.

But here there arises a problem in assigning Galileo's precise role in the discovery of the inertial principle. In the *Dialogue*, Galileo sometimes spoke as though the inertial motion of bodies leaving the earth's surface was itself circular, causing historians to question whether he himself fully understood the rectilinear character of this motion, and some have gone so far as to say that Galileo believed the planetary motions to be perpetuated by a sort of circular inertia. These, I believe, are misconceptions that require some further comment.

The passage on projectile motion in the *Dialogue* and the derivation of the parabolic trajectory in the *Two New Sciences* show that Galileo as a physicist knew inertial motion to be rectilinear. Nevertheless, Galileo as a propagandist, when writing the *Dialogue*, stated that rectilinear motion cannot be perpetual, though circular motion may be. In the same book he ascribed some special properties almost metaphysically to circles and circular motions. Now the passages in question occur mainly in the opening section of the *Dialogue on the Two Chief World Systems, Ptolemaic and Copernican*, and they should be construed in the light of the purpose for which that book was written. It was not written to teach physics or astronomy, but to weaken resistance to the Copernican theory, and it was very effective in doing so. For Galileo was not only an outstanding scientist, but a first-rate polemicist and a writer of exceptional literary skill and psychological insight. He knew when he wrote the *Dialogue* that strong opposition could be expected from the professors of philosophy, most of them convinced Aristotelians. It was for that reason, I believe, that in the opening section of his book he deliberately conceded (or appeared to concede) to the philosophers everything he possibly could without compromising his one objective. This is still the best way to proceed if you wish to espouse an unpopular view. Accordingly, when I read the metaphysical praise of circles in the *Dialogue*, I do not conclude with most historians that its author was unable to break the spell of ancient traditions; rather, I strongly suspect that he was up to something again, as usual. This suspicion is confirmed when I read his other books and his voluminous surviving correspondence, and find nowhere else any trace of metaphysics about circles. On the contrary, Galileo often scoffs at such ideas, and in a book published in 1623 he expressly denied that any geometrical form is prior or superior to any other, let alone is perfect, as Aristotle had claimed for the circle.[12] It is not likely that he had changed that opinion by 1632, as he was nearly sixty years of age before he published it in the *Assayer* of 1623.

All that Galileo wanted to accomplish in the *Dialogue* was to induce his readers to accept the idea set forth by Copernicus, and Copernicus had placed the planets, including the earth, in circular orbits around the sun. Galileo probably knew better, having read at least the preface of Kepler's *Astronomia Nova*. But it was hard enough for him to get acceptance in Italy of any motion for the earth, and for his immediate purpose it would have been fatal to argue for an elliptical orbit. It was far better strategy for him to ennoble the circle, using arguments extracted from Aristotle himself, and to argue that circular motion was as suitable to the earth as to the heavens, if he wanted to win over or even neutralize any philosophers. And I can see in my mind's eye some of them starting to read the *Dialogue* for no other purpose than to find and answer his hostile arguments against Aristotle, and then in the first forty or fifty pages finding themselves so much at home as to wonder whether there might not be some merit in the other ideas of so sound a writer. And that, in my opinion, is precisely what Galileo was up to when he composed those opening pages, balancing his criticisms of Aristotle's physics with extracts from his metaphysics about circles. Galileo certainly did not state or believe that the celestial motions would perpetuate themselves merely by being circular in form, but this does not mean that he was averse to letting philosophers believe that if they wished to. If Galileo had held such an opinion, he would not have hesitated to declare it; I see little point in looking for hidden beliefs behind the words of a man who spent his last years under arrest for disdaining to conceal his opinions. And the fact is that in the *Dialogue* itself, Galileo declared that he had no opinion about the cause of the planetary motions, but went on to say that if anyone could tell him the cause of gravitation, he could then give a cause for those motions.[13]

To me, the amusing thing is that in saying that no rectilinear motion can be perpetual, Galileo seems to have been on sounder ground than are many of those who criticize his having said it. Galileo's modern critics seem still blissfully unaware that in the last fifty years a question has arisen about the meaning, the nature, even the existence of straight lines in the physical universe. Perhaps the general law of inertia is tautological; perhaps it is our only definition of "straight motion"; certainly it has lost the absolute physical character with which those critics invest it still. Nor do they seem to realize that physicists now question the infinite extent of the universe, an attribute which would be a necessary condition for "perpetual straight motion" in the sense in which Galileo denied its possibility. For Galileo has been criticized both for his failure to declare the universe to be infinite, and for his denial of the possibility of perpetual straight motion, as if this impugned the modernity of his physics.

In conclusion, I freely grant that Galileo formulated only a restricted law of inertia. Perhaps this too is a tribute to his modernity as a physicist, for there is an advantage in refusing to generalize beyond the reach of your available experimental evidence; it is that four hundred years later, your restricted statement will still be true, while the speculations of your more daring colleagues may have gone out of date. Galileo's restricted law of inertia, applying only to heavy bodies near the surface of the earth, was in a sense all that was needed or justified in physics up to the time of Newton's discovery of the law of universal gravitation. Any speculation by Galileo about the behavior of bodies in interstellar space would at his time have been essentially meaningless metaphysics – the very sort of philosophizing that he had undertaken to replace with a science of physics. And in any event, Galileo's restricted law of inertia enabled him and his disciples in a few decades to advance physics further than had been done in eighteen centuries following the death of Archimedes. That characterizes it indeed as a true revolution in physical thought.

NOTES

1 Stillman Drake, *Discoveries and Opinions of Galileo* (Doubleday & Company, Inc., New York, 1957), p. 237.

2 See Ref. 1, p. 84.

3 *Dialogue Concerning the Two Chief World Systems*, translated by Stillman Drake (University of California, Berkeley, and Los Angeles, California, 1962), p. 54.

4 Sir Isaac Newton, *Principia*, edited by F. Cajori (University of California, Berkeley and Los Angeles, California, 1947), p. 21.

5 See Ref. 3, p. 51.

6 I. Drabkin and S. Drake, *Galileo on Motion and on Mechanics* (University of Wisconsin, Madison, Wisconsin, 1960), p. 72.

7 See Ref. 6, p. 68.

8 See Ref. 6, p. 171.

9 See Ref. 6, p. 171, n.

10 See Ref. 1, pp. 113–114.

11 Kepler, who was a contemporary of Galileo's, uses the word *inertia* to denote the supposed tendency of bodies to come to rest.

12 See Ref. 1, p. 263.

13 See Ref. 3, p. 234.

2

The Concept of Inertia

The formal introduction of the concept of inertia into modern physics is due to Sir Isaac Newton, who wrote in his *Mathematical principles of natural philosophy*:

> Law I. Every body perseveres in its state of rest, or of uniform motion in a right line, unless it is compelled to change that state by forces impressed thereon.
>
> Law II. The alteration of motion is ever proportional to the motive force impressed; and it is made in the direction of the right line in which that force is impressed.[1]

In a *scholium* to the section containing these laws, Newton added:

> By the first two Laws and the first two Corollaries, *Galileo* discovered that the descent of bodies observed the duplicate ratio of the time, and that the motion of projectiles was in the curve of a Parabola.[2]

Until the beginning of the present century, historians of science usually followed Newton's lead and attributed to Galileo the origin of the concept of inertia which lies at the base of modern physics. That attribution, however, has in recent decades been subjected to further scrutiny and scholarly criticism, and is no longer literally accepted. We may distinguish two main lines of research and criticism which have led to its modification, of which one is primarily historical and the other is conceptual or philosophical.

The historical researches of Pierre Duhem with regard to scientific thought in the Middle Ages brought to light a large body of writings which had been gen-

Reprinted from *Saggi su Galileo Galilei* (Florence: Barbera, 1967), 3–14. All reasonable effort has been made to contact the copyright holder of this article.

erally ignored by earlier historians of science. Among these were the important writings of John Buridan, Nicholas Oresme, and others, in which the theory of *impetus* was introduced to account for projectile (and certain other) motions, in opposition to the Peripatetic doctrine of *antiperistasis*. In the light of such documents, Duhem and his followers came to regard the concept of inertia as a mere logical extension of medieval impetus-theory. That theory in its turn was then related to still earlier speculations concerning the possibility of impressed forces, going back at least to the time of John Philoponus and perhaps to that of Hipparchus. Such researches tend to support a general theory of the history of ideas in which a fundamental continuity of human thought is assumed, and under which the efforts of the historian are directed principally toward the tracing of possible connections between and influences of all writers mentioning an idea in any form or for any purpose upon any later writers employing similar ideas for analogous or even quite different purposes. In that theory of history, a virtual necessity is felt for some influence by each writer upon his successors in order to account for the recurrence of fundamental ideas. Opposed to this is the theory of revolutionary ideas, in which the contribution of a single writer may be regarded as the origin of an entirely new concept. These two opposing theories are largely the expressions of an individual historian's temperament rather than of demonstrable historical facts, and so is much of what has been written in recent decades concerning the history of the concept of inertia, particularly with respect to Galileo's role in its development.

The most objective summary of the relation of medieval impetus-theory to inertia in the sense of modern physics has been given by Miss Annaliese Maier in her *Zwei Grundprobleme der scholastischen Naturphilosophie*:

Die Situation its also um 1600 die, dass die Impetustheorie von der offiziellen scholastischen Philosophie für sich in Anspruch genommen wird. Das schliesst natürlich nicht aus, dass vereinzelt immer noch die aistotelische Lehre Anhänger findet; und es schliesst auch nicht aus, dass von andere Seite, die ihrerseits der Schulphilosophie ablehnend gegenübersteht, gleichfalls die Impetustheorie vertreten wird. Und hier wird sie in der alten Weise ausdrücklich der aristotelischen Auffassung entgegengestellt und dieser vorgezogen. Unter denen, die die Impetustheorie auch noch im 16. Jahrhundert so aufgefasst haben, und die übrigens den anderen gegenüber, die sie zur aristotelischen Lehre stempeln wollen, in der Minderzahl sind, gehören ... Telesio, ... Bruno, ... Benedetti ... und schliesslich Galilei. Bei ihnen hat die Impetusthreorie noch durchaus die polemische Spitze gegen den Aristotelismus. Diese Reihe allein war es, die von Duhem und denen, die seine Ergebnisse übernahmen, berücksichtigt wurde. Die Folge war natürlich ein nicht ganz zutreffendes Bild von der tatsächlichen geistesgeschichtlichen Situation. Man sah in der Impetustheorie des ausgehenden 16. Jahrhunderts ein Wiederaufleben des grossen revolutionären Gedankens des 14. Jahrhunderts, der nun erst jetzt sein volles Gewicht und seine volle Wirkung gegen den Aristotelismus erhalten haben soll, und man nahm weiter an, dass aus dieser Impetustheorie sich die neue Mechanik in gerader Linie entwickelt habe. Aber so haben die Dinge nicht gelegen.[3]

Miss Maier goes on to say that whereas in impetus-theory each motion impresses an inhering moving force on the body moved, the inertial concept sees in uniform motion, as in rest, a state that is conserved so long as it is undisturbed; whereas according to impetus-theory the moved body has an inclination to return to rest and thus opposes the inhering force, according to the inertial concept the inclination in the case of uniform motion is to continue with no resistance on the part of the body; and whereas impetus-theory would allow for an acting force in circular as well as straight motion, the inertial concept postulates only a continuance of the latter. This last point, she says, is of secondary importance; the essential differences between the concepts of impetus and inertia are the two first named, for these are in direct conflict with the Aristotelian principle that everything moved requires some mover; "und dieser Gegensatz ist allerdings so stark, das der neue Gedanke sich nicht *aus* dem alten sondern nur *genen* ihn entwickeln konnte."[4]

We may safely take Miss Maier's summary as definitive with respect to the historical aspect of the relation between impetus-theory and the inertial concept. Grounded as it is on the most thorough analysis of the writings of the scholastics and of modern students of this problem, it comes as a welcome conclusion to a series of controversies that related rather to the theory of history than to the ostensible subject matter, and it is unlikely to be seriously modified by further investigations. But as to Miss Maier's comments on the conceptual or philosophical aspects of the matter, though I thoroughly agree with them, it must be admitted that they are by their nature less apt to receive universal acceptance. The late Professor Alexandre Koyré, for instance, took quite a different position; and though his main writings on the subject preceded those of Miss Maier, I do not believe that he was inclined to modify them after hers appeared. Here, for example, is an indication of one fundamental difference of opinion:

Or, ... si la dynamique de Galilée est, dans son fond le plus profond, archimédienne et tout entière fondée sur la notion de la pesanteur, il en résulte que Galilée ne pouvait pas formuler le principe d'inertie. Aussi ne l'a-t-il jamais fait. En effet, pour pouvoir le faire, c'est-à-dire pour pouvoir affirmer la persistance éternelle non point au movement en général, mais du mouvement *en ligne droite*, pour pouvoir se représenter un corps, abandonné à lui-même et *privé de tout support*, comme demeurant au repos ou continuant à se mouvoir *en ligne droite* et non *en ligne courbe*, il eût fallu qu'il pût concevoir le mouvement de la chute comme un mouvement non point naturel mais au contraire ... comme causé par une force exterieure.[5]

Aussi, au cours du *Dialogue* l'*impetus* se trouve-t-il identifié au moment, au mouvement, à la vitesse ... glissements successifs qui, insensiblement, amènent le lecteur à concevoir le paradoxe du mouvement se conservant tout seul dans le mobile, d'une vitesse "indélébilement" imprimée au corps en mouvement.

En principe, le privilège du mouvement circulaire est battu en bréche: c'est le mouvement en tant que tel qui se conserve, et non le mouvement circulaire. En principe. Mais, en fait, le *Dialogue* ne va pas plus loin. Et quoiqu'on l'ait dit, jamais nous ne glissons, ni ne glisserons jusqu'au principe d'inertie. Jamais, dans les *Discours* pas plus que dans le *Dialogue*, Galilée n'affirmera la conservation éternelle du mouvement rectiligne.[6]

Clearly, to Professor Koyré it would by no means be acceptable to say that the third of Miss Maier's distinctions between impetus-theory and the inertial concept is of secondary importance; for him, the limitation of the inertial concept to uniform rectilinear motions was every bit as important as the recognition of continuance in a state of rest or motion by a body otherwise undisturbed. Now, opinions will always differ concerning the aspects of any concept which are to be considered essential and those which are secondary or subordinate. My opinion, like Miss Maier's, is that the essential aspect of the concept of inertia is that of motion and rest as states of a body which are indifferently conserved. If the disagreement ended there, all would be well. But Professor Koyré went on to argue at length that because Galileo asserted that truly straight motions are impossible in nature, *only* circular motions could for him be really perpetuated, and hence that the inertial concept was inextricably linked in Galileo's own mind with privileged circular motions. The prevalence of this view is illustrated by the following passage from a work by Professor E.J. Dijksterhuis:

> The situation is thus as follows: according to the Galilean law of inertia proper, a particle that is free from external influences (note that gravity is not included among them) perseveres in a circular motion having the centre of the earth for its centre. Over short distances this motion is considered rectilinear; subsequently the limitation to short distances is forgotten, and it is said that the particle would continue its rectilinear motion indefinitely on a horizontal plane surface if no external factors interfered. Thus what might be called the circular view of inertia of Galileo gradually developed into the conception that was formulated in the first law of Newton.[7]

The view thus expressed by Professors Koyré and Dijksterhuis is based on a deduction about Galileo's personal beliefs which I consider ill-founded. More than one reconstruction is possible of what went on in Galileo's mind, if we examine from the beginning all his writings concerning the inertial concept. Professor Koyré's evidence that Galileo himself believed circular motion to be privileged, and to be the only motion inertially perpetual, is drawn principally from the *Dialogue* of 1632. The passage cited from Professor Dijksterhuis follows his discussion of a passage from the *Discorsi* of 1638, preceded immediately by comments on the speculation in the *Dialogue* concerning the absolute path of an object falling freely from a tower to the surface of a rotating earth. This famous speculation is the subject of discussion by Professor Koyré in the work previously cited, as well as in a later monograph.[8] Now, it is true that Galileo implicity assumed in that passage that the terrestrial rotation shared by an object at rest on the top of a tower would be continued during the fall of the object when released. To the best of my knowledge, however, that assumption is never again made by Galileo anywhere else in his writings; certainly not in the *Discorsi*, his chief work on physics. Moreover, this particular passage is described by Galileo himself, both in the *Dialogue* and in subsequent correspondence, as an amusing and bizarre speculation rather than a serious conclusion (*G.G.*, VII, 192; XVII, 89). And elsewhere in the *Dialogue*, speaking of objects released from circular motion, he explicitly stated that the tendency of their motion is along a straight line tangent to the circle at the point of release.[9] Hence it seems to me rash to draw a conclusion about Galileo's inner beliefs concerning inertial motion from the particular passage in question.

It is true that many references are to be found in the *Dialogue*, particularly in its opening section, to special properties of circular motions. In particular, Galileo asserts there that the motions of heavenly bodies cannot be straight, or anything but circular. A selection of such passages, removed from their context, would indeed form a powerful argument in favor of the view that Galileo shared the metaphysical views of Aristotle concerning circles and circular motions, including the idea that such motions were especially appropriate to the heavenly bodies and were self-perpetuating. And most modern writers follow Professor Koyré in attributing to Galileo the idea that the motions of the heavenly bodies, and particularly of the planets, could be explained as inertially perpetuated. Yet Galileo, in the *Dialogue* itself, expressly denied that he could explain the planetary motions, any more than he could explain the fall of heavy bodies (*G.G.*, VII, 261). In the same work he also rejected Aristotle's assumption that circular motions were more appropriate to the celestial bodies than to terrestrial bodies or to the earth itself (*G.G.*, VII, 61). Indeed, the basic purpose of the *Dialogue* was to induce people to attribute a circular motion to the earth. In view of the circumstances in which the *Dialogue* was written, and considering Galileo's need at that time to neutralize the hostility of convinced Aristotelians who dominated the philosophical thought of the epoch, I think it probable that his preliminary endorsement of certain Aristotelian views concerning circles and circular motions is better explained in terms of polemic strategy than in terms of Galileo's inner convictions. At any rate, it is inappropriate to deduce his inertial views from his cosmological discussions. Rather, let us try to trace their development from the very beginning.

It is well known that Galileo's first physical researches dealt with the behavior of bodies placed in water, and that about the year 1586 he composed his first scientific treatise, *La bilancetta*. Those studies in turn aroused his interest in questions relating to the speed of descent, and led him to attempt to formulate a theory of motion, in opposition to that of Aristotle, founded on the Archimedean principle. This "buoyancy" theory of motion, in which not weight but relative density was invoked to determine the speed of natural motion of a body in any medium, was set forth in a treatise *De motu*, composed about 1590 during Galileo's brief professorship at Pisa. Although he did not publish that treatise, he intended to do so, and some parts of it are preserved in more than one form, while the manuscripts include numerous changes, additions and marginal notations which assist in the tracing of his thought. The precise order in which the parts were written, and even the order in which they were intended to appear in the final work, is partly conjectural. The chapters of special interest for our present purpose occur in the manuscript directly after those which present in detail Galileo's theory of weights and speeds; they deal respectively with motion on inclined planes, circular motions, and the motions of projectiles. The definition of natural and forced motion, essential to all these chapters, occurs in connection with the second of the above topics, so we may commence with that.

"Let us see," writes Galileo, "whether motion about the center of the universe is forced motion or not: for example, the motion of a marble sphere at the center of the universe. ... Now we have natural motion when bodies as they move approach their natural places, and forced motion when they recede therefrom. This being so, it is clear that such a sphere rotating with its center at the center of the universe moves with a motion that is neither natural nor forced" (*G.G.*, I, 304).

He goes on the remark that even a sphere of non-uniform density would have this un-Aristotelian kind of motion if its center of gravity were at the center of the universe. Moreover, a sphere with its geometric center at the center of the universe and its center of gravity elsewhere, moving about its geometric center, would still have a motion that was neither natural nor forced, since its center of gravity would be neither approaching toward nor receding from its natural place. Again, the same would hold in theory for any homogeneous sphere rotating about its center at a place remote from the center of the universe; but here Galileo took care to remark that in practice, such motion is forced motion because of friction between the axis and the supports required to hold the sphere away from the center of heavy bodies, and also because of resistance by the air. Finally, he says, a heterogeneous sphere rotating with its center removed from the center of the universe will be moving with a motion sometimes natural and sometimes forced; that is, at times its center of gravity will be approaching and at other times receding from the center of heavy bodies. Here Galileo noted that on the whole, the latter is a forced motion, since the impetus of approach will not effect recession to the same extent. Although he promised to explain this elsewhere, his explanation is not to be found in the surviving manuscripts.

In the discussion summarized above, the starting-point of any reconstruction of Galileo's thought-processes leading to his inertial concept, he described two kinds of motion which would not fit into the Aristotelian dichotomy of natural and forced motions – one which is "neither natural nor forced" and one which is "sometimes natural and sometimes forced." In connection with the former, he raised the question whether such motion would be perpetual, but deferred its discussion to another place and did not take it up explicitly again. Later in the treatise, however, he stated that forced motions continually diminish, and probably that is the reason which he intended to assign for the impliedly non-perpetual motion of a heterogeneous sphere with its center removed from the center of heavy bodies. One might find in this a further implication that motion which was "neither natural nor forced" would already be perpetual in Galileo's mind. A similar implication may be contained in his remark that a homogeneous sphere similarly located will in practice be impeded by friction (and by the air), but that this friction would in theory be obviated by reducing the supports of the axis to indivisible points.

In the chapter summarized above, Galileo did not assign specific names to his two kinds of non-Aristotelian motion; that is, motion "neither natural nor forced" and motion "sometimes natural and sometimes forced." But the concepts were applied also in his chapter on motion along inclined planes, though in the manuscript that chapter precedes the other. It is a chapter of great importance in the development of Galileo's mechanical investigations, for here he first made use of the argument that a horizontal plane may be considered as the limiting case of the inclined plane, a conception associated later with the presentation of his inertial concept in the *Dialogue* and the *Discorsi*. In *De motu*, this reasoning appears as follows:

A body subject to no external resistance on a plane sloping however little below the horizon will move downward in natural motion without the application of any external force. This can be seen in the case of water. And the same body on a plane sloping upward, however little above the horizon, does not move up except by force. And so the conclusion remains that on the horizontal plane the motion of a body is neither natural nor forced. But if its motion is not forced, then it can be made to move by the smallest of all possible forces (*G.G.*, I, 299).

This is immediately followed by a demonstration of the latter proposition, to which was subsequently appended a marginal addition which helps greatly in the reconstruction of Galileo's progress toward an inertial concept:

From this it follows that mixed motion [except circular] does not exist. For the forced motion of heavy bodies is away from the center, and their natural motion is toward the center, but a motion partly upward and partly downward cannot be compounded there-from, unless perhaps we might say that such a mixed motion takes place on the circumference of a circle around the center of the universe. But such a motion will be better described as "neutral" rather than "mixed," for "mixed" partakes of both, "neutral" of neither (*G.G.*, I, 300).

Having completed this addition, Galileo returned to the beginning and struck out the words "except circular" which are enclosed in brackets above. Thus it would appear that this process of thought was somewhat as follows: "Among the rotations of spheres, there are cases in which the motion is neither natural nor forced (any sphere centered at the center of the universe, and any homogenous sphere wherever centered), and there are also cases in which the motion is partly forced and partly natural (heterogenous sphere with center remote from center of universe). Now, on the horizontal plane there are also motions of the first type, and they are not necessarily limited to rotations. Let us give the name 'neutral' to motions which are neither natural nor forced, and call the others 'mixed' motions. Then 'neutral' motions will include 1) rotation of spheres with centers at the center of the universe, 2) rotations of all homogeneous spheres, and 3) any undisturbed motion on a horizontal plane. 'Mixed' motions will include 1) rotation of heterogeneous spheres with centers remote from the center of the universe, and 2) motions of bodies upon circles centered at the center of the universe." At this point, Galileo perceived that the last-named group did not really belong where he had put it, since bodies moving in that manner, though seeming to move alternately "up" and "down," would in fact be neither approaching toward nor receding from the center of the universe. Accordingly he went back to the beginning of the note and struck out the circular exception he had had in mind at first. And since he considered the sole remaining case of "mixed" motion (heterogeneous spheres rotating away from the center of the universe) to be essentially in forced motion, he abandoned "mixed" motions as a separate classification. "Neutral" motions, which remained in contradiction to Aristotle's dichotomy, included both the conservation of angular momentum and horizontal motions of supported bodies, and these were invariably associated thereafter in Galileo's writings.

Galileo's chapter on projectile motion in *De motu*, immediately following the chapters mentioned above, is simply a presentation of the impetus or impressed-forced theory of his time, with a refutation of the Peripatetic arguments for antiperistasis. Comment on this fact will be made later.

About three years after writing *De motu*, Galileo composed a treatise on mechanics for the use of his students at the University of Padua. In that treatise he included the remark that on a horizontal plane, though a body would not move spontaneously, the slightest force would suffice to set it in motion. He also observed that bodies have no resistance to transverse motion except in proportion to their removal from the center of the earth. The treatise on mechanics underwent several revisions, reaching its final form about the year 1600. It contains the following passage:

On a perfectly horizontal surface, a ball would remain indifferent and questioning between motion and rest, so that the least force would be sufficient to move it, just as any little resistance, even that of the surrounding air, would be capable of holding it still. From this we may take the following conclusion as an indubitable axiom: That heavy bodies, all external and accidental impediments being removed, can be moved in the horizontal plane by any minimal force (*G.G.*, II, 180).

From these two propositions Galileo can scarcely have failed to perceive that horizontal motion would continue perpetually if unimpeded. That essential step in his progress toward the concept of inertia, taken no later than 1600, was not explicitly stated until several years later. Yet Galileo must have been teaching it in his private classes, for his pupil Benedetto Castelli, who had left Padua some time before, wrote to Galileo in 1607, mentioning "your doctrine that although to start motion a mover is necessary, yet to continue it the absence of opposition is sufficient." (*G.G.*, X, 170).

Galileo's ideas on these matters were first published in 1613, when in his *Letters on sunspots* he argued for the sun's axial rotation, and as a preliminary to that argument he wrote:

I have observed that physical bodies have an inclination toward some motion, as heavy bodies downward, which motion is exercised by them through an intrinsic property and without need of a special external mover, whenever they are not impeded by any obstacle. And to some other motion they have a repugnance, as the same heavy bodies to motion upward, wherefore they never move in that manner unless thrown forcibly upward by an external mover. Finally, to some movements they are indifferent, as are heavy bodies to horizontal motion, to which they have neither inclination ... nor repugnance. And therefore, all external impediments being removed, a heavy body on a spherical surface concentric with the earth will be indifferent to rest or to movement toward any part of the horizon. And it will remain in that state in which it has been placed; that is, if placed in a state of rest, it will conserve that; and if placed in movement toward the west, for example, it will maintain itself in that movement. Thus a ship ... having once received some impetus through the tranquil sea would move continually around our globe without ever stopping ... if ... all extrinsic impediments could be removed (*G.G.*, V, 134–135).

If the essential core of the inertial concept lies in the ideas of a body's indifference to the states of motion and rest, and its perpetual continuance in either state if undisturbed, then the above passage contains its first published announcement. It is of great interest to note that it appeared in connection with an argument for the conservation of angular momentum – a fact which on the one hand tends to corroborate the reconstruction of Galileo's intellectual progress toward his inertial concept that has been presented here, and on the other hand illustrates the reasons for which so many scholars have been imbued with the idea that Galileo's conception of inertia was confused by a metaphysical preoccupation with circular motions. In any event, we have now reached a point at which the essential idea is complete, and we may pause to review the implications of the foregoing reconstruction.

Contrary to a common opinion of historians, Galileo was not put on the track of his inertial concept by the impetus-theory which had preceded it. For impetus had been conceived as a device for the explanation of projectile motion, and in Galileo's first writings on that subject he showed himself still satisfied with that explanation. Even though he had already formulated the idea of "neutral" motions, and had extended that idea to the case of motion on a horizontal plane, he did not rewrite or marginally annotate his discussion of projectile motions, nor did he indicate in any way that he saw any application of the neutral-motion concept to projectiles when he wrote this treatise *De motu*. The essential germ of inertia – namely, indifference of a body to the state of rest or of uniform motion – came to him from an entirely different set of reflections, concerning the nature of certain rotations, and it was only after many years that he came to apply it to the analysis of the motion of a projectile.

The case of Galileo and the concept of inertia illustrates an interesting and recurrent historical situation. If one were to say that historically, the concept of inertia arose from the failure of all previous attempts to account for certain observed phenomena, and that inertia was accepted as the upper limit of duration of an impressed force which in impetus-theory had been assumed to decay from the moment of its receipt by the moved body, there might be good reason to make that assertion. In effect, we should be saying only that men had been generally dissatisfied with the previous theory, and had thus been prepared for the acceptance of a more general rule, under which the impressed force did not inherently decay. But if we were to say biographically of Galileo, who first came forward with an inertial concept, that he must have arrived at it by experiencing such a dissatisfaction himself, and then seeking in all directions for a better explanation than impetus-theory, we should be grossly mistaken. That is one way in which the concept might have been developed, but it is not in fact the way in which it was first hit upon. The concept of inertia (like the concept of vectors) had its origin in purely theoretical considerations. Only much later was inertia found to have a practical application to projectile motions. That application, as made by Galileo and by his pupil Bonaventura Cavalieri, is too well known to need repetition here. It brings to mind, however, a question that disturbed Professor Koyré, a question which I hope may now be finally laid to rest, together with the paradox that gave rise to it.

Si – ainsi que nous croyons l'avoir montré – Galilée n'a pas formulé le principe d'inertie, comment se fait-il que ses successeurs et disciples ont pu croire le trouver dans son oeuvre? Si, ainsi nous croyons l'avoir également démontré, Galilée, non seulement n'a pas conçu, mais encore n'a pas pu concevoir le mouvement inertial en ligne droite, comment se fait-il, ou, mieux, comment s'est-il fait, que cette conception, devant laquelle s'est arrêté l'esprit d'un Galilée, a pu paraitre facile, évidente, allant de soi, à ses disciples et successeurs? ... Aussi n'ont-ils pas eu tout à fait tort d'attribuer à Galilée une découverte qu'il n'avait pas faite, et de trouver dans son oeuvre ce qui, sans doute, n'y était pas expressément, mais y était "en germe?"[10]

Galileo's "failure to formulate the principle of inertia," in the sense of the foregoing passage, is a question of formal criteria, not one of significant fact. The "inability to conceive of inertial movement in straight lines" is an illfounded conjecture concerning Galileo's mental processes. Let us select a passage from the *Dialogue*, not from its opening section which was designed to lull the fears of Galileo's Peripatetic opponents, but from the heart of the book, in which the motion of a stone flung from a whirling stick is under discussion:

SIMPLICIO: The impressed motion, I say, is undoubtedly in a straight line.
SALVIATI: But what straight line? ...

SIMPLICIO: It can be no other than that line which touches the circle at the point of separation ... .

SALVIATI: ... That is, the projectile acquires an impetus to move along the tangent to the arc described by the motion of the projectile at the point of its separation from the thing projecting it. ... And you have said that the projectile would continue to move along that line if it were not inclined downward by its own weight, from which fact the line of motion derives its curvature ... .[11]

Such are the passages that were read by Galileo's pupils and his immediate successors. The lack of an explicit statement of the law of inertia for rectilinear motion of bodies freed from gravitational influence, which led Professor Koyré to construct a fascinating and paradoxical portrait of Galileo's scientific character and philosophical beliefs, is merely a formal lack in Galileo's writings. Such an explicit statement, sought in vain by historians, is not required by physicists. Newton did not hesitate to attribute a knowledge of the law of inertia to Galileo, nor did Einstein doubt that he should christen inertial frames of reference "Galilean."

Whoever believes in the relevance of formal historical criteria to actual scientific revolutions would do well to consider the case of Descartes, whom Professor Koyré correctly credits with the first precise formulation of the law of inertia, but whose physics was largely mistaken and almost completely sterile. Perhaps I may be permitted a counter-paradox in conclusion: before the discovery of Newton's general law of gravitation, all conjectures about the behavior of bodies in free space were in a sense nothing more than idle metaphysics of the kind which Galileo particularly abhorred.

NOTES

1   I. Newton: *The mathematical principles of natural philosophy*, tr. A. Motte, London, 1729, I, p. 19. The concept of inertia had been introduced into philosophy by Descartes in 1644.

2   I. Newton: [cit. *n.* 1], I, p. 31.

3   A. Maier: *Zwei Grundprobleme der scholastischen Naturphilosophie*, Roma, 1951, pp. 304–305.

4   A. Maier: [cit. *n.* 3], p. 306.

5   A. Koyré: *Études Galiléennes*, III, *Galilée et la loi d'inertie*, Paris, 1939, pp. 246–247.

6   A. Koyré: [cit. *n.* 5], p. 228.

7   E.J. Dijksterhuis: *The mechanization of the world picture*, tr. C. Dikshoorn, Oxford, 1961, p. 352.

8 A. Koyré: [cit. *n.* 5], pp. 261–262: A. Koyré: *A documentary history of the problem of fall from Kepler to Newton*, in *Transactions of the American Philosophical Society*, n. s., 45, 4 (1955), pp. 332 sq.

9 *G.G.*, VII, 218, 219, 220, 222; see also below.

10 A. Koyré: [cit. *n.* 5], p. 282.

11 *G.G.*, VII, 218–220 (extracts).

3

Galileo's Experimental Confirmation of Horizontal Inertia: Unpublished Manuscripts (Galileo Gleanings XXII)

I

More than a decade has elapsed since Thomas Settle published a classic paper in which Galileo's well-known statements about his experiments on inclined planes were completely vindicated.[1] Settle's paper replied to an earlier attempt by Alexandre Koyré to show that Galileo could not have obtained the results he claimed in his *Two New Sciences* by actual observations using the equipment there described. The practical ineffectiveness of Settle's painstaking repetition of the experiments in altering the opinion of historians of science is only too evident. Koyré's paper was reprinted years later in book form without so much as a note by the editors concerning Settle's refutation of its thesis.[2] And the general literature continues to belittle the role of experiment in Galileo's physics.

More recently James MacLachlan has repeated and confirmed a different experiment reported by Galileo – one which has always seemed highly exaggerated and which was also rejected by Koyré with withering sarcasm.[3] In this case, however, it was accuracy of observation rather than precision of experimental data that was in question. Until now, nothing has been produced to demonstrate Galileo's skill in the design and the accurate execution of physical experiment in the modern sense.

In the circumstances, it has become unfashionable to support the view of the earliest historians of science that Galileo was the father of experimental science. Reading his published works without a preconceived theory of medieval-Renaissance continuity, these early writers saw them as imbued with a spirit of concern for empirical facts, and they construed this to be evidence of experiments conducted with care though not reported in detail. Modern historians dis-

Reprinted from *Isis* 64 (1973), 291–305, by permission of the University of Chicago Press. Copyright 1973 by the History of Science Society, Inc.

trust such conclusions; they insist upon having documents in evidence. Thus the view that Galileo ever resorted to careful experimentation remained merely an opinion (and a professionally unpopular one) even after Settle's conclusive demonstration that Galileo *could* have obtained the results he claimed in print with the equipment he there described. This was sufficient to refute Koyré's position he could *not* have done so, but it remained insufficient to prove that Galileo *did* obtain those results. (The same skeptical historians, however, believe that to show that Galileo could have used the medieval mean-speed theorem suffices to prove that he did use it, though it is found nowhere in his published or unpublished writings.)

Historians in the twentieth century have probably felt confident that no evidence of precise physical experimentation on Galileo's part was likely ever to show up, after his notes on motion had been edited and published by Antonio Favaro. Indeed, those who have scorned the opinions of pioneer historians have not thought it worth while first to search those manuscript notes for themselves. Not even the customary cautionary phrases used by earlier scholars – "so far as is presently known" or "in the absence of documents to the contrary" – are to be found in modern monographs making light of Galileo's interest and ability in experimentation.

Now, it happens that among Galileo's manuscript notes on motion there are many pages that were not published by Favaro, since they contained only calculations or diagrams without attendant propositions or explanations. Some pages that were published had first undergone considerable editing, making it difficult if not impossible to discern their full significance from their printed form. This unpublished material includes at least one group of notes which cannot satisfactorily be accounted for except as representing a series of experiments designed to test a fundamental assumption, which led to a new, important discovery. In these documents precise empirical data are given numerically, comparisons are made with calculated values derived from theory, a source of discrepancy from still another expected result is noted, a new experiment is designed to eliminate this, and further empirical data are recorded. The last-named data, although proving to be beyond Galileo's powers of mathematical analysis at the time, when subjected to modern analysis turn out to be remarkably precise. If this does not represent the experimental process in its fully modern sense, it is hard to imagine what standards historians require to be met.

The discovery of these notes confirms the opinion of earlier historians. They read only Galileo's published works, but did so without a preconceived notion of continuity in the history of ideas. The opinion of our more sophisticated colleagues has its sole support in philosophical interpretations that fit with preconceived views of orderly long-term scientific development. To find manuscript evidence that Galileo was at home in the physics laboratory hardly surprises me. I should find it much more astonishing if, by reasoning alone, working only from fourteenth-century theories and conclusions, he had continued along lines so different from those followed by profound philosophers in earlier centuries. It is to be hoped that, warned by these examples, historians will begin to restore the old cautionary clauses in analogous instances in which scholarly opinions are revised without new evidence, simply to fit historical theories.

In what follows, the newly discovered documents are presented in the context of a hypothetical reconstruction of Galileo's thought. The principal documents are reproduced in facsimile, so that others may arrange and interpret them as they like.[4] Other, minor, notes still unpublished also bear on my reconstruction, and still others hint at phases of Galileo's experimental activity not directly involved in the matter of chief concern here.

II

As early as 1590, if we are correct in ascribing Galileo's juvenile *De motu* to that date, it was his belief that an ideal body resting on an ideal horizontal plane could be set in motion by a force smaller than any previously assigned force, however small. By "horizontal plane" he meant a surface concentric with the earth but which for reasonable distances would be indistinguishable from a level plane. Galileo noted at the time that experiment did not confirm this belief that the body could be set in motion by a vanishingly small force, and he attributed the failure to friction, pressure, the imperfection of material surfaces and spheres, and the departure of level planes from concentricity with the earth.[5]

It followed from this belief that under ideal conditions the motion so induced would also be perpetual and uniform. Galileo did not mention these consequences until much later, and it is impossible to say just when he perceived them. They are, however, so evident that it is safe to assume that he saw them almost from the start. They constitute a trivial case of the proposition he seems to have been teaching before 1607 – that a mover is required to start motion, but that absence of resistance is then sufficient to account for its continuation.[6]

In mid-1604, following some investigations of motions along circular arcs and motions of pendulums, Galileo hit upon the law that in free fall the times elapsed from rest are as the smaller distance is to the mean proportional between two distances fallen.[7] This gave him the times-squared law as well as the rule of odd numbers for successive distances and speeds in free fall. During the next few years he worked out a large number of theorems relating to motion along inclined planes, later published in the *Two New Sciences*. He also arrived at the rule that the speed terminating free fall from rest was double the speed of the fall itself. These theorems survive in manuscript notes of the period 1604–1609. (Work during these years can be identified with virtual certainty by the watermarks in the paper used, as I have explained elsewhere.[8])

In the autumn of 1608, after a summer at Florence, Galileo seems to have interested himself in the question whether the actual slowing of a body moving horizontally followed any particular rule. On folio 117^r of the manuscript just mentioned, the numbers 196, 155, 121, 100 are noted along the horizontal line near the middle of the page (see Plate 1). I believe that this was the first entry on this leaf, for reasons that will appear later, and that Galileo placed his grooved plane in the level position and recorded distances traversed in equal times along it. Using a metronome, and rolling a light wooden ball about 4¾ inches in diameter along a plane with a groove 1¾ inches wide, I obtained similar relations over a distance of 6 feet. The figures obtained vary greatly for balls of different materials and weights and for greatly different initial speeds.[9] But it suffices for my present purposes that Galileo could have obtained the figures noted by observing the actual deceleration of a ball along a level plane. It should be noted that the watermark on this leaf is like that on folio 116, to which we shall come presently, and it will be seen later that the two sheets are closely connected in time in other ways as well.

The relatively rapid deceleration is obviously related to the contact of ball and groove. Were the ball to roll right off the end of the plane, all resistance to horizontal motion would be virtually removed. If, then, there were any way to have a given ball leave the plane at different speeds of which the ratios were known, Galileo's old idea that horizontal motion would continue uniformly in the absence of resistance could be put to test. His law of free fall made this possible. The ratios of speeds could be controlled by allowing the ball to fall vertically through known heights, at the ends of which it would be deflected horizontally. Falls through given heights thereafter would consume the same time from table top to floor, and the distances moved horizontally would be proportional to the speeds acquired at table top.

Equipment suitable for an experimental test of this kind was drawn by Galileo on folio 175 (Plate 2). A very steeply tilted plane is shown, provided with a curved deflector at the bottom which would serve to arrest all vertical motion and put the ball into new free fall with its previous motion converted horizontally. This device was probably not found satisfactory in practice and was replaced by a plane tilted about 60° resting near the edge of a table on which the ball rolled briefly before falling to the floor. An inked bronze ball was presumably used which would leave a small and distinct spot at the point of impact, permitting precise measurement of the horizontal travel. Duplication of the experiment shows that proper control is very difficult with the steeper plane, both as to tracking and as to direction of subsequent motion, while analysis of the recorded data fixes the slope used with considerable accuracy.

The results of Galileo's experiment are entered on folio 116^v, where the height of the table used is given as "828 points." Another fragment (fol. 166) makes it possible to deduce that the "point" was the unit distance marked along the inside scale of Galileo's proportional compass. Thomas Settle, who was in Florence when I found this reference, kindly measured for me the units on Galileo's own instrument, preserved at the Science Museum there, and found that 180 points measure 169 millimeters. Hence Galileo's table was of about the standard modern height, nearly 80 centimeters.

The vertical line above the table shows that the ball was dropped through distances of 300, 600, 800, and 1,000 points, though the last-named does not show in the reproduction, Plate 3. The entries on the horizontal show that one other drop was added, of 828 points, in which the initial and final vertical falls were equal. Along the horizontal are entered the measurements from the edge of the table to the point of impact on the floor, and each of these is accompanied by another entry marked *doveria*, "it should be." The difference between actual and expected results is also recorded.

Galileo's calculations of the expected distances were also carried out on folio 116. For the reader's convenience I have transcribed in Plate 4 the essential parts, which are blurred in the reproduction. These calculations were made according to the rule that the squares of the speeds acquired in vertical fall to the table top are proportional to the distances fallen from rest, or $V_2^2:V_1^2::S_2:S_1$. Since all horizontal distances were traversed in the same time (that of free fall through 828 points), they are proportional to the speeds acquired in the initial fall; hence, designating the horizontal distances by the letter D, we have $D_2^2:D_1^2::S_2:S_1$. Taking D_1 as 800 and S_1 as 300, these being the empirical data from the shortest drop, Galileo had $D_2^2:800::S_2:300$, whence each expected horizontal distance in other drops was obtained by a mean proportional, equivalent to extracting the square root of ($800^2S/300$). For the first case, in which S_1 was exactly doubled, Galileo multiplied 800 by its double and took the square root, as seen at the top of folio 116^v. In all other cases he multiplied the initial fall by 800, divided this product by 300, multiplied the quotient by 800, and extracted the square root of the product. His results are summarized in the following table, in "points" of about 17/18 millimeter.

Galileo thus found very good agreement between actual and calculated distances run horizontally after deflection. The agreement with modern theory using the rate of acceleration due to gravity is also quite good, as will be seen below. Analysis of the data shows that the rate along the plane was about 630 centimeters/second^2, indicating a tilt of 64° if no slipping took place. As will

Fall to table	Table to floor	Horizontal projection	Galileo's computed expectation by his mean-proportional rule
300	828	800	Used as standard comparison
600	828	1,172	1,131
800	828	1,328	1,306
828	828	1,340	1,330 (Galileo's calculation, 1,329)
1,000	828	1,500	1,460

be seen later, Galileo may have employed a plane of that slope, though it seems likely that he would have used a tilt of 60° and that there was slippage of about 10 per cent, which would produce the same rate of acceleration. Since we are concerned only with the overall precision of his experiment, and cannot establish the precise slope from the data, the ensuing analysis assumes an angle of 64° with perfect rolling. Converting all data into centimeters, the fall through 828 points (77.7 cm) from table to floor is found to require 0.4 second. The first fall is treated as exactly given, since Galileo used this fall for his own standard; on this basis, the discrepancies from calculated expectancies do not exceed 1 inch in 3 feet.

Vertical drop	Time along 64° incline	Terminal speed	Travel in 0.4 sec	Actual (fol. 116)	Difference	
28.1 cm	0.299 sec	187.9	75.2 cm	75.1 cm	—	—
56.3	0.423	266.1	106.4	110.0	+3.6 cm	+3.4%
75.1	0.489	307.1	122.8	124.6	+1.8	+1.5
77.7	0.497	312.6	125.0	125.8	+0.8	+0.6
93.9	0.546	343.6	137.5	137.1	−0.4	−0.3

In making these computations, account has been taken of the reduction of the rate of acceleration by a factor of 5/7 for rolling as against free fall, because of the rotational moment of inertia of the ball. The loss of energy of a bronze ball striking a wooden table is not considered, because Galileo took the first pair of data to establish his ratios, and therefore for comparison that pair is treated here as theoretically correct. On that basis the indicated experimental errors bear out the idea of a carefully conducted test.

Galileo, on the other hand, was confronted with an apparent discrepancy from another result, to be expected in connection with his special test for an initial drop of 828 points. This value is not shown on the vertical line; it seems to have been added as an afterthought with the purpose of testing his double-distance rule by letting the ball fall initially through the same vertical height that it was to fall after deflection. According to his rule, deduced in 1607–1608 by reasoning from one-to-one correspondence of speeds in vertical motion and along inclined planes of the same height, the ball should travel twice as far along the horizontal as along the incline during a second time equal to that of the accelerated motion. For a 60° plane the descent is 1.15 times the vertical, or 952 points against 828; hence in another such time the ball should go about 1,900 points horizontally. But it went only 1,340 in a time equal to that of vertical fall, and since Galileo knew that the times of fall through equal height were as the lengths of plane and vertical, this implied only 1,531 in place of 1,900 for travel during a second equal time. Thus, although the distance of 1,340 fitted properly into the mean-proportional pattern, it did not verify the double-distance rule by a wide margin.

We know that the basic source of trouble is the factor of 5/7 for rolling as against free fall (or frictionless slide), but Galileo did not know this. Accordingly, he was obliged to look for a possible explanation in the only obvious place to him – in the disturbance occasioned by impact with the table top. (In later years he mentioned more than once the loss of motion in angular deflections.) It was this, I believe, that led him to a refinement of the foregoing experiment of which we have a record on folio 114 (Plate 5). The idea was simple: in order to circumvent the effect of deflection, the ball was simply to be allowed to roll off the end of an inclined plane through a fixed height at various speeds whose ratios were known. The experiment was duly performed and the empirical data written down; but having got these, Galileo found himself quite unable to make the appropriate calculations to get new *doveria* figures. He had been easily able to compound a horizontal with a vertical motion, but he did not see how to deal with an oblique impressed speed in composition with free fall. It is this that we shall now proceed to do.

It is fair to assume that the data are again given in "points." Galileo's rough sketch indicates an angle of about 25° for this experiment; in fact, 26° turns out to account very well for the data. When I first attacked this fragment I did not assume any particular purpose for the experiment, but I did assume that Galileo would have allowed the ball to roll through some set of integral multiples of a distance along the plane. Since 30° is a very reasonable angle to use, easy to set up precisely and easy to calculate with, I asked my colleague James MacLachlan to put to a computer the question of whether the data would conform to rolls through integral multiples of a length followed by falls though a common height. He did this and reported that for 30° this was possible except for one distance – the middle one. Recalling then the special case of 828 points above, I suggested that the misfit might be the distance for a roll from a height along the plane that was equal to the height of the final drop. This immediately led to the gratifying result that the basic lengths of roll in this case, omitting the middle one, were exactly related as the vertical heights in the first experiment, with the addition of two more; that is, they stood in the ratios 1:2:3:6:8:10.

Good results were obtained for a 30° plane, assuming a final drop of about 48.5 centimeters, as I reported at the annual meeting of the History of Science Society in 1972. Subsequently, however, a very striking result was obtained for a 26° plane, which MacLachlan found to give the best fit with Galileo's data. A modification of the best fit (unlike the previous calculations) gives a rational and psychologically plausible set of measured rolls using Galileo's "point" as the unit. It is a rather striking coincidence, also, that the angle 26° is the complement of the 64° angle which gives the appropriate rate of acceleration for the experiment recorded on folio 116. It seems possible, at least, that Galileo had at one time a rigid triangular frame, grooved along the hypotenuse, which he used in experiments of this kind. In any event, for a 26° plane followed by a free fall through 450 "points," the horizontal distances for rolls along the plane have been computed and compare with the data recorded by Galileo as follows:

Distance of roll, in "points"	Horizontal motion expected	Galileo's data (fol. 114)	Per cent short of prediction
200	256	253	1.17
400	339	337	0.59
600	395	394	0.25
900	454	451	0.66
1,200	499	495	0.80
1,600	543	534	1.66
2,000	579	574	0.86

The remarkable accuracy of these data, considering that the unit is less than 1 millimeter, may cause some readers to wish to check my computations. Converting Galileo's data into centimeters, and taking sin 26° as 0.4384, cos 0.8988, I obtained $a = 306.9$ centimeters/second^2. The final drop is 42.25 centimeters; t is the time of roll and T the time of drop, obtained by the formula $V_v T + (g/2)T^2 = D$; V is the terminal velocity of roll, V_v and V_h being its vertical and horizontal components. The theoretical values of the drop D range from 41.26 to 42.04; these were adjusted to 42.25 in obtaining the value of T. A different algorithm used on the computer gave virtually identical results.

At the time when Galileo performed this experiment he had not yet deduced the parabolic trajectory of projectiles or worked out his theorems concerning it. Shortly afterward, probably within a month or two, he had reached

Roll	t	V	V_v	V_h	T	Horizontal motion		Per cent short
						Theoretical	Fol. 114	
18.78	0.350	107.4	47.08	96.52	0.249	24.04	23.75	1.2
37.56	0.495	151.9	66.52	136.5	0.233	31.82	31.64	0.6
56.33	0.606	186.0	81.53	167.2	0.222	37.08	36.99	0.2
84.50	0.742	227.8	99.86	204.7	0.208	42.66	42.34	0.7
112.67	0.857	263.0	115.3	236.4	0.198	46.83	46.48	0.8
150.20	0.990	303.7	133.1	273.0	0.187	51.05	50.14	1.7
187.80	1.110	339.6	148.9	305.2	0.178	54.34	53.89	0.8

the rule of vector addition for composition of uniform motions that appears as Theorem II in the Fourth Day of his *Two New Sciences*. But even after that time, he would have been unable to carry out the computations necessary for the analysis on his recorded data on folio 114 without a knowledge of the factor of 5/7 that relates rolling motion and free fall. Since in all his work he compared only ratios and did not attempt to compute individual speeds, his published results were not affected by this, though workers like Marin Mersenne, who attempted to apply them indiscriminately to falling and to rolling bodies, and who compared speeds rather than ratios of speeds, encountered seeming contradictions.

Deduction of the parabolic trajectory followed immediately on the experiment recorded on folio 116, as suggested by the little diagrams on folio 117. The discovery is the heuristic aspect of Galileo's experiment; it was, of course, a merely accidental byproduct of a carefully executed experiment designed for quite a different purpose. It just happened that Galileo carried out his work in confirmation of uniformity of horizontal motion using a table about 32 inches high and made the longest drop from about 6 feet above floor. This was as high as he could reach while making sure that the ball was on the plane at 1,000 points above the table. This height permitted him to stoop after releasing the ball and thus to observe fairly accurately its paths after deflection, which are drawn on folio 116 as approximately parabolic. Returning to folio 117, we see that this same sheet, which I believe led him to the experiment on folio 116, was next used for some drawings of parabolas and one of a complete trajectory, with tangents.

III

On February 11, 1609, Galileo wrote to Antonio de'Medici at Florence, including a drawing of some parabolas of differing amplitudes, lying between parallel lines. He wrote in response to a request concerning his activities; and since he had been in Florence during the preceding summer, it may be assumed that the line of research described was new and had not been in progress when the two men had last met. In part, Galileo wrote:

I am also about some questions that remain to me concerning the motion of projectiles, among which are many that bear on artillery shots. Only recently I have found this: that putting the cannon on some elevated place above the plane of the field, and aiming it exactly level, the ball leaves the cannon, driven by much or little gunpowder, or even just enough to make it leave the mouth; and yet it always goes declining and descending to the ground with the same speed, so that all balls will arrive on the ground at the same time for all level shots, whether the shots are very long or very short, or even if the ball merely emerges from the mouth and falls plumb to the plane of the field. And the same happens with shots at an elevation; these are all completed in the same time provided that they are lifted to the same vertical height. Thus, for example, the shots AEF, AGH, AIK and ALB, contained between the same parallels CD and AB, are all completed in the same time; and the ball consumes the same time in traversing line AEF as AIK, or the others – and in consequence, in their halves; that is, the parts EF, GH, IK and LB are all made in equal times, corresponding to level shots ...

from E, G, I, and L, respectively.[10]

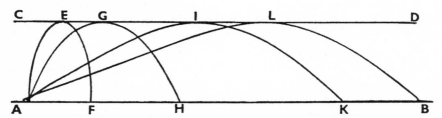

Antonio Favaro believed from this letter that early in 1609 Galileo was already in possession of the parabolic trajectory. I once considered the drawings to be insufficient evidence of that, and even shared with other historians the suspicion that Galileo may not have had the derivation of the trajectory as late as 1632, when he complained of Bonaventura Cavalieri's publication of it. But there is no longer any doubt that Galileo knew the trajectory and its theoretical derivation early in 1609, when he wrote out the proofs of three propositions that later appeared, almost unchanged, in the *Two New Sciences* at the beginning of the Fourth Day. These, and several other theorems on the parabolic trajectory, are written on paper with a distinctive watermark (rhinoceros) that is never found in Galileo's correspondence after 1609. All these notes are, moreover, written in a distinctively small and cramped hand that was probably associated with an illness which afflicted him in that year.

IV

In conclusion, I should like to stress a few points related to my initial remarks concerning Galileo's place as an experimentalist. Having tried these experiments myself, I know that to obtain results as accurate as Galileo's would require more care and more elaborately prepared equipment than I myself used. I doubt that Galileo went to the trouble of preparing such equipment and then used it only for the experiments represented by the documents reproduced here. Also, it takes a certain amount of previous experimental practice before one even becomes aware of the devices required to obtain data as reliable as is found in these records. Had physical experiment been common in Galileo's day, he might have just borrowed some equipment for a single brief session with it, as I did. But accurate experimentation then was anachronistic, as Galileo's Platonist admirers never tire of pointing out. The correct conclusion is not that Galileo was an armchair speculator, but that Galileo's physics was anachronistic. That is just what the pioneer historians of science felt from the spirit of his published books. If we do not sense the same thing, it is because we are too firmly wedded to certain philosophical and historical theories. The present example should stand as a warning; for if the Florence flood had destroyed Galileo's manuscript notes in 1966, as it came within a few feet of doing, then the prevailing sophisticated myth of Galileo's essential resemblance to medieval physicists would forever have held the field.

That Galileo did not publish detailed data concerning his inclined plane experiments, and did not even mention these others which confirmed his restricted inertial concept for horizontal motions of heavy bodies near the earth's surface, need not surprise anyone. None of his opponents would have been convinced by a parade of data, for they were interested not in physical laws but in the causes of things, and causes are not revealed by experiments. As a matter of fact, his law of free fall was rejected by Descartes as only an approximation, and two other physicists ostensibly demonstrated that the *true* law which lay behind the merely sensible appearances of Galileo's odd-number rule was a quantum-jump increase of distances traversed in the progression of the natural numbers.[11] As for interested laymen, Galileo had no need to parade his data; it sufficed for him to assert that he had made careful tests and found his rules to be true. What surprised even me in the manuscripts described above was the degree of precision that Galileo required in order to convince himself before he made those assertions. I had always imagined that he was content to have made the sort of simple observations described in his books in support of his restricted inertial principle.

If the publication of experiments that influenced others is the criterion by which we should judge who was the father of experimental physics, then Marin Mersenne was the man. But Mersenne was inspired to a large degree by his reading and questioning of Galileo's results, as MacLachlan has amply shown in his as yet unpublished doctoral thesis.[12] Mersenne and his contemporaries showed by their behavior, and sometimes by their statements, that Galileo had started the trend. Hence it may be better to call Galileo the grandfather of experimental science than to revere him as a Platonist organizer of fourteenth-century mathematical speculations.

NOTES

1 T.B. Settle, "An Experiment in the History of Science," *Science*, 1961, *133*: 19–23.

2 Alexandre Koyré, "An Experiment in Measurement," *Proceedings of the American Philosophical Society*, 1953, 97: 222–237, reprinted in Alexandre Koyré, *Metaphysics and Measurement: Essays in Scientific Revolution* (Cambridge, Mass.: Harvard University Press; London: Chapman & Hall, 1968), pp. 89–117.

3 In this issue of *Isis*, James MacLachlan, "A Test of an 'Imaginary' Experiment of Galileo's," pp. 374–379; cf. Koyré, *Metaphysics and Measurement*, pp. 83–84.

4 The manuscript leaves reproduced herein are from Vol. 72, MSS Galileiani, preserved at the Biblioteca Nazionale di Firenze, with whose kind permission they are included here.

5 I.E. Drabkin and Stillman Drake, *Galileo on Motion and on Mechanics* (Madison: University of Wisconsin Press, 1960), p. 68; *Le opere di Galileo Galilei*, ed. Antonio Favaro, 20 vols. (Florence, 1890–1909), Vol. II, pp. 301; hereafter cited as *On Motion* and *Opere*.

6 *Opere*, Vol. X, p. 170.

7 See Stillman Drake, "Galileo's Discovery of the Law of Free Fall," *Scientific American*, May 1973, *228* (No. 5): 84–92.

8 Stillman Drake, "Galileo Gleanings XXI – On the Probable Order of Galileo's Notes on Motion," *Physis*, 1972, *14*: 55–68.

9 It is perhaps worth noting that on a gently inclined grooved plane the large wooden ball rapidly overtook a smaller steel ball simultaneously released in front of it. A mistaken extrapolation of this result to free fall may account for Galileo's otherwise puzzling statements around 1590 (in support of an experiment described by Girolamo Borro) that a wooden ball at first outruns a lead ball; cf. *On Motion*, pp. 105, 106, 109.

10 *Opere*, Vol. X, p. 229.

11 Cf. letters of Descartes to Mersenne, Oct. 11, 1638, *Opere*, Vol. XVII, p. 390, lines 109–111; Jan. 29, 1640, Vol. XX, p. 612; G.B. Baliani, *De motu naturali gravium* ... (Genoa, 1646), pp. 110–111; H. Fabri, *Tractatus physicus de motu locali.* ... , ed. P. Mousnier (Lyons, 1646), pp. 88–90, 98–108, passim.

12 J.H. MacLachlan, "Mersenne and Galileo: Ideas in Motion" (Harvard University, June 1971).

4

Galileo's Discovery
of the Parabolic Trajectory

Galileo, in the last part of his final work *Discourses on Two New Sciences*
(1638), proved that the trajectory of a projectile traveling through a nonresisting
medium is a parabola. The proof is simple and straightforward, once it is known
that the vertical distance the object falls from rest is proportional to the square
of the time elapsed and that the projectile's horizontal velocity will remain uniform. As a matter of fact, a similar proof had been published six years earlier in
a book on conic sections by Bonaventura Cavalieri, a pioneer of the calculus
who knew Galileo and who had studied mathematics under one of Galileo's
pupils, Benedetto Castelli. Galileo was indignant when he first learned of Cavalieri's publication; he wrote to a friend that the proof was the fruit of studies he
had begun 40 years earlier and that the least he deserved was the courtesy of
first publication. No one knew better than he, he said, how hard it had been to
make the discovery and yet how easy it was to work out the proof when the
shape of the trajectory was known.

Cavalieri was much distressed when he learned of Galileo's wrath, and he
wrote at once to say that he had credited both Galileo and Castelli in his book, a
copy of which he sent to Galileo. He added that everyone knew the discovery
was Galileo's and that he himself had believed Galileo had already published it.
Galileo was satisfied and went out of his way in *Discourses on Two New Sciences* to praise Cavalieri as a new Archimedes.

Some historians of science have naturally wondered how much of this
sequence of events could be believed. Was it really possible that a man who had
made such a discovery would never have mentioned it in print over a span of

Written with James MacLachlan. Reprinted from *Scientific American* 232 no. 3 (March 1975), 102–
110. Reprinted with permission Copyright © 1975 by Scientific American, Inc. All rights reserved.

three decades? On the other hand, would Cavalieri have said that he had believed the result to have been already published by Galileo if the discovery, and not just the proof, was really his own?

These questions can now be answered on the basis of Galileo's notes on motion, preserved at the National Central Library in Florence in Volume 72 of the Galilean manuscripts. It is certain that Galileo discovered the parabolic trajectory no latter than 1608 and proved it mathematically early in 1609, although he did not mention it in print until 30 years later. The papers that support this conclusion were not published with Galileo's collected works; they are reproduced here for the first time.

As we have mentioned, two laws must be known in order to derive the parabolic trajectory. One is the law of free fall, which states that the distance an object falls from rest is proportional to the square of the time elapsed, and which Galileo had discovered in 1604 by a combination of luck and mathematical reasoning [see "Galileo's Discovery of the Law of Free Fall," p. 248 in this volume]. The other law needed was a restricted principle of inertia that gives a relation between the motion imparted to a body and the behavior of the body after it is free from the initial impulse. Such a principle was precisely what Galileo was testing when he discovered the parabolic trajectory. Luck played no part this time. Galileo's discovery of the parabolic trajectory was a case of serendipity: the discovery of something other than what the seeker had set out to find.

To a historian of science the interesting thing about Galileo's discovery is the remarkable extent to which he made use of experimental methods in science that we now take for granted but that were not standard procedure in the 17th century. It is difficult to tell from Galileo's published works how much use he really made of careful experiments. Those that he did describe as having been actually carried out have been repeated in recent years, and they work very well. Galileo did not, however, describe many such experiments, and he did not give his results in numerical form, as became the custom among his successors. It is only in his private notes that traces of his experimental data survive.

Naturally we must be cautious about interpreting those data. In the case of Galileo's discovery of the parabolic trajectory, however, one set of seven three-digit numbers (253, 337, 395, 451, 495, 534 and 573) exists that he could not have obtained in any way other than by careful measurement. We shall reconstruct the steps that led to those numbers and then describe an experiment we have conducted that nearly duplicates them. Although we may be wrong about details, the evidence in favor of our overall picture seems to be convincing.

As early as 1590 Galileo had proved mathematically that in theory a ball on a level plane will be set in motion by any force, no matter how small. By 1607 he was teaching his student Castelli that once any motion has begun it needs no force for its continuation. It was accordingly reasonable for him to believe that on a level plane a ball would continue rolling indefinitely with a uniform motion if it was not impeded.

During the years 1602 through 1608 Galileo developed many theorems concerning motion in vertical free fall, along inclined planes and in combinations of the two. It was during this period that he experimented with an actual level plane. In the document labeled *f. 117* (Plate 1) in Volume 72 of the Galilean manuscripts a series of diminishing numbers is seen along a central horizontal line. The numbers probably represent the distances traveled by a ball in successive equal times after it had been given an initial push along a groove on a level plane. Obviously it would have been unsafe for Galileo to conjecture from such data alone that the ball was slowing down entirely because of friction and other external influences. It might have been that the initial impetus simply "got tired" and would have done so even on an ideally flat and frictionless plane. Hence it would have been natural for Galileo to wonder what would happen in the absence of contact between the ball and the plane.

Galileo's own law of free fall and his knowledge of how it is modified on inclined planes had put into his hands a means of answering that question. We know that a body heavy enough not to be much influenced by the friction of the air takes the same time to fall a given distance whether or not its motion is straight down. Its horizontal progress during the time it falls should therefore depend only on its horizontal speed at the moment it begins to fall. Then if one could control the ratios of different horizontal speeds, it would be possible to test how uniformly the horizontal motion is continued in the absence of friction. The ratios of the terminal speeds at the end of motions from different heights along a fixed inclined plane could be calculated by using any single actual measurement and the law of free fall. Such a set of calculations is found together with Galileo's measurements in the document labeled *f. 116* (Plate 3 and transcription in Plate 4).

Galileo rolled a ball from different heights down a grooved plane supported on a table and recorded the heights above the table with the numbers 300, 600, 800 and 1,000. He deflected the ball horizontally at a height above the floor recorded as 828. For each of the four original heights he marked the point at which the ball hit the floor. The distances from the edge of the table were 800, 1,172, 1,328 and 1,500. For a reason that we shall explain below he also rolled the ball from a height of 828 above the table, and he recorded the horizontal distance traversed as 1,340.

After obtaining these measurements he used his law of free fall to relate the height of the plane to the horizontal distance traversed. The first pair of measurements, 300 units of height to 800 units of horizontal distance traversed, he took as a standard ratio. On the basis of the standard ratio and his times-squared law of free fall he calculated the horizontal distances the ball should have traversed. The calculated distances for the heights of 300, 600, 800 and 1,000 came out as 800, 1,131, 1,306 and 1,460. He also calculated a horizontal advance of 1,329 or 1,330 for the added case when the height of the ball on the plane above the table (828 units) was equal to the drop from the table to the floor (828 units). The fact that the calculated horizontal distances so nearly agreed with the measured distances established the restricted principle of inertia that Galileo had been seeking and that he needed. Such a principle can be stated concisely: In the absence of any appreciable resistance the horizontal motion is uniform and of indefinite duration.

In a third document, *f. 175* (Plate 2), which is clearly associated with these experiments and which bears the same watermark as a letter Galileo wrote dated late in 1608, he drew a steep inclined plane equipped with a curved deflector designed to convert all the ball's motion into horizontal motion before the drop from the table to the floor. In practice he could not have used such a steep plane because it is very difficult to make the ball roll along it uniformly or even to make it always travel in the same straight line. In fact, our analysis of Galileo's data shows that his plane could not have been steeper than 64 degrees because the ball's acceleration implied by the data is less than the acceleration a ball would attain even if it rolled perfectly on such a steep plane. Moreover, it is quite difficult to get a perfect roll even on slopes much shallower than 64 degrees.

One might reasonably ask how we can determine the acceleration of the ball in Galileo's experiment, since we have said nothing so far about his units of distance. It happens that in other notes Galileo gave the length of certain lines in what he called *punti*, or points. These measures turn out to agree very nearly with the small units engraved on the linear scale of Galileo's proportional compass preserved at the Museum of the History of Science in Florence. On that instrument the unit is 169/180 millimeter, which we have used even though the drawings in some of his notes put one *punto* at 58/60 millimeter.

In *f. 116* (Plate 3) Galileo wrote *pū. 828 altezza della tavola*, or "828 points height of the table." On the basis of this remark we have taken the height of his table as being 777 millimeters in order to attain the times, terminal velocities and rates of acceleration stated in his measurements. Thus the table of Galileo's experiments was about the standard height of modern tables, and the highest point from which he released the ball was about five feet eight inches above the floor. That is reasonable for a man of average height who wanted to be sure he had the ball exactly on the mark before he released it. Moreover, the same arrangement would have allowed Galileo to stoop and observe the ball in flight. He sketched the paths in his diagram, and we can hardly doubt that their form suggested the parabolic trajectory to him. Galileo knew a great deal about parabolas, having begun his scientific career with a study of the centers of gravity of solids with parabolic shapes.

Once the idea of the parabolic trajectory had suggested itself to Galileo, it was easy for him to derive it theoretically. If we look back at *f. 117* (Plate 1), we see that he began at once to design a parabolic trajectory and to sketch little parabolas with steps under them, each of equal horizontal length, corresponding to the height of the curve above the ground for each equal unit of horizontal distance traversed. On the other side of the page there are some related calculations. A later group of pages, written early in 1609, contains the basic theorems of the parabolic trajectory that he did not publish until 30 years later.

When we first analyzed Galileo's data in 1972, one of us (Drake) believed the inclined plane Galileo used for the experiment recorded in *f. 116* was probably tilted at an angle of 64 degrees to the table, the steepest angle that could be accounted for. Now, however, we believe that, for that experiment as well as for the continuation of the work we shall discuss below, Galileo employed a plane at an angle of only 30 degrees to the table. A plane at that angle is easy to set up with considerable precision, and it also lends itself to easy computation. All the experiments were probably made with a bronze ball covered with ink running in a groove and then quickly deflected horizontally; when the ball struck the floor, which was presumably marble, it would leave a well-defined ink spot. From such a spot it would be possible to measure the distance to a point directly under the edge of the table to an accuracy of within about a millimeter. Galileo's drawing of the ball's path in *f. 175* (Plate 2) shows he was aware that the ball would rise after it was deflected, so that it would be above the level of the table for a short time before it again reached that level. How he took this rise into account experimentally is not recorded.

We have mentioned that Galileo included a drop of 828 *punti* to the table. His reason for adding this test is easy to guess. Galileo had proved mathematically, probably in 1604 or 1605, that a body deflected horizontally after it has been accelerated from rest for a certain time interval will, in another such time interval, traverse a distance double the distance it moved under the initial uniform acceleration. He had proved this double-distance rule for inclined planes as well as for vertical descent. Naturally Galileo wanted to test this result as well as the uniformity of horizontal motion. He supposed he could do so by having the ball start initially from exactly the same vertical height above the table as it would subsequently drop from the table to the floor. Such an arrangement, he thought, should make it easy for him to confirm his double-distance rule.

The result, however, must have surprised and disturbed him. First he computed the expected distance for this case exactly as he had computed the others, using the standard ratio and the times-squared law, and the result fitted with the others very well, giving a horizontal distance of 1,330 *punti* traversed as against the 1,340 *punti* actually measured. The double-distance rule still had to stand on its own feet, so to speak. According to Galileo's theory of inclined planes, the times consumed by a ball descending vertically and along any plane of the same height were proportional to the ratio of the height of the plane to the length of the plane. Now, that would be true if there were free, frictionless fall in both descents, as if a heavy block of ice were sliding on a hot track. The phenomenon of rolling, however, is a different matter. For Galileo's bronze ball two-sevenths of the energy from the attraction of gravity went into rotating the mass, leaving only five-sevenths to go into the motion of translation down the plane. Hence the ratio of the time it takes a ball to roll down a plane 828 *punti* high to the time it takes the ball to fall 828 *punti* is not the same as the ratio of the length of the plane to the height of the plane. Galileo did not know this, and so his separate calculations according to the double-distance rule would have led him to expect the ball to advance 1,656 *punti* horizontally (twice the vertical height of 828 *punti*), as against the actually observed advance of 1,340 *punti*.

That is a pretty serious discrepancy. Galileo could not have reasonably accounted for it except by supposing some of the ball's motion must be lost as a result of its sudden horizontal deflection at the end of the incline. That is probably why he devised a new experiment in which there would be no deflection and thus no artificial change in the direction of motion. The record of this experiment is found in the document *f. 114* (Plate 5), although it could not have been reconstructed without the information already gained from *f. 116* (Plate 3). In this experiment the ball was allowed simply to roll off the end of the inclined plane and drop to the floor. Data are given for the respective horizontal distances traversed after the ball rolled seven different lengths, but there are in this case no relevant calculations. On the face of it, it seems all we can say about this experiment is that the plane seems to have been at an angle of about 30 degrees and that that is the end of the matter.

Fortunately for us it is far from the end. We can further relate the experiment shown in *f. 114* to the experiment shown in *f. 116*. Our analysis indicates that Galileo's data are consistent with the horizontal distances traversed by a ball

rolling six different times along a plane from heights having the ratios of
1:2:3:6:8:10, plus one ratio that is not expressible in whole numbers. These are
precisely the same as the ratios of the vertical heights of 300, 600, 800 and
1,000 shown in *f. 116*, plus heights of 100, 200 and 450. Considering the complexity of the analysis, the agreement is not likely to be coincidence.

Next, it is highly probable that when Galileo selected the heights from which
the ball was to be rolled in the experiment recorded in *f. 114*, he would have
used round hundreds of *punti* for the sake of convenience, and that he would
have done the same for the terminal drop. Our analysis shows that his recorded
data are indeed consistent with the assumption that the lengths of the rolls along
the plane, in *punti*, were 200, 400, 600, 900, 1,200, 1,600 and 2,000; the final
vertical drop to the floor was 500 *punti*. The same calculations show that the
terminal speeds reached by the ball as it rolled from the four vertical heights
originally mentioned in *f. 116* are very nearly the same as the terminal speeds
calculated for four of the seven above-mentioned rolls we have assumed for the
results in *f. 114*. We can accordingly consolidate both experiments by supposing the same tilt of 30 degrees was always used for the inclined planes, an
assumption that is quite plausible. On the basis of our deductions we have compiled a table of data showing Galileo's experimental results.

The acceleration of a body falling under the influence of gravity at the latitude of Padua is 980.7 centimeters per second per second. Assuming that the
ball rolled perfectly without slipping and touched the plane at only one point as
it rolled, its acceleration down a plane at an angle of 30 degrees would be 350
centimeters per second per second. Its acceleration as it rolled along a groove in
the plane, however, would be less, and the rate of acceleration is related to how
much of the ball is above the top of the groove and how much is below. We
shall be noting some experimental observations of this effect, but for our
present purpose we assumed that in Galileo's experiment the ball accelerated at
the rate of 320 centimeters per second per second. If in the second experiment
the angle of Galileo's plane deviated very slightly from 30 degrees when it
rested on a different table only 500 *punti* above the floor instead of 828 *punti*,
such a departure would be sufficient to account for the consistently negative
deviations of Galileo's measurements from modern theory.

The experiment recorded in *f. 114* is of much more use to historians of science than it turned out to be to Galileo himself. After he had obtained the
seven measurements of the horizontal distances traversed he was unable to calculate how they compared with theory. It had been easy for him to combine
uniform horizontal motion with accelerated vertical motion, but he did not see
how he could combine a uniform "impetus" along a slope with vertical free
fall. In due course he did learn how to determine the impetus of a projectile at

Length of roll on plane (punti)	Vertical height (punti)	Time (seconds)	Terminal speed (punti per second)	Horizontal distance traveled in air (punti) f. 116 theory	Galileo	Percent difference from theory	Horizontal distance traveled in air (punti) f. 114 theory	Galileo	Percent difference from theory
200	100	.343	1,168				261	253	-3.2
400	200	.484	1,651				344	337	-1.9
600	300	.593	2,022	805	800	-.7	398	395	-.8
900	450	.727	2,477				456	451	-1.2
1,200	600	.839	2,860	1,139	1,172	+2.8	499	495	-.8
1,600	800	.969	3,303	1,315	1,328	+1.0	542	534	-1.4
1,656	828	.986	3,360	1,338	1,340	+.2			
2,000	1,000	1.083	3,692	1,470	1,500	+2.0	574	573	~0

MODERN THEORY AND GALILEO'S RESULTS for the experiments in *f. 116* and *f. 114* are compared in this table. In both cases the plane was inclined at an angle of 30 degrees. The final drop for the experiment in *f. 116* was 828 *punti* (77.7 centimeters) and the final drop for the experiment in *f. 114* was 500 *punti* (46.9 centimeters). At an angle of 30 degrees the distance the ball rolls down the plane is twice the vertical height of the plane above the table. Consistently negative departures of Galileo's results from modern theory for experiment in *f. 114* may be due to a slight difference in width of the groove, diameter of the ball or the slope of plane.

any point along its path, and after that he could have interpreted the data in
f. 114. Even though he preserved this scrap of paper with its experimental data,
however, he appears not to have returned to analyze it. In a way, that is fortunate for historians of science. If Galileo had been able to compute the figures
we have been discussing, someone might now say that he had not conducted
the experiments at all and had simply calculated the answers. As things stand
there is no way to explain the series of numbers 253, 337, 395, 451, 495, 534
and 573 except by acknowledging that Galileo was a very careful experimentalist indeed, working with an inclined plane that was at an angle only imperceptibly different from 30 degrees and measuring distances to an accuracy of a
millimeter.

Up to this point our interpretation of *f. 116* and *f. 114* as recording carefully
planned experiments has depended entirely on paper work based on some diagrams and numbers. We have shown that Galileo's numerical data fit well with
modern analysis that takes into account how much the linear acceleration is
reduced by the energy involved in the ball's rolling, a factor of which Galileo
was not aware. A random variation of the data in *f. 116* around the expected theoretical result makes it appear even more likely that the numbers were obtained
experimentally and not calculated hypothetically. Still, if we could physically
duplicate Galileo's experiments and not just reconstruct them on paper, the evidence would be even stronger. In short, what we have developed up to this point
is itself a theory – not a physical theory but a theory of Galileo's procedures. It
remains to test that theory as we would any physical theory, namely by actual
experiment.

We obtained a plank that measured two inches by four inches by six feet and
planed it smooth on one side. Two grooves were cut straight along its length.
The plank was then mounted at an angle of 30 degrees from the horizontal
along the edge of a level table that was 77.7 centimeters high. In the first trials
we rolled a ball down one of the grooves and allowed it to strike a horizontal
board before it took off in free flight. Although the ball bounced slightly, with
the result that it stayed in the air longer than it would have if it had simply fallen
from the edge of the plane, it nevertheless landed considerably short of the horizontal distances reported by Galileo for the same roll.

In an attempt to remedy the situation we installed a curved deflector at the
base of the plane to convert the ball's motion from the 30-degree slope to the
horizontal more smoothly. The results now obtained followed the pattern of
Galileo's results more closely, although our measured distances were consistently somewhat greater than his. The ball fell on a sheet of carbon paper on top
of a paper strip that was supported on a sheet of hard plastic resting on a smooth

plank. The carbon paper marked the point of impact on the paper strip, and the distances from the edge of the table could be measured quite accurately.

When we duplicated the experimental shown in *f. 114*, no curved deflector was used, and a box supported the sheet of plastic and raised the point of impact to a vertical distance of 500 *punti* from the edge of the table instead of 828. Once again the measured distances followed the same pattern as Galileo's in *f. 114*, although again they were consistently slightly longer.

We used a steel ball four centimeters in diameter and rolled it in a groove 1.5 centimeters wide. We also tried rolling a second ball only 2.5 centimeters in diameter down a groove one centimeter wide. The results of the trials showed that the width of the groove significantly influences the rate of the ball's acceleration along the plane. The maximum acceleration of a ball rolling perfectly along a flat plane at an angle of 30 degrees is about 350 centimeters per second per second. By way of comparison, the four-centimeter ball in the 1.5-centimeter groove had an acceleration of about 325 centimeters per second per second; the 2.5 centimeter ball in the same groove had an acceleration of about 280 centimeters per second per second. The difference between the accelerations of the two balls was less for the narrower groove.

Our analysis of Galileo's data shows that 320 centimeters per second per second is a good intermediate value for his two experiments. The acceleration implied by the data in *f. 116* was larger and that implied by the data in *f. 114* was smaller. A good intermediate value for our two experiments is 330 centimeters per second per second, and the acceleration was somewhat greater for our duplication of the experiment in *f. 116* and somewhat less for our duplication of the experiment in *f. 114*. Thus our measured accelerations deviated from our intermediate value in the same way that Galileo's did from his intermediate value. Our experiments further indicate that Galileo did use a curved deflector at the end of his plane for the results shown in *f. 116*. Hence the diagram of his apparatus in *f. 175* was not entirely the fruit of his imagination but reflected some experimental design.

Two conclusions seem reasonable to us as a result of our experiments. First, the data we recorded are sufficiently similar to the data recorded by Galileo to verify the hypothesis that he experimentally obtained sets of numbers measured to three or four significant figures. Second, the experimental effects of the width of the groove provide a plausible explanation for the accelerations implied by Galileo's data being less than the theoretical acceleration of 350 centimeters per second per second that would hold only for a sphere rolling perfectly on top of a flat plane.

When Galileo published his discussion of the parabolic trajectory in 1638, he did not refer to any experiments. All he could derive was an ideal law that excluded the sources of actual variations. Hence he wrote: "I mentally conceive of some movable projected on a horizontal plane, all impediments being put aside. ... Equable motion on this plane would be perpetual if the plane were of infinite extent, but if we assume it to be ended, and [situated] on high, the movable (which I conceive of as being endowed with heaviness), driven to the end of this plane and going on farther, adds on to its previous equable and indelible motion that downward tendency which it has from its own heaviness. Thus there emerges a certain motion, compounded from the uniform horizontal and the naturally accelerated downward." That motion he showed to be parabolic.

Now that we have seen *f. 116* it is apparent that Galileo was describing as a mental conception something he had carefully observed with his own eyes 30 years earlier. The first historians of science jumped to the conclusion that that was what he had done. Recent historians of science, critical of their predecessors, have jumped instead to the conclusion that Galileo worked from pure mathematics without empirical evidence; faith in ideal Platonic forms rather than attention to physical detail, they say, opened the way to modern science. As far as Galileo is concerned, the earlier historians came closer to the truth. What they lacked in philosophical insight they made up for in common sense. To the conclusions of the recent historians we reply in the words of Salviati, who spoke for Galileo in his *Dialogue on the Two Great Systems of the World* (1632).

"What you refer to is the method Aristotle used in writing his doctrine, but I do not believe it to be that with which he [originally] investigated it. Rather, I think it certain that he first obtained it by means of the senses, experiments and observations, to assure himself as much as possible of the truth of his conclusions. Afterward he sought means to make them demonstrable. That is what is done for the most part in the demonstrative sciences. ... And you may be sure that Pythagoras, long before he discovered the proof for which he sacrificed a hecatomb, was sure that the square on the side opposite to the right angle in a right triangle was equal to the squares on the other two sides. The certainty of a conclusion assists not a little in the discovery of its proof."

5

Galileo's New Science of Motion

Galileo asserted that in the last two days of his *Discorsi* of 1638 he was present-ing a very new science of a very old subject, motion. I see much justice in that claim. It may seem eccentric today to support Galileo's own appraisal of his work on motion and to see him as having introduced a new method, though that was the opinion of pioneer historians such as Montucla, Andres, and Whewell. Even as late as 1909, despite the doubts already raised by Raffaello Caverni, Angelo Valdarnini was still able to write:

Nessuno vorrà certo negare o contendere a Galileo il merito insigne d'aver usato il vero metodo sperimentale comprensivo nello studio dell'universo, e di aver esposto qua e là nelle sue Opere scientifiche le norme fondamentali del metodo sperimentale. Ma non scrisse il Galileo un apposito trattato sul Metodo Scientifico.[1]

Thus it is only during my own lifetime that the opinion of historians has been totally reversed, following the researches of Pierre Duhem into medieval manu-scripts and the studies of Alexandre Koyré which presented Galileo's work as chiefly the culmination of mathematical reasoning in a Platonic tradition, in which experiments were of no importance, and were more frequently imagined than performed. Yet often the seemingly best established conclusions have had to be modified in the light of new evidence, and the historian must be wary of taking superficial resemblances for essential continuities in the progress of ideas. Balance requires a sort of Devil's Advocate not only for Galileo, but for the great pioneers of our discipline.

Reprinted from *Reason, Experiment, and Mysticism in the Scientific Revolution*, eds. M.L. Righini Bonelli and William R. Shea (Canton, MA: Watson Publishing International, 1975), pp. 131–56.

It was through an excessive attention to printed books alone, and neglect of manuscripts, that pioneer historians were misled into believing that medieval physics had produced nothing of lasting value. Similarly, historians of the present era have been misled into deprecating the novelty of Galileo's science of motion through excessive attention to his final published book, and neglect of his manuscript notes. Unlike his great contemporary, Johann Kepler, Galileo did not publish an account of the bypaths and blind alleys into which he wandered before he arrived at his correct results. Instead, he published in deductive order an array of theorems and problems, stemming from a single definition and a single postulate concerning the motion of free fall. We are thus led along step by step, and we may come to feel that we are following in Galileo's own tracks, as if he had reached his conclusions in the same order as that in which he presented them to his readers many years later. And though we know how improbable it is that anyone ever worked so logically and so unfalteringly in practice, this impression has in fact led several modern historians into a certain trap.

Thus it has come about that Galileo's work on naturally accelerated motion is often treated as if he had started on it from the definition given in the *Discorsi*, and the first theorem which follows directly from it. That same theorem is then treated as indistinguishable from the Merton Rule, though no use was made of any mean-speed concept in Galileo's statement or proof; and, taking this theorem to be a legacy of the Middle Ages, some make the history of the physics of free fall appear as logical and as unfaltering as was Galileo's own final presentation of his science of motion. This simplistic view is, however, even farther from the truth than an older view, now rejected by every historian, that Galileo's science of motion began from some elementary experiments with falling bodies. Yet it has appealed so strongly to historians with a philosophical bias that some have gone so far as to question whether Galileo ever made any experiments, at least in the modern scientific sense of the word.

The simplistic historical position will not hold up, because however Galileo started on his studies of motion, he certainly did not begin with the correct definition of uniform acceleration. This is evident from his celebrated letter to Paolo Sarpi late in 1604; for no matter how we interpret the principle adopted in that letter, it was certainly not a mere legacy of the Middle Ages. Whether we take it to be a blunder on Galileo's part, or a sophisticated semantic manoeuvre to simplify the proof of a result already arrived at in another way, we cannot say that Galileo *began* his work on free fall with the correct medieval definition of uniformly difform motion or with any form of mean-speed analysis.

In the search for Galileo's procedures in reaching his new science of motion, then, we cannot rely to any large degree on the methodology he chose in presenting that science to the public. No more can we discover how Isaac Newton reached the propositions in his *Principia* from a study of the geometric proofs he offered in their support. Nor is there any real puzzle about Galileo's methodology in the *Discorsi*; its pattern is the classic procedure of Euclid and Archimedes, though supplemented by scholia and interspersed with conversational discussions of the meaning and applicability of the demonstrated conclusions. The real puzzle over Galileo's method concerns the logic of discovery, not that of proof. In the discovery of his 38 propositions on free fall and descent along inclined planes, a role was played by mathematics, another role was played by experiment, and still another was played by chance, or luck. Errors were made and corrected; wrong leads were followed and abandoned. To reconstruct the historical process we must turn to other documents – to treatises Galileo withheld from publication, to his letters, and above all to his notes and working papers. From these we may discover Galileo's methodology in the science of motion; or rather, we can reconstruct it, for it was doubtless largely unconscious on his part in the early stages, and unconscious methodology is perhaps a contradiction in terms.

The task of reconstruction is difficult, but not impossible. It is difficult because Galileo's surviving notes are undated, and were intended only for his own eyes. But it is not impossible, because the notes of any one man must have a chronological sequence, and those of a man like Galileo must also have had a psychologically plausible, if not a logically necessary, concatenation. The problem, then, is to find an order for these seemingly chaotic notes that is internally plausible, and is also consistent with everything else we know about Galileo and his work.

Fortunately, some benchmarks are provided by dated documents or by others that can be placed in time with reasonable accuracy. One of these, of course, consists of the lectures *De motu* that certainly belong to the Pisan period, and probably to the year 1590. Another is Galileo's syllabus on mechanics, probably brought to its final form around 1600. Certain letters, notably in 1602, 1604, 1609, and 1611, also give us definitive evidence about Galileo's state of knowledge regarding particular problems of motion in those years.

Characteristic forms of handwriting at various times also provide clues, as do the watermarks in the paper used for the notes. Clues from watermarks are particularly valuable because the papers Galileo used at Pisa, at Padua, and at Florence came from different makers; and in a few cases, the same paper was used for dated letters and for notes on motion. Copies made by pupil-assistants at Florence of theorems proved earlier at Padua establish still further details of order.

Employing these kinds of evidence, I have been able to reconstruct a pattern of Galileo's progress in the study of motion which, while always subject to correction and refinement, seems to me likely now to hold up well in its main out-lines. From this I shall select representative procedures followed by Galileo in reaching his principal results.

Galileo's demonstration of equilibrium conditions on inclined planes, in *De motu*, by reduction to the principle of the lever, led him to a proof that a body should be moved on the horizontal plane by a force smaller than any previously assigned force, and he noted that such motions should be called "neutral" rather than natural or forced. In attempting to deal with *motion* on inclined planes, however, he assumed that speeds should be proportional to effective weights, on the Aristotelian basis that weight is the cause of motion. From this it fol-lowed that the speeds of a given body along two planes of equal height should be inversely proportional to the lengths of the planes. Such ratios, Galileo noted, were not borne out by experiment, which was perhaps the reason that his treatise *De motu* was ultimately withheld from publication.[2]

In 1602, Galileo wrote to Guidobaldo del Monte in answer to the latter's objection to his statement that the times of fall along chords to the bottom of a vertical circle were equal.[3] He declared that he had a proof of this proposition, and also of another, that the time of fall was longer along a chord than along its two conjugate chords, but said that he had been unable to prove the isochronism of descents along circular arcs, which was his objective. Since Galileo did not yet have the law of free fall in 1602, it is reasonable to ask the nature of his proofs concerning chords. These appear not to survive, but in fact one can rigor-ously prove the equality of times along chords to the bottom of a vertical circle from the earlier, incorrect, assumption that speeds along different planes of the same height are inverse to the lengths of the planes.[4] Since only overall speeds are concerned, acceleration need not be considered. The shorter time along con-jugate chords follows also from this proof. These two theorems are easily con-firmed by experiment, by placing boards of different length in a hoop, and noting whether balls released simultaneously along them reach the bottom together. Quite possibly, Galileo discovered this interesting theorem in that way, and then found a proof from his old false assumption.

In the same letter, Galileo described the use of long pendulums in the study of descent along arcs of vertical circles. Such observations probably made him question his idea in *De motu* that acceleration is only of brief duration, at the very beginning of motion. At any rate, by mid-1604 at the latest, he had found a law relating ratios of speeds in descent to the distances from rest, finding that the acquired speeds when plotted against distances fell along a parabola. On the discovery document, *f. 152r* (Plate 6, with transcription in Plate 7) of volume 72, among the manuscripts preserved at Florence, Galileo started with the assumption that as the distances increase as the natural numbers, overall speeds grow as the triangular numbers, a rule suggested by medieval sources which I will discuss later. But Galileo found that for continuous growth, this led to contradictory speed ratios, and he accordingly amended it by introducing a mean proportional of distances from rest, and by using this in the ratios of time.[5] Thus the times-squared law emerged from an arbitrary mathematical device intended to reconcile conflicting ratios of speeds.

Galileo's next move, naturally enough, was to test whether the new rule held good also for inclined planes, and this again was a purely mathematical exercise. The sheet on which it was carried out, *f. 189r* (Plate 8), implies all but one of the rules he would need for the comparison of motion along inclined planes in general, from rest or after a given fall, though this was not immediately apparent to him.[6] On this same sheet there is a sketch of the parabola mentioned in the last line of the previous discovery, and there is also another parabola converted into a triangle by extending the abscissae. Again an arbitrary mathematical device – this time, that of squaring – enabled Galileo to simplify his procedure by adopting a new definition. This particular conceptual change, however, was destined to produce a semantic obstacle for historians, who had only the demonstration that resulted and not this unpublished preliminary step. When Galileo wrote to Paolo Sarpi in October, 1604, and drew up a proof for him, he did not consider the speeds acquired as squared, but simply adopted the squaring as part of the meaning of the word *velocità*, which had not previously been assigned any physical definition. Thus, he offered the rule that *velocità* were proportional to distances from rest, adducing experience of machines that act by striking as his example. His wording in 1604 thus made it seem that Galileo literally asserted a rule of proportionality of speeds acquired to distances fallen, in the sense of those words today, though there is no instance in any of his notes of application of that rule, even in the years 1604–09. Early in 1609 he adopted our ordinary meaning of "velocity"; meanwhile he had worked out most of the essential theorems found in the Third Day of the *Discorsi*.

With the law of free fall in hand and mathematically extended to descent along inclined planes, where it could be verified experimentally, Galileo returned to his earlier project of seeking a rule for times along arcs of a vertical circle. This time his method was one of detailed numerical calculation. Starting with the chord of a quadrant, on *f. 166r* (Plate 9), Galileo drew the two equal conjugate chords, and then the four, and the eight, obtained by successive bisections.[7] He proceeded to compute the times in straight and broken fall along these, in various combinations. In so doing, he arrived at the calculational technique that was needed for the comparison of times along any set of slopes, beginning at any height. But after writing some vain conjectures about circular fall across his elaborate diagram, Galileo seems to have lost interest in this quest. Instead, he decided to write out and prove theorems about motion in the vertical, on inclined planes, and along broken lines. These notes, made at Padua in 1607–08, have been published by Favaro.[8] The work at this time was entirely geometrical, with relatively few numerical examples and none that suggest experimental data.

Toward the end of 1608, however, there is clear evidence of Galileo's use of precise experiment in the modern scientific sense. In the autumn of that year, it occurred to him that his mathematics of free fall had put into his hands a means of testing his longstanding belief that a body in horizontal motion would continue uniformly if unimpeded by external resistances. This belief had originated with his proof in 1590 that a body on a horizontal plane should be capable of being set in motion by a vanishingly small force. In his *Mechanics*, about 1600, the same idea had been further developed, and by 1607 he was teaching that though force was necessary to commence motion, absence of resistance sufficed to conserve it.[9] Contrary to Peripatetic tradition, but in accordance with ideas in the ancient pseudo-Aristotelian *Questions of Mechanics*, Galileo held that different tendencies to motion could exist in the same body, and that these would give rise to a single resultant motion.[10]

Given all this, Galileo had needed only some way of knowing the ratios of speeds of a body leaving a horizontal surface and falling freely through a fixed distance. Those ratios could now be known by dropping a body from rest through various heights and then deflecting it horizontally. The apparatus is drawn on *f. 175v* (Plate 3 with transcription in Plate 4), and the experiment is recorded on *f. 116v* (Plate 2). A table 828 *punti*, or about 80 cm high, was used, and horizontal travel during this drop was recorded for initial drops to the table of 300, 600, 800, and 1000 *punti*. Using the initial drop and its related horizontal advance as a standard, Galileo computed the others, which were in very good agreement with the actual measured distances of advance during fall.[11]

This part of the experiment alone marks a methodological epoch in the study of motion. All known earlier studies of motion (including Galileo's) appear to have been based either on more or less casual, even if careful, observations, or else on pure reasoning, logical or mathematical, from explicit or implicit definitions. This appears to have been true of ancient studies, including those of Aristotle and of Archimedes, as well as of medieval studies by Bradwardine, the "calculators," and Buridan. Galileo's own *De motu* indicates that he had tested some theoretical ratios of speed, but that they had not been vindicated by experiment. The present case is rather different. Ratios of speeds that had already been tested and found to conform to experience were now applied to a physical theory of compound motion in order to test a physical theory about one of its component motions. The test was successful, meaning that more could safely be built on these theories.

But this was not all. Galileo next added a drop of 828 *punti* to the table, equal to the drop after horizontal deflection. The purpose was evidently to confirm his double-distance rule, derived in 1607 by reasoning from one-to-one correspondence of uniform and uniformly accelerated speeds.[12] But in this test, the horizontal travel fell far short of what Galileo expected. The reason was that he did not know of the factor of 5/7 that applies to the acceleration of a rolling body as against one falling freely. Galileo had no way of knowing this, so he attributed the loss of motion to the impact at the moment of deflection. Accordingly, he devised a new experiment, shown on *f. 114v* (Plate 5), in which the ball simply rolled off the end of an inclined plane. His recorded data permit reconstruction of the entire experiment, and though he was unable at the time to compute the horizontal progress for an oblique impetus, we can do so. The result shows him to have been remarkably accurate in recording his results. Using 980 cm/sec^2 for the value of acceleration in free fall, and 5/7 of this for rolling, the greatest discrepancy from theoretical values is about 3%. Descents along the slope were 200, 400, 600, 900, 1200, 1600, and 2000 *punti*, the final drop for a 30° plane being 500 *punti*. The devising of such an experiment is itself methodologically significant.

The accuracy of these recorded data suggest that Galileo had a good deal of practice as an experimentalist before 1608. This contradicts the prevailing opinion, drawn only from Galileo's printed books and our knowledge of the customary practices of physicists up to his time. Even more interesting is the care that Galileo took to assure himself of the validity of his basic assumptions, in view of the fact that he did not later describe these experiments for his readers.

There is still another aspect of this work that is of methodological importance, and that is that Galileo's experiments in 1608 led on immediately to a new discovery. The height of the table was such as to permit him to stoop after releasing the ball, and observe its path from table to floor. These paths he drew in on *f. 116* (Plate 3), with fair accuracy; and he noticed them to be approximately parabolic. On *f. 117r* (Plate 1), which bears the same watermark and which contains notes of a preliminary observation that led to the experiment, we find drawings of parabolas and of a parabolic trajectory with its tangent. Galileo turned at once to the demonstration of the parabolic trajectory, on various sheets that can be dated early in 1609 and which are associated with a letter written early in February of that year. It was in relation to this analysis that Galileo finally adopted the physical definition of "velocity" that we still use. His reasoning is embodied in the third proposition of the Fourth Day of the *Discorsi*, which is virtually unchanged from the form in which it was written early in 1609.[13]

I have given but a sketchy account of the chronology and nature of Galileo's early work on motion, up to the point at which the main body of propositions had been reached that were later to appear in the *Discorsi*. However elementary the content, the pattern is that of modern scientific investigation. First, a causal hypothesis was adopted that turned out to be wrong; namely, that speed of fall depends on effective weight. The incorrect ratios derived for speeds on inclined planes were tested, found defective, and so reported. *De motu* was withheld from publication, but the problem was not forgotten. In time, two theorems were found that agreed with experience and that could be proved from the false assumption. This time the circumstances of testing drew attention to the role of acceleration, previously neglected. Search for a rule of acceleration revealed contradictions in the traditional law. A mathematical device, introduced to reconcile ratios of speed in continuous acceleration, led to the law of free fall. This law was found consistent also with descent on inclined planes, but in a form inconvenient for relating speeds to simple distances from rest. Accordingly, speed was redefined in a manner consistent with observations of a different kind. There followed a long train of mathematical theorems, in which the concept of speed seldom appeared, being replaced by relations of distances and times. The new law permitted precise testing of an earlier physical hypothesis related to inertial motion. This test succeeded, and led incidentally to a new discovery. Mathematical development of the new phenomenon required a re-definition of speed. At this point, Galileo's new science of motion was completed in all essential respects except the decision as to the best postulational basis for its formal deductive presentation. If none of the elements in this pattern are new – causal hypothesis, mathematical expression, experimental test, new discovery, reconciliation of concepts by new definition, and so on – these elements were nevertheless combined in a manner for which it is hard to find earlier examples in the study of motion, and perhaps in any area of physics.

Nevertheless, Galileo did not disdain to borrow from earlier sources anything that seemed to him to fit in with his conception of exact science. Most notable is his adoption, as his first theorem on uniform motion, of the first proposition of Archimedes in his treatise *On spiral lines*. This theorem is of capital importance with regard to the principal respect in which Galileo's science of motion departed mathematically from medieval physics. Its proof by Archimedes depended solely on the Eudoxian theory of proportion, which was lost in the standard Latin text of Euclid during the Middle Ages and was restored only in the sixteenth century. It was that theory of proportion alone that permitted Galileo to deal with acceleration as continuous in the modern mathematical sense, the importance of which fact I shall presently stress. Borrowing also Aristotle's unchallenged definitions of "equal speed" and "greater speed," Galileo applied the same Eudoxian definition to the proof of his second theorem on uniform motion. This appears to be novel, as do his other four theorems which ingeniously used the notion of compound ratio also used by Archimedes in lieu of the later concept of "function."

From medieval tradition, Galileo at first adopted a rule of free fall related to impetus theory, though he found this inconsistent with continuous acceleration and abandoned it, as I have already indicated. It is also possible that he found in Swineshead's "Calculations" the first clue to his own fruitful use of one-to-one correspondence betwen infinite aggregates, in its applications to proportions, though this is dubious on other grounds. It is highly improbable that either Galileo's method or his results in the science of motion are *founded* on medieval mathematical physics, for reasons I shall now set forth.

Antiquity offers no mathematical treatment of free fall or of uniform acceleration, but writers of the fourteenth century afford both. The odd thing, in the light of the high quality of both the mathematics and the physics of that century, is that free fall and uniform acceleration were then dealt with quite separately, and not in conjunction. I shall try to show that our surprise at this has arisen from a basic oversight on our part, and from a misinterpretation of medieval physics as a whole that has arisen in the process of singling out certain parts of it precisely because, in the light of later knowledge, they seemed to be harbingers of Galileo's science of motion. But in fact those parts not only belonged to a different physics, founded on the search for causes, but also depended on a different mathematics, essentially arithmetical and discrete rather than geometrical and continuous.

First, let us consider the medieval treatment of uniform acceleration. This arose in the classification of every possible type of motion, as Father William Wallace has shown, rather than in the quest for a mathematics of free fall.[14] It is no accident that Galileo's treatise in the *Discorsi* is entitled "On naturally accelerated motion" and not, in the medieval pattern, "On uniform acceleration." Galileo's purpose here, as elsewhere in science, was to limit the scope of his inquiries to separate and well-defined areas, and not to seek a general theory of the universe. This is an extremely important part of his scientific methodology, expressed in his *Dialogue* when he wrote that there is no phenomenon in nature, not even the least that exists, which is such that even the most profound theorist will ever attain a complete understanding of it. Earlier, in the *Assayer*, he had remarked that the best philosophy will offer the fewest promises and will teach very little as certain.

The concept of a mean speed as a single speed chosen from an infinite aggregate to represent all speeds, so important to medieval analysts of acceleration, is nowhere to be found in Galileo's writings, or even in his private notes. Nor was the mean proportional, so essential to Galileo's treatment, given any particular meaning in the Middle Ages. Medieval physics dealt always with bounded motions, which implied a mean speed. Galileo considered unbounded motions, within which ratios may be found between speeds, times, or distances at any two points, but for which the concept of a mean speed is superfluous. It is noteworthy that even Galileo's earliest triangles for accelerated motion were open at the bottom, as was that used in the proof of the times-squared law in the *Discorsi* near the end of his life. It is likewise noteworthy that Oresme, who first used the triangle for acceleration, implied that the area represented total motion, or distance traversed, whereas Galileo never implied any meaning for such an area other than that of a kind of overall speed. These conceptual differences are related to the medieval search for causes as against the quest for laws, and to the assumption of discreteness as against that of continuity in physical phenomena.

Next, as to the medieval treatment of free fall as such, which was a part of Buridan's impetus theory, I was formerly inclined to agree with Alexandre Koyré that Galileo, in his *De motu*, simply adopted bodily the impetus theory of the fourteenth century, and that he subsequently abandoned it only because it did not lend itself to precise mathematical formulation. This, however, is not so. The very ingenious explanation of acceleration in free fall offered by Jean Buridan was not even mentioned in *De motu*, though it had recently been expounded again by Benedetti. In its place, Galileo offered an inferior theory, so it cannot be maintained that he adopted impetus theory bodily. But more than that, it cannot be maintained that Buridan's impetus theory of free fall resisted precise mathematical formulation. Such a formulation had in fact been given to it by Albert of Saxony, and this was accepted by Oresme, by Leonardo da Vinci, and probably by all physicists up to the second half of the sixteenth century. Galileo himself attempted to use it when he first seriously sought a rule of acceleration in free fall, probably early in 1604. Albert's formula, I believe, has simply been misunderstood by modern historians, through neglect of medieval physics as a whole. To understand it, let us review the physics of impetus.

Buridan's explanation of acceleration in free fall was the first fully rational account ever to be given, and the best to appear up to the time of Newton. It is perhaps still the best explanation for those who insist on causes, and are not content with laws. Buridan regarded impetus as a kind of force impressed in a body and remaining there except as reduced either by external resistances or by an internal tendency on the part of the body to some contrary motion. In the case of the heavenly spheres, neither external resistence nor tendency to any motion except rotation was present, whence an initial impetus given to them would be conserved forever. In terrestrial heavy bodies, however, there was always a natural tendency downward, which tendency would normally conflict with and reduce the impressed impetus, and ultimately would bring the body to earth.

Buridan, however, was an excellent physicist, and he did not fail to see that there was one unique case in which impetus would be conserved undiminished in a heavy body, in the absence of any external resistance. This was the case of a heavy body hurled straight down, the natural tendency being also in that direction. For the explanation of acceleration in free fall, only a single additional postulate was required; namely, that impetus could be imparted not only by violent projection, but also by natural motion itself. And thus Buridan wrote that one must imagine that in the first movement from rest, a body is moved only by its heaviness; but along with that motion, it gains an impetus, so that in the next movement it is moved by that impetus together with its heaviness, and hence moves faster. With the faster motion it receives still stronger impetus, and so on, the speed increasing to the end.[15]

Under such a theory it is evident that during the first movement from rest, when only a single cause – the heaviness – is acting, the speed is uniform; and in the second movement it is likewise uniform, though faster, since only the first impetus is acting along with the heaviness, and so on. Buridan's explanation in fact implies a sort of quantum theory of speeds, increasing discretely though very rapidly. This accords well with Aristotelian causal principles, as well as with Aristotle's own remark that though the motion of projectiles appears continuous, it is not really so, only that motion which is imparted by an unmoved mover being truly continuous. Gassendi was later to support a theory of successive discrete impulses in his book entitled *On the Motion Impressed by a Moved Mover*, and I shall presently consider another post-Galilean impetus theory. In the Middle Ages, a quantum physics of acceleration was also mathematically appropriate because (as previously mentioned) the theory of proportion was arithmetical and discrete, in the absence of one Eudoxian definition in the medieval Euclid.

In strict Aristotelian physics, there could be no first *mathematical* instant of motion from rest, for at such an instant a body would be both in motion and at rest, in violation of the law of contradiction. But there could be a first *physical* instant of motion, meaning by this a duration shorter than any previously assigned time. In short, the successive "degrees" of speed, which were always represented by integers, were contiguous rather than continuous.

With this in mind, let us now consider the mathematization of Buridan's theory of free fall by Albert of Saxony. Albert's rule stated that when a body has fallen a certain distance, it has a certain speed; and when it has moved twice that distance, it has twice that speed, and when it has gone triple the first distance, it has triple the first speed, and so on. Historians of our time, thinking in post-Galilean and algebraic terms, have naturally supposed this to mean that the distances of which Albert spoke were to be taken *from rest*, in which case the rule would amount to putting speeds acquired proportional to distances from rest. Such a rule was indeed put forth in 1584 by Michel Varro, and it was shown by Galileo to be untenable and to imply an instantaneous motion.[16] Indeed, this rule contradicts the concept of rest itself. But Albert of Saxony would have been perfectly capable of perceiving that contradiction, and his rule was in my opinion quite different. The distances of which he spoke were meant to be taken successively, so that in the second distance the body went twice as far and twice as fast as in the first, implying an equal time, and so on. Thus in Albert's rule, the successive distances, speeds, and times, all went up as the natural numbers (and any cumulative values progressed as the triangular numbers). It was this rule that was accepted by Oresme, Leonardo, and others down to the time of Galileo, who indeed tried to apply it in the opening lines of *f. 152r* (Plate 6 with transcription in Plate 7), but mistakenly. He found that it could not apply to continuous acceleration, and moved on to the times-squared law. But there was nothing internally contradictory in Albert's rule; in an Aristotelian world in which bodies fell with different speeds, proportional to their weights, a rule of this kind might well apply.

This account of Albert's mathematical rule for fall clears up an otherwise perplexing problem of historians; that is, why medieval writers did not associate free fall with uniformly difform motion. In the latter, it was known that the distance traversed in the second of two equal times from rest must be three times that covered in the first. But Albert's rule for free fall would make the distance in the second time only double that of the first; hence free fall could not be a case of uniform acceleration. This account also clears up a second puzzle; namely, why in all the commentaries on physics after Albert, no one is known to have raised the question whether, in free fall, the speeds are proportional to the distances fallen or to the times elapsed. In the usual modern interpretation of Albert's rule, such a question would have been bound to arise long before Galileo.

Any lingering doubts concerning this interpretation of impetus mathematics are dispelled if we consider a suggestion of G.B. Baliani in 1646 that behind Galileo's odd-number rule for distances in equal times, as observed by experiment, lay the natural-number rule for vanishingly small times. Baliani noted that if we take three observable equal times, and the distances traversed, and if we divide each of these distances into ten parts in which the spaces increase as the integers, then 55 such tiny units of space are covered in the first large equal time, 155 in the second, and 255 in the third; but ignoring the last digit, these distance-quanta are 5:15:25 or as 1:3:5, very nearly. Now, if we had made the division by one hundred instead of ten, said Baliani, the figures would have been 5050, 15050, 25050, very close to 1:3:5; and since physical instants are millions of times smaller than observable times, Galileo's odd-number rule for large distances is implied by a natural-number law for infinitesimal distances.[17]

In the same year Honoré Fabri published at Lyons a very detailed treatise on the impetus theory of free fall, in which similar ideas were developed at length. Among other things, Fabri argued against Galileo's position that in order to reach any speed from rest, a body must have passed through every possible smaller speed, a position also rejected by Descartes.[18] Fabri noted that the same body, falling freely, and along an inclined plane, goes faster vertically. Hence, he said, it must start faster; and therefore there must exist some speeds along the incline that are not to be found in vertical fall. Such arguments help us to understand medieval impetus theory, with its implicit assumption of infinitesimal successive motions, as against Galileo's continuity theory of motion, made possible by the restoration of Euclid, Book V. Not only medieval physical thought, but the elaborate medieval arithmetical theory of proportion continued to be taught well into the seventeenth century, and conservative scholars were not swept away by Galileo's new science of motion.

Even more fundamental was Fabri's contention that his physics was better than Galileo's because it could provide a cause for acceleration in fall, while Galileo's could not. Here we have a truly methodological debate, though Galileo was not alive to reply. His view was that whatever laws were confirmed by precise experiment should be assumed to hold beyond the boundaries of observable phenomena, much as Newton was to give it as a rule of science that properties found in all bodies accessible to experience were to be attributed to all bodies whatever. Ultimately, this procedure would eliminate causes in favor of laws, and we all know what Galileo said about inquiries into the cause of acceleration in fall. Whether or not we happen to agree with the substitution of laws for causes as a program of physics, it will be admitted that such a program had not been suggested before Galileo, and that it has received a good deal of support since his time. Fabri argued, on the other hand, that his physics and Galileo's were both consistent with the observable phenomena, while only his could discover something ultimate behind those phenomena, and different in form from their laws.[19]

In this sense there is perhaps something methodologically new even in Galileo's ultimate presentation of his science of motion to his readers. Previously, I outlined the novelties in Galileo's procedures in reaching his conclusions, but said that his presentation of them was simply that of Archimedes, interspersed with scholia and interpretative conversations. I am not unwilling to stand by that, and it was certainly Galileo's own view, expressed in a letter which I shall cite presently. But it is worth mentioning that if that very form of presentation was not an open and explicit attack on the Aristotelian conception of physics as an understanding of nature in terms of causes, it at least implied that everything of lasting value in physics could be presented in the form of precise laws, experimentally confirmed, and their deductive consequences. It is my view that Galileo's reason for presenting part of his last work in Latin, within an Italian dialogue, was to mark out those parts which he himself considered to be established irrefutably, and to which he wished his own name, rather than that of any interlocutor, to be forever associated. And it was precisely his new science of motion, consisting of definitions, postulates, and theorems, that was published in Latin, for scholars everywhere. Not even the theorems in his new science of strength of materials, which remained to some degree tentative, were so distinguished from the discursive and speculative parts of his last great book.

Those who look upon Galileo's science of motion as in effect the mere addition of some new results to medieval physics, carried on logically from earlier beginnings should, I think, explain historically why the times-squared law and the odd-number rule were not deduced much earlier and applied to inclined planes and projectile motions. One common explanation is a fancied inhibition of scientific thought by the rise of humanism. Now, no intellectual movement has ever been so strong as to seduce every individual thinker for two centuries away from all other pursuits, so this is not so much an explanation as an ad hoc reason – what Galileo called "reaching for sky-hooks." But in fact it is so far from the truth that the very opposite is true. It was a humanist, Bartolomeo Zamberti, who, moved by his abhorrence of everything touched by the Arabs and his Renaissance love for Greek antiquity, first made accessible in print a correct Latin translation of Euclid's fifth book. But Zamberti did not explain its true significance. The historical event that made possible Galileo's new science of continuous acceleration in free fall was therefore not Zamberti's Latin Euclid of 1505, but Nicolò Tartaglia's Italian Euclid of 1543, in which the real meaning of Eudoxian proportion theory was explained, and the specific correction of the errors in the medieval commentary of Campanus was set forth. But since the universities did not use Italian texts, and especially such colloquial Italian as Tartaglia's, intended to educate the layman, it was not until the 1570s that Eudoxian proportion theory began to invade the universities. The Latin commentaries of Federico Commandino and of Christopher Clavius date from that decade; and thus Galileo, who entered the University of Pisa in 1581, was among the first generation of students who could reasonably be expected to put aside the medieval theory of proportion and to think in terms of ratios of continuous magnitudes in the modern sense. And by the time Galileo published his work, there was a new generation of mathematicians capable of understanding and advancing it, though there were also still men like Fabri and Descartes to oppose his new science of motion. I mention this because it seems to me that datable events are preferable to intellectual and social theories in the explanation of new modes of thinking in physical science.

In conclusion, here is Galileo's own account of the methodology used in his ultimate publication, written in a letter to Baliani in 1639:

I have treated the same material [as you], but somewhat more at length and with a different attack; for I assume nothing but the definition of that motion with which I wish to deal, and whose events I wish to demonstrate, in this imitating Archimedes in his *Spiral lines*, where he, having explained what he means by motion made in a spiral – that is, that it is composed of two equable motions, one straight and the other circular – goes on immediately to demonstrate its properties. I explain that I wish to examine the properties that are found in the motion of a moveable which, leaving from a state of rest, goes moving with speed always growing in the same way; that is, that the acquisitions of speed grow not by jumps, but equably with the growth of time, so that the degree of speed acquired, for example, in two minutes of time, shall be double that acquired in one minute, and that acquired in three minutes, and then in four, is triple and then quadruple that which was acquired in the first minute. And premissing nothing more, I go on to the first demonstration in which I prove that the spaces passed by such a moveable are in the squared ratio of the times; and I go on to demonstrate a goodly number of other events. You touch on some of these, but I add many more – and perhaps more marvellous [ones], as you will see from my dialogue on this matter, already published two years ago at Amsterdam, though none have come to me except page by page, sent for corrections and for the making of an index of important matters ...

But getting back to my treatise on motion, I argue *ex suppositione* about motion, so that even though the consequences should not correspond to the events of the natural motion of falling heavy bodies, it would little matter to me, just as it derogates nothing from the demonstrations of Archimedes that no moveable is found in nature that moves along spiral lines. But in this I have been, as I shall say, lucky; for the motion of heavy bodies and its events correspond punctually to the events demonstrated by me from the motion I defined. ...[20]

From this, it seems to me that Galileo's own view of his method in the science of motion was that which was expressed only much later, by Heinrich Hertz, in these words:

We form for ourselves images or symbols of external objects, and the form which we give them is such that the necessary consequents of the images in thought are the necessary consequents in nature of the things pictured. In order that the requirement be satisfied, there must be a certain conformity between nature and thought. Experience teaches us that the requirement can be satisfied, and hence that such a conformity does exist.[21]

NOTES

1 A. Valdarnini, *Il metodo sperimentale da Aristotele a Galileo* (Asti, 1909), p. 63.

2 Galileo, *Opere* I, p. 301; tr. I.E. Drabkin (Madison, 1960), p. 70. (*Opere* refers to the *Edizione Nazionale* of Galileo's works, ed. Antonio Favaro.)

3 *Opere* X, pp. 97–100.

4 Cf. S. Drake, "Mathematics and Discovery in Galileo's Physics," *Historia Mathematica* I (1974), pp. 135–137. (Cited below as *Historia*.)

5 Cf. S. Drake, "Galileo's Discovery of the Law of Free Fall," *Scientific American*, 228 (1973), n.5, p. 89.

6 *Historia*, pp. 139–143.

7 *Historia*, pp. 143–148.

8 *Opere* VIII, pp. 371–423.

9 *Opere* I, pp. 299–300; II, p. 179; X, p. 170.

10 Pseudo-Aristotle, *Questions of Mechanics*, n.1 (Loeb edition), pp. 337–339.

11 Cf. S. Drake, "Galileo's Experimental Confirmation of Horizontal Inertia," *Isis* 64 (1973), pp. 291–299.

12 *Opere* VIII, pp. 383–384; cf. also p. 243. A similar approach was used in Galileo's *Dialogue*, *Opere* VII, pp. 255–256; tr. S. Drake (Berkeley, 1953), pp. 228–229.

13 *Opere* VIII, pp. 281–282; tr. S. Drake (Madison, 1974), pp. 230–231.

14 W. Wallace, "The Enigma of Domingo de Soto," *Isis* 59 (1968), pp. 384–401.

15 Cf. M. Clagett, *The Science of Mechanics in the Middle Ages* (Madison, 1959), pp. 560–561.

16 M. Varro, *De motu tractatus* (Geneva, 1584).

17 G.B. Baliani, *De motu gravium solidorum et liquidorum* (Genoa, 1646), pp. 110–111.

18 P. Mousnier (ed.), *Tractatus physica de motu locali ... ex praelectionibus Honorato Fabry* (Lyons, 1646), pp. 88–90; cited below as Fabri. For the opinion of Descartes, see *Opere* XVII, p. 390 and XX, p. 612.

19 Fabri, pp. 98 ff.

20 *Opere* XVIII, pp. 11–13.

21 H. Hertz, *Principles of Mechanics*, tr. Jones (Dover, 1956), p. 3.

6

Galileo's 1604 Fragment on Falling Bodies (Galileo Gleanings XVIII)

I

The first attempted derivation by Galileo of the law relating space and time in free fall that has survived is preserved on an otherwise unidentified sheet bound among his manuscripts preserved at Florence. It is undoubtedly closely associated with a letter from Galileo to Paolo Sarpi, dated 16 October 1604, which somehow found its way into the Seminary of Pisa, where it is still preserved. Those two documents, together with the letter from Sarpi to Galileo which seems to have inspired them, are translated in full below. Sarpi's letter, dated 9 October 1604, suggests that recent oral discussions of problems of motion had recently taken place between the two men. It reads as follows:

In sending you the enclosure, it occurs to me to propose to you a problem to resolve, and another that seems to me paradoxical.

We have already concluded that no heavy body can be thrown upward to a given terminus without some force, and consequently some speed. We are agreed (as you finally affirmed and discovered) that it will return downward by the same degrees by which it went upward. There was some objection about the musket-ball; here the power of the firing clouds the force of the objection. But let us say: a strong arm that shoots an arrow with a Turkish bow will send it clear through a board; but if the arrow should descend from the height to which the arm with the bow can send it, it would make but a small entry [into the board]. I think the objection may be trivial, but I don't know what to say.

The paradox: if there are two movables of different material, and a force smaller than they are capable [of receiving,] either of them will receive it; if the force is communicated

This article is reproduced with the permission of the Council of the British Society for the History of Science. It was first published in *BJHS* 4 (1969), 340–58.

to both, they will receive it equally. For example, say that gold can take a force of 20 and no more and silver 19 and no more; if both are moved by a power of 12, both will receive [the full twelve]. So it seems, because the whole force is communicated, the moving body is capable of receiving it, and thus the effect is the same. But it appears not, because it would follow that two movables of different materials driven by equal force would go to the same point with the same speed. If one should say that the force of 12 will move the silver to the same point but not with the same speed, why not? Especially if both are capable [of receiving] even more than that which 12 can communicate to them?

I do not require that you answer, but just not to send this page blank, which had a peripatetic desire to be filled with these characters, I wanted to content it, as the agent does to prime matter. So here I make an end, and salute you.[1]

Galileo replied one week later, saying:

Thinking again about the matters of motion, in which, to demonstrate the phenomena [*accidenti*, events] observed by me, I lacked a completely indubitable principle to put as an axiom, I am reduced to a proposition which has much of the natural [*naturale*, physical] and the evident; and with this assumed, I then demonstrate the rest; that is, that the spaces passed by a naturally falling body are in squared proportion to the times, and consequently the spaces passed in equal times are as the odd numbers from one, and the other things. And the principle is this: that the natural movable goes increasing in speed with that proportion with which it departs from the beginning of its motion; as, for example, the heavy body falling from the end A along the line ABCD, I assume that the degree of velocity that it has at C, to the degree it had at B, is as the distance CA to the distance BA, and this continuing, at D it has a degree of velocity greater than at C according as the distance DA is greater than CA

I should like your reverence to consider this a bit, and tell me your opinion. And if we accept this principle, we not only demonstrate (as I said) the other conclusions, but I

believe we also have it very much in hand to show that the naturally falling body and the violent projectile pass through the same ratios of speed. For if the projectile is thrown from the point D to the point A, it is manifest that at the point D it has a degree of impetus able to drive it to the point A; and likewise the degree of impetus at B suffices to drive it to A, whence it is manifest that the impetus at the points D, C, and B go decreasing in the proportions of the lines DA, CA and BA; whence, if it goes acquiring degrees of speed in the same [proportions] in natural fall, what I have said and believed up to now is true.

As to the experiment of the arrow, I believe that in falling it will acquire like force to that with which it was shot, as we may discuss orally with other examples, since I have to be there [in Venice] before the end of the month. Meanwhile I beg you to think a little about the aforesaid principle.

As to the other problem you propose, I believe that the same movables both receive the same force, but that it does not work the same effect in both; as, for example, the same man, rowing, communicates his force to a gondola and a rowboat, both of them being capable [of receiving] even greater [force]; but the same effect does not follow in both as to speed or length or interval through which they move.

I write but darkly; this little suffices rather to satisfy the duty of replying than that of solution, putting this off until I talk to you soon face to face.[2]

The fragment associated with this correspondence has been translated previously, but not in full, and usually with a specific mistranslation discussed below, which has led to much confusion. In the ensuing translation, I have rendered the word *velocità* as "velocity" when it refers to momentary velocity at a point, and as "speed" when it refers to the effective velocity over an interval of changing velocities. The distinction will be found useful in an understanding of Galileo's reasoning.

I suppose (and perhaps I shall be able to demonstrate this) that the naturally falling body goes continually increasing its velocity according as the distance increases from the point from which it parted; as, for example, the body departing from the point A and falling along the line AB: I suppose that the degree of velocity at the point D is as much greater than the degree of velocity at the point C, as the distance DA is greater than CA; and so the degree of velocity at E to be to the degree of velocity at D as EA to DA, and thus at every point of the line AB it is to be found with degrees of velocity proportional to the distances of the same points from the end A. This principle appears to me very natural, and one that responds to all the experiences that we see in the instruments and machines that work by striking, in which the percussent makes so much the greater effect, the greater the height from which it falls; and this principle assumed, I shall demonstrate the rest.

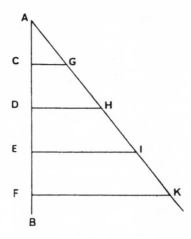

   Draw the line AK at any angle with AF, and through the points C, D, E and F draw the
parallels CG, DH, EI and FK. Since the lines FK, EI, DH and CG are to one another as
FA, EA, DA and CA, therefore the velocities at the points F, E, D and C are as the lines
FK, EI, DH and CG. So the degrees of velocity go continually increasing at all the points
of the line AF according to the increase of the parallels drawn from all those same
points. Moreover, since the speed with which the moving body has come from A to D is
compounded of all the degrees of velocity it had at all the points of the line AD, and the
speed with which it has passed over the line AC is compounded of all the degrees of
velocity that it has had at all the points of the line AC, therefore the speed with which it
has passed the line AD has that proportion to the speed with which it has passed the line
AC which all the parallel lines drawn from all the points of the line AD over to AH have
to all the parallels drawn from all the points of the line AC over to AG; that is, the square
of AD to the square of AC. Therefore the speed with which it has passed the line AD has
to the speed with which it has passed the line AC the square of the proportion that DA
has to CA. And since speed to speed[3] has contrary proportion to that which time has to
time (because it is the same to increase the speed as to diminish the time), therefore[4] the
time of motion through AD to the time of motion through AC has a proportion that is the
square root of that which the distance AD has to the distance AC. The distances, then,
from the beginning of motion are as the squares of the times; and, dividing, the spaces
passed in equal times are as the odd numbers from unity, which corresponds to what I
have always said and [have] observed with experiences; and thus all the truths agree with
one another.
   And if these things are true, I demonstrate that the speed in forced motion goes
decreasing in the same proportion with which, in the same straight line, natural motion
increases. For let the beginning of forced motion be the point B, and its end the point A.

And since the projectile does not pass the point A, therefore the impetus it has had at B was such as to be able to drive it to the point A; and the impetus that the same projectile has in F is sufficient to drive it to the same point A; and the same projectile being at E, D, and C, it is found joined with impetuses capable of driving it to the same point A, neither more nor less; therefore the impetus goes evenly falling away according to the diminution of the distance of the moving body from the point A. But according to the same proportion of the distances from the point A, the speed goes increasing when the same heavy body shall fall from the point A, as assumed above and compared with our other previous observations and demonstrations; therefore what we wished to prove is manifest.[5]

I am inclined to believe that the sequence of events was as follows. Oral discussions had taken place shortly before the above correspondence, in which Galileo had asserted that in free fall, the successive spaces from rest passed over in equal times were in the proportion of the odd numbers starting from unity; that a body would rise to the same height from which it had fallen, and in so doing would pass in reverse through the same degrees of speed; and the distances fallen from rest in unequal times were in the ratio of the squares of the times. Sarpi questioned whether a musket ball would acquire in descent anything like the original force of the upward shot, and wished to see a proof of Galileo's space and time relationships. Galileo, thinking the matter over again at Padua, was unable to find any truly unquestionable principle to assume, but tentatively accepted one that he thought was physically reasonable, and for which he hoped eventually to find a demonstration; this assumption was that the velocities increased in proportion to the distances fallen. In saying that he might be able to demonstrate this, he had in mind not a mathematical demonstration, but some means of physically measuring velocities. Meanwhile he supported the assumption only on the basis that what could be observed in pile-drivers and the like showed that force, and hence speed, did increase with height of fall. Granted the assumption, he proceeded to demonstrate the special phenomena he had already observed himself; that is, that in natural acceleration the spaces passed from rest in equal times grew as the odd numbers from one.

This demonstration, written out at Padua, he intended to show to Sarpi when they met at Venice later in October. First, however, he wanted Sarpi to consider the assumption and to find out whether he believed that the desired consequences followed from it. Whether or not this was actually done, we do not know. By a curious coincidence, Sarpi's letter to Galileo was written on the very date that the nova of 1604 was first sighted at Verona, and Galileo's interest in that event may have prevented his planned visit to Venice. Or it may be that further oral discussions with Sarpi are the reason that the correspondence on motion with him ends here as suddenly as it began.

II

What has been published on the 1604 fragment by Duhem,[6] Koyré,[7] Hall[8] and Humphreys[9] is interesting, but with respect to the reconstruction of Galileo's reasoning it is unsatisfactory. The trouble lies partly in our habit of translating Galileo's words into modern mathematical symbolism. That symbolism inevitably involves equations, and it is hard not to read equations algebraically. Hence we easily see a fallacy, or fallacies, that escaped Galileo entirely; we see them, moreover, so clearly and evidently as to be puzzled that they escaped him. We know that we are not dealing with the productions of a mathematical tyro, and we know that this particular fragment was taken seriously by Galileo at the time, since it is written out in a fair hand and was not destroyed or cancelled. Hence all that we know is that so far we have not succeeded in precisely following his own thought. That thought has to have been such as to have appeared to him, at least, free of contradiction.

Another trouble in our retrieval of Galileo's train of thought arises from inadequate attention to the precise Euclidean concepts that Galileo inherited. An example may be helpful. We read in Euclid, Book V, Definition 13: "*Inverse ratio* means taking the consequent as antecedent in relation to the antecedent as consequent."[10] So familiar are we with the modern concept of inverse proportion that we may easily be tricked into thinking that that is what Euclid was defining, as if his word "as" meant "to." Euclid meant only that if A/B is a ratio, then B/A is called its inverse ratio. Hence if $A/B = C/D$, then $B/A = D/C$, which seems to us a mere tautology. The latter relation was proved by Theon as a corollary to Book V, Proposition 4. It was included as Euclid's in sixteenth-century texts, but I have not found any authentic use of the term "inverse proportion" in Euclid's *Elements* as a relation between ratios. It is not a concept for which any necessity can be found in pure geometry, though it is enormously useful in physics or in any other application of mathematics. It is therefore anachronistic and certainly misleading to translate Galileo's phrase "contrary proportion" in the 1604 demonstration as "inverse proportion," in the modern sense, and to draw conclusions from the phrase itself.

Galileo's clear understanding of the role of mere definitions in geometry is amusingly illustrated when, attempting to rectify Euclid's incomplete definition of "compound proportion," and encountering an objection from Simplicio, he has Salviati reply: "But here there are no theories, or demonstrations, inasmuch as this is the simple imposition of a name. If you do not like the word 'compounded,' and name this 'decomposed,' or 'stuck together,' or 'confused' [proportion], or whatever else you like, then at least allow me this: that when later we shall have three magnitudes of the same kind and I call the proportion 'decomposed,' or 'stuck together,' or 'confused,' I shall mean the proportion that exists between the extremes of those magnitudes, and no other."[11] Similarly, those who want to say "inverse proportion," when Galileo writes "contrary proportion" are bound by his use of the term, and not by its present meaning, still less by a definition in Euclid, applying to a form of ratio, that Euclid himself never went on to apply to any form of proportion.

The reason for the difference between pre-Galilean and modern uses of "inverse proportion" is that in mathematics, proportionality can always be arranged to suit the user, whereas the same freedom does not exist when mathematics is applied to physical reality. Or rather, we prefer not to exercise that freedom. Thus when we say that the weight of a body is inversely proportional to the square of its distance from the earth, we could just as well have said that the lightness of the body is directly proportional to the square of that distance. This would seem to us a clumsy way of talking; but in Galileo's day, it is exactly the way men did talk. Some find it strange that Galileo should have gone on using expressions like "degree of tardity," and explain this by saying that he was still rooted in the meaningless physical distinctions of Aristotle. But it would be equally tenable, and more illuminating, to say that Galileo's terminology often reflected his mathematics, which was that of the Euclidian theory of proportion.

Under that theory, on which Galileo lectured at Padua, and on which he composed the work from which the foregoing quotation was taken, it was impossible to conceive of velocity as the ratio of space traversed to time elapsed. No ratio whatever was allowed to exist between two things different in kind. That is of the utmost importance in any reconstruction of Galileo's thought. To him, velocity was not a ratio, but a "degree of speed," something that was measured by a pure number, just as time and space were measured. The numbers measuring degrees of speed could be added and subtracted, just like those measuring time or space. Ratios could be formed of velocities, and such ratios could be compared with ratios of spaces, or ratios of times, or of numbers, or of lines, or of areas, or ratios of anything else; but both terms of any ratio had to represent measures of the same kind of thing.

The 1604 fragment concerns four definite physical concepts that are now seen as mathematically related in a simple way, and their relation is so intimately know to us that we have a hard time thinking of them in any other way than that which we know to be correct. The concepts are (1) terminal speed, (2) average speed, (3) distance traversed from rest, and (4) time elapsed, all (of course) in uniformly accelerated motion of a freely falling body. We see quite plainly that there are really only two variables, the distance traversed and the time elapsed. We know that the latter is the independent variable, of which the former is a function, as usually expressed in the form $s = 1/2\ at^2$, while either type of velocity is entirely derived from this by the definition $v = s/t$. It is only by an effort that we can realize the absurdity to Galileo of all these expressions, as indeed of any kind of algebraic equation or of any ratio between quantities of different kinds.

Thus if we want to understand the 1604 fragment, we must first make the effort of thinking of the four concepts mentioned above as four entirely different kinds of entity; or better, of four different qualities or properties a body was capable of exhibiting: a quality of instantaneous speed, or "intensity of speed"; the body's "quantity of motion" or actual speed over an interval during which the first quality varied; a property of traversing space, and one of existing through a period of time. Galileo's problem was the establishment of a mathematical relationship between those qualities, if possible, or between pairs of them, as exhibited by a freely falling body. It was a conviction of his, an act of faith at the outset, that the task could be accomplished in such a way that any measurements conducted on observable phenomena would not contradict the mathematical relationship.

For this purpose he asked that a principle be granted, one which at the time appeared natural and reasonable to him. That it was wrong, as he later recognized, should not affect our attempt to follow his original reasoning. We know that in theory at least, any result desired can be made to follow from a false assumption, so we should not be surprised that Galileo's result was correct. Still less should we assume that Galileo must have made further errors in order to get a true conclusion out of a false assumption.[12] If he did, we should correctly identify and analyse them. But it is one thing to discover internal errors in a man's reasoning, and quite another to discover that he made certain assumptions that we would not accept. If Galileo did bad mathematics at the age of forty, it is going to be hard indeed to understand his later works. But if he merely made unwarranted assumptions in the process of relating physics to mathematics, there is no occasion for surprise, and by studying his train of thought we may learn something about the process by which physics actually became mathematical.

In short, the historian's first task, as I see it, is to perceive how it was possible for Galileo to arrive at the correct relationship of space and time in uniform acceleration without his having immediately become aware of the falsity of his initial assumption. For we must recall that later, when he did become aware of it, it was easy for him to modify it and finally to replace it by the correct assumption, which he eventually published. It is also important to note that Galileo was not convinced, even in 1604, of the correctness of his first assumption, but only of its apparent support by gross observations of percussion effects. In contrast to this, he was quite confident of the truth of his conclusions – that in free fall from rest, the spaces covered in equal times grow as the odd numbers from unity, and that the over-all spaces are as the squares of the elapsed times.

Now it is hardly possible that from a principle he described as merely proba- ble, by means of a demonstration that was not purely mathematical,[13] Galileo had arrived for the first time at the correct conclusions in which he expressed such confidence. He says, in fact, that he has long asserted the truth of those results, but has just now sought a demonstration of them. We have Galileo's own word for it that the usual method in the demonstrative sciences is first to make certain of the truth of a proposition by sensible experiences, and then to seek a proof for it.[14] There is no sound reason for believing that this was not his procedure in the present instance. I believe that it was, and that the 1604 dem- onstration is entirely *ad hoc*. If that is so, then it raises grave doubts concerning the prevailing view of Galileo's early physical thought and of its relation to the writings of his medieval predecessors. I shall presently reconstruct Galileo's *ad hoc* approach to the 1604 demonstration and his subsequent correction of it. But first it is necessary to show that he might indeed have been completely con- vinced of the odd-number rule and the times-squared law without reference to any demonstration at all, or at most with reasoning of an entirely different kind which he is known to have called merely "a probable reason."

III

There are at least two ways in which Galileo could have arrived at his knowl- edge of and belief in the odd-number rule and the times-squared relationship. Both of these have been neglected in previous discussions of the 1604 frag- ment. I think that both probably entered into the matter, in the order set forth below.

The first way is by means of observation – and I hasten to add that I do not mean by Galileo's having performed, as early as 1604, anything like the elabo- rate experiments he described in 1638 as having been employed in confirming his laws of falling bodies. It would be simply anachronistic, both with respect to the history of experimental physics in general and to the known procedures of Galileo, to suppose that he carried out careful and controlled observations in order to arrive at the times-squared law, though it is compatible with both that he made such observations to confirm the law before its eventual publication. Experiments at least as well conceived as those he later described would be nec- essary for a direct test of the times-squared rule, let alone for its discovery in that form.[15] But a far less elaborate observation might have led to it indirectly.

The times-squared rule follows immediately from the odd-number rule for successive spaces in equal times. Now, Galileo had known since 1590 that speeds on inclined planes were far from following his rule in *De motu*, the rule that equated "speeds" with positional weights.[16] He can hardly have known that without his having tried to confirm the false rule by some experiments, at least by crude ones. In *De motu*, his theory was that acceleration was evanescent and unimportant. Hence the rule he had deduced was so far from the truth that even the crudest test would contradict it. I believe that he made such a test, from which he learned only that his rule was wrong.

By 1602, however, he was making observations on pendulums and on fall along concave surfaces in which the importance of acceleration was almost inescapable. At that point it would be perfectly possible, and not psychologically implausible, for Galileo to have rolled a heavy ball down a long gentle incline to see whether acceleration continued. In so doing, he might well note, or have someone else mark, its place at each pulse beat. Using the first space as a basis of comparison, by means of a string, the successive spaces might accord with the 1–3–5–7 relationship well enough to suggest the rule – which, as I have said, would immediately suggest a square law to any mathematician of the epoch. So it is not entirely impossible that Galileo was put on the track of the times-squared law by observations. All that is unthinkable is that he carefully designed an experiment to discover that law, such as the one he later designed to test it for diverse slopes.

The other possibility of Galileo's source of knowledge will be more plausible to some scholars as the sole source, though I think a rough observation probably started him thinking about it. It consists in his having reasoned about the problem along the following lines. If the spaces traversed in equal times during natural acceleration increase uniformly, they form an arithmetic progression. But this would be a special kind of arithmetic progression, in which the ratio of the first two terms must also be the ratio of their sum to the sum of the second two terms, and the ratio of the sum of the first three terms to the sum of the second three terms, and so on; because the original unit chosen for time is perfectly arbitrary and might equally well have been doubled or tripled, etc. Is there such a progression? It is not 1, 2, 3, 4, because 1 plus 2 is not one-half of 3 plus 4. But the very next arithmetic progression, 1, 3, 5, 7, ... , meets the test; 1 plus 3 is one-third of 5 plus 7, and so on. No other arithmetic progression will serve, though of course there are many arithmetically non-uniform series of numbers that will. An example of these is 1, 7, 19, 37, 61 ... , the differences of consecutive cubes. But the growth of that series is not uniform.

In this way, by merely assuming that the spatial increases are uniform in natural acceleration, Galileo might have arrived, by simple reasoning, at the conviction that spaces increase for equal times as the odd numbers beginning from unity. As a matter of fact, Galileo gave this argument, at least in part, to Baliani in 1615. Writing to Castelli in 1627, Baliani mentioned " ... a proposition that Sig. Galileo told me as true but without adducing the demonstration for me; and it is that bodies in natural motion go increasing their [successive] velocities in the ratios of 1, 3, 5, 7 etc. and so on *ad infinitum*; but he did adduce a probable reason for this, [namely] that only in this proportion [do] more or fewer spaces preserve always the same ratio. ... However, I have demonstrated it with very different principles ... "[17]

Baliani does not mention whether Galileo also gave him the times-squared rule at the same time. Probably he did, since the two rules are essentially identical, and no mathematician could fail to recognize the fact. It was the latter rule that Baliani published in 1638, six years after its publication in Galileo's *Dialogue*, and for which he attempted to claim priority over Galileo shortly after the latter's death. Baliani did not know of the 1604 fragment, however, and claimed only that his researches had begun in 1611, three years before he entered into correspondence with Galileo.

It may be added in passing that if Galileo was as thorough a Platonist as many now pretend, it is rather curious that he did not adduce this remarkable property of the series 1, 3, 5, 7 ... in the *Dialogue* or the *Discorsi* as an instance of the mysterious reign of numbers over nature. The argument is easy for a lay reader to understand, and the facts are somewhat remarkable. But Galileo regarded it as only a probable reason, not as a demonstrative proof, and left it unpublished. Christian Huygens appears to have been the first to deduce it analytically, in one of his earliest demonstrations.[18]

IV

In recent years there has arisen a widely prevailing belief that in 1604 Galileo was familiar with the Merton Rule, and used it. This view has its basis in three grounds. First, there is the general continuity principle in the history of ideas; it is said that the doctrine of the latitude of forms developed in the Middle Ages had become an integral part of university education before Galileo entered Pisa. Second, there is a philosophical treatise in Galileo's own hand, dating from his student days (1584, to be precise), that mentions the doctrine and the names of several medieval authors who discussed it. Third, there is a formal resemblance between the diagram used in the 1604 fragment and the diagrams employed previously in published discussions of the Merton Rule. Taken together, these grounds seem very convincing in favour of the prevailing view, particularly when it is added that the Merton Rule seems to lie at the very basis of Galileo's ultimate presentation of his law of falling bodies in 1638. When these grounds are taken separately, however, it may be seen that each alone is highly provisional and may ultimately be rejected.

As to the first, there is no question that the doctrine of the latitude of forms, originating in the fourteenth century, had an intensive and consecutive development up to the beginning of the sixteenth century. The mean speed theorem was printed and reprinted several times between 1490 and 1515. After that time, however, it was not reprinted again during the sixteenth century in Italy. After 1530, there was in Italy a general drift of interest away from medieval writings toward those of classical antiquity on the one hand, and toward contemporary authors on the other. During the half-century that elapsed between the last printing of the Merton Rule in Italy and Galileo's matriculation at Pisa, mere lip-service to great medieval writers may have replaced the serious study of their works.

As to the second point, there is good reason to believe that the treatise in Galileo's hand was not his own production, but was either a set of dictated lectures or his copy of a manuscript treatise composed by another. The virtual absence of changes and the nature of the relatively few corrections, coupled with the vast number of authorities cited, suggest such an origin. In any event, the treatise is by no means conclusive proof that Galileo himself had read the works of all the thirty or forty authorities mentioned in it.

As to the formal resemblance in diagrams, this is not matched by any resemblance in the demonstrations based on them, nor does any representation of mean speed, essential to Merton Rule diagrams, appear in Galileo's diagram. Diagrams virtually identical with those of the letter to Sarpi and the 1604 fragment will be found in Michael Varro's *De motu tractatus* of 1584, a work devoid of any connection with the Merton Rule. The use of a line to represent distances and of triangles to represent proportionality is not in itself sufficient evidence on which to base a case for a common source. Differences in aim and viewpoint militate against such a source.

Essential to the ideas of the Merton School writers is the concept that in uniform acceleration, "the motion as a whole will be as fast, categorematically, as some uniform motion according to some degree [of velocity] contained in the latitude being acquired, and likewise, it will be as slow."[19] The determination of such a degree – a single value by means of which an over-all uniform change might be represented – constituted the Merton Rule, or mean-degree theorem, which stated that to the midpoint in time corresponded the mean degree in uniformly difform change. To represent a set of changing velocities, medieval writers took a single velocity, chosen from within that set, and the same in kind with every member of that set.

It is evident to us now that the comparison of two uniformly accelerated motions, or of two segments of a single such motion, could have been most simply carried out by utilizing the ratio of two means, each representing one of the motions and each being by definition the same kind of entity, capable thereby of forming a ratio. But that was not the procedure adopted by Galileo in 1604, after he had become convinced that acceleration was an essential and continuing phenomenon of actual falling bodies. The Merton Rule directly related instantaneous velocities, mean velocities, and times elapsed. Galileo related instantaneous velocities to spaces traversed, spaces traversed to sets of such velocities, elapsed times to such sets by their "contrary" relation to spaces traversed, and thereby, finally, times elapsed to spaces traversed. He did not assume the existence of a mean speed within the set, or attribute any property to a midpoint, temporal or spatial.

Even if we assume that as early as 1584, Galileo was familiar with the Merton Rule sufficiently to know that it applied to uniform acceleration in which velocity increased proportionally to time, and that it made the mean speed correspond to the midpoint in time, we cannot reasonably maintain that he still remembered that rule in 1604. At that time he was convinced that the spaces traversed in equal times from rest were as the odd numbers commencing from unity, and that the successive distances from rest were as the squares of the times of fall. In attempting a mathematical proof of those convictions, he appears not to have tried the Merton Rule. Had he done so, the desired proof would have emerged at once. He told Sarpi that he had been unable to find an unquestionable principle on which to base a proof, and therefore had recourse to one that was merely physically probable; namely, that velocities increased in proportion to spaces traversed. This sounds as if he had tried to think of others, and makes it unlikely that among them he had remembered the Merton Rule and tried in vain to apply it.

The central role of the mean-speed concept for medieval mathematicians of motion and its total absence in the 1604 fragment are highly significant for the history of science. We merely miss the point when we attempt to make the Merton Rule the historical, as well as the logical, predecessor of the law of falling bodies. It is here as with impetus and inertia; medieval studies had prepared the way for recognition and acceptance of Galileo's physics, but they did not put it in his hands. It is a mistake to suppose that we can divine a man's ideas without paying attention to his precise words. I am not sure that Galileo ever used the expression "mean speed" in his life, though every medieval writer on the Merton Rule did. Nevertheless, English translators of the *Two New Sciences* have put the expression in his mouth when he did not use it. For example, in the crucial first theorem on accelerated motion, they have him say " ... a uniform speed whose value is the mean of the highest speed and the speed just before acceleration began,"[20] whereas Galileo said only, " ... by a uniform motion whose degree of speed is as one-half to the highest and last degree of speed ... " (*motu aequabili ... cuius velocitatis gradus subduplus sit ad summum et ultimum gradum velocitatis*).[21] The distinction may not be as trivial as it appears. What Galileo consistently presents only as a ratio, the medieval writers (and Galileo's translators) presented also as some kind of entity. There are reasons for the difference.

Generally, medieval writers were concerned with a specific change from a definite *terminus a quo* to a definite *terminus ad quem*, in accordance with the Aristotelian concept of change. Philosophically, the determination of a single measure of overall change was the solution of their problem. The Merton Rule accomplished such a determination. It solved the problem not only for uniform acceleration from rest (change from its beginning), but also for change from any subsequent point to a definite *terminus ad quem*. But it should be noted that medieval writers solved this latter problem as a separate one, noting that such intermediate motions had their own means, but disclaiming the possibility of a general rule of proportionality. For like reasons they did not seek rules of proportionality for any change that had no *terminus ad quem*. Such a change was for Aristotelians a contradiction in terms; in the case of accelerated motion, it would lead to infinite speed. Proportionality was used by medieval writers to determine relations within a finite change, and hence tended to be confined to converging series.

Galileo cared little or nothing for the determination of a mean value as such, but he was deeply interested in proportionality in every form. It was his key to the discovery of physical relationships. (The mean proportional, not the arithmetic mean, became his speciality.) When he became convinced that the spaces in free fall progressed as 1, 3, 5, 7 ... *ad infinitum*, he sought a general relationship of velocities, spaces and times. It is perfectly true that the Merton Rule would have afforded a simple and direct path to the solution of that problem; it is also true that Galileo's ultimate solution at a later date coincided with that way and embraced that rule. Yet it is neither necessary nor even probable that he achieved either his first (incorrect) or his ultimate solution by means of applying the Merton Rule, in the sense of his having relied upon a past tradition for either solution. It is not necessary because there is more than one way to arrive at the same truth. It is not probable because of the chronological order in which his physical conclusions first appear, as well as because of fundamental differences in concepts and methods between mean-speed and velocity-ratio determinations as outlined above.

v

The entirely *ad hoc* basis of the 1604 demonstration, as I see it, suggests the following reconstruction of Galileo's reasoning.

I assume the truth of Galileo's statement to Sarpi, repeated in the fragment, that he already knew the spaces of fall in equal times from rest to be as the numbers 1, 3, 5, 7 ... [22] Since the times were equal, the speeds of fall through these successive spaces were likewise necessarily as the numbers 1, 3, 5, 7 ... , and for the same reason, the corresponding cumulative times were as 1, 2, 3, 4 ... The cumulative distances fallen at the ends of these time-intervals were as the numbers 1, 4, 9, 16 ... , to which numbers the terminal velocities were made proportional by hypothesis in this demonstration.

Let us suppose that with these sets of numbers before him Galileo drew first a scale-diagram in which the vertical line of fall was divided as 1:3:5, and the terminal velocities at these divisions were represented by horizontal lines, cut by an oblique line to preserve the assumed proportionality. The resulting diagram looked like this:

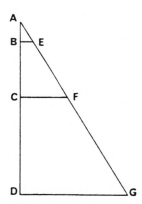

The triangles ABE, ACF and ADG offered a convenient representation for the speeds of fall from rest, justified in the demonstration by the "compounding" argument. It also assured Galileo that the ratios of compounded speeds would always be a ratio of squares, whether the terminal velocities happened to be squared or not, as they were in the numbers before him – a fact that helped to conceal from Galileo the error he made in generalizing this diagram for equal spaces. Since Galileo was concerned only with ratios, he did not care how the "compounding" took place, provided only that it took place in the same way for any interval commencing from the common origin. Of this he was confident,

not only from his belief in the uniformity of nature, but from the evident inapplicability of mathematics to the whole problem if it did not hold true. His
attention was not on mean speeds, but on ratios of growth of speeds.

The ratio of compounded speeds having been established as a ratio of squares
of distances, a corresponding ratio of times was next sought. Galileo knew,
from the numbers before him, that this was a ratio of square roots of distances.
In a single sentence, by invoking "contrary proportion," he boldly produced this
conclusion. His wording is so compact as to suggest either that he suddenly
confused uniform motion with accelerated motion, or that he used a merely
arbitrary definition as a step in a proof. Alternatively, he may have omitted a
line of argument (as obvious) that we have failed to perceive. Here are his
words, immediately after the establishment of compounded speeds as related to
squares of distances:

> Since speed to speed has contrary proportion to that which time has to time (because it
> is the same to increase the speed as to diminish the time), therefore the time of motion
> through AC to the time of motion through AB has a proportion that is the square root of
> that which the distance AC has to the distance AB.[23]

The initial statement is usually explained by saying that Galileo seized on a
rule for uniform motion over a single fixed distance, which rule relates in
inverse proportions any velocity to any time. If there was such a rule in 1604, I
think it must have been Galileo's own, which makes it unlikely that he would
be unaware of its assumption of uniform motion. But apart from that, the concept of a fixed distance seems not to fit his words here. He is speaking of speed
to speed, time to time, motions through AB and AC, and spaces AB and AC.
"Contrary proportion" of speed and time can hardly mean a simple or inverse
ratio between them, or it would follow immediately that the same ratio (or its
inverse) as that which Galileo had just deduced for speeds and distances – a
ratio of squares – would apply to times and distances. Yet it is not a ratio of
squares, nor of inverse squares, that is invoked in the same sentence for times
and distances; rather, that ratio is said to be one of square roots.

Clearly Galileo's "contrary proportion" was not an isolated ratio of any kind,
but one chosen to illustrate the linkage of two sets of numbers having a certain
kind of contrariety to which he wished to direct attention. This may appear to be
a capricious use of "proportion," but historically it was not. The use of that word
was not confined to ratios, but embraced arithmetic and harmonic "proportions"
– what we would call "series" – as well. Galileo appears to have been thinking of
the series of squares as a proportion, to which the square roots are contrary.
Euclid himself gave a strange general definition of proportion, one that he never

employed and that has caused commentators some perplexity: "Proportion is a certain relation of two magnitudes of the same kind with respect to quantity."[24] Galileo would have been quite within his rights to say that "speed has to speed a certain kind of contrary relation with respect to quantity as compared with that which time has to time," in accelerated motion. The contrariety was that of squares to square roots of a common measure, distance from rest.

Galileo's reasoning is obscured by the wording of his parenthetical justification for the general statement that ratios of times and spaces were "in contrary proportion." It is the word "diminish" that makes him appear to have reverted to uniform motion. In acceleration, the spaces, times and speeds are all continually increasing; how, then, can he speak of time as "diminished"? If we credit him with thinking only of accelerated motion here, it is evident that he meant to compare the growth of speed with the *growth* of time, just as he related the growth of velocity with the growth of distance fallen. He speaks throughout of ratios, not of points, and all ratios are of pairs taken from a common origin. It is the same, he sees, for the growth of speed to carry the body over ever-increasing distances, as for ever-decreasing times to elapse during the body's passage over those distances. The proportion of speed must be contrary to the proportion of time to time. The proportion of speed to speed is a ratio of squares; but the contrariety cannot be one of inverse squares, or the time of fall from rest would be shorter for a longer distance. (If the speeds for the respective distances of 1 and 2 from rest have been shown to be as 1 to 4, the times for those falls from rest, if "inverse," would be as 4 to 1; that is, four units of time would have to elapse before one unit had elapsed.) Nor could it be the contrariety of sides to squares; for under Galileo's initial assumption that is the ratio of distances from rest to speeds; and if the times were in the ratio of the spaces, the motion would be uniform and not accelerated. There remained only the contrariety of squares and square roots, which complied with all the required conditions, making the distances outstrip the times as the speeds outstripped the distances. Besides (as Galileo knew all along), this "contrary" relation agreed with actual fall.

If Galileo reasoned this way, he can hardly be blamed for thinking that his conclusion was really demonstrated, and not merely pulled out of the hat by an arbitrary definition of "contrary" as applied to a relation existing between two proportions. It is hard even to blame him for having been so concise in his wording; there was no need for him here to describe the three possible ways in which the relation could be "contrary," and to show two of them to be contradictory. For his own use, this demonstration would serve; to Sarpi, it could be explained if necessary; for the general reader, it could be embodied in a discussion that would eliminate the alternatives. Assuming the reconstruction I have set forth, and the postulated scale-diagram, we could say that no error had been made up to this point, except the initial assumption. The error came with the redrawing of the diagram.

For the scale-diagram shows only those points on the line of fall that are distant from rest by multiples of a unit of time. Thus restricted, it simultaneously represents the correct assumption. So long as Galileo made no real use of the space-proportionality assumption except in drawing the scale-diagram, it was powerless to lead him astray. The incompatibility of that assumption with the initial state of rest was cloaked in the diagram by his having constructed the diagram to scale, for equal units of time. The lines representing instantaneous velocities play no role in his conclusion, and serve the same purpose in his "compounding" argument regardless of the meaning of the vertical line. As he later confessed in the *Two New Sciences*, Galileo thought at the time that it did not matter whether velocities were said to be proportional to space or to time, so the ambiguous use of the vertical line (for equal times and growing spaces) did not concern him.[25]

Thus it came about that in drawing a fresh diagram for the 1604 demonstration, he made the space-intervals equal, in accordance with the initial assumption. In so doing, he failed to notice that time was no longer represented in the diagram at all. Satisfied that the reasoning he had previously used was perfectly general and self-evident, he wrote it out and put it aside.

The manner in which Galileo discovered and corrected the error in his 1604 assumption will be left to the next paper in this series. This requires the introduction of some fragmentary notes associated with a projected Latin treatise on motion by Galileo. The correct translation and interpretation of his published rejection of space proportionality in 1638 can then be shown to confirm the foregoing reconstruction and to be confirmed by those notes. It will appear that the sequence of Galileo's thought was one of an attempt to build further on his own results, and not (as is now believed) of his having recalled a theorem of his schooldays that he failed to remember or apply correctly in 1604, or his having turned back in frustration to the study of medieval writings after the age of forty, in the hope of gaining new light. And indeed, such assumptions are as out of keeping with what is known of Galileo's character and habits as with the whole spirit of the seventeenth century in Italy. There is no good reason to assume that Galileo switched his allegiance from Euclid, or put more faith in the writings of past centuries than in his own efforts, or that he was dissatisfied with his own demonstration at the time. And in fact his later fragmentary notes, like his *Dialogue* and *Discorsi*, are as innocent of a mean-speed concept in the Mertonian sense as was his attempted demonstration of 1604. But the oversight – the illicit generalization of the diagram – was a fatal mathematical error that rendered inapplicable the purported proof that accompanied it. The moment that equal times ceased to be implicit in the diagram, Galileo's reliance on the known law betrayed him into thinking that his reasoning had been general.

Particularly striking to me is Galileo's failure to mention the double-distance rule in his letter to Sarpi or in the fragment of 1604. He did adduce the odd-number rule and the times-squared rule. The double-distance rule is at least as striking as those, and as easy to state. Its absence suggests that it was still unknown to Galileo in 1604. It would in fact be harder to find or to verify by observation than the odd-number rule, yet it would be inescapable from a knowledge of the Merton Rule. Conversely, the odd-number rule was not developed in the printed medieval works, except with respect to the initial ratio of 1:3 for the spaces traversed in the first two equal times. The times-squared law, so far as I know, was not developed in them at all, perhaps because it implied an infinite velocity and made no use of a *terminus ad quem*.

NOTES

1 Galileo, *Opere* (Ed. Naz.), x, 114.
2 *Opere*, x, 115.
3 It is evident that *velocità* here means speed over an interval, and not instantaneous velocity, into which time does not enter.
4 In the autograph, Galileo first wrote *la linea*, which he cancelled and then wrote *il tempo*. The difference is one of thought rather than one of a plausible mistaken copying. The argument as we have it is thus probably not the result of successive revisions on paper, but a first record of an argument thought out. Possibly Galileo started to write "the line AD has to the line AC a subduplicate ratio to the speeds and a duplicate ratio to the times of motion over the distances AD and AC," but decided instead to omit speeds and relate the times and distances directly.
5 *Opere*, viii, 373–374.
6 *Études sur Leonard de Vinci* (Paris, 1955), iii, 562–566.
7 *Études Galiléennes* (Paris, 1939), ii, 21–25.
8 "Another Galilean Error," *Isis*, 50 (1959), 261–262.
9 "Galileo, Falling Bodies and Inclined Planes," *British Journal for the History of Science*, iii (1966–67), especially pp. 230–244.
10 This is the wording of Heath's Euclid. Commandino used the word "converse" instead of "inverse," in which he was followed by the first English translation of 1570. Tartaglia used "contrary." All these, of course, merely name the inversion of a ratio, or of both ratios in a proportionality. In 1638, Galileo used either the phrase "taken contrarily" or "taken permuted" after the second ratio named in a proportionality, when he wished to assert inverse proportionality in the modern sense.

11 *Opere*, viii, 361. Intended as an addition to the *Discorsi*, Galileo's treatment of the theory of proportion was first published by Viviani in 1674.

12 Thus from the false assumption that all women are Greek, one might argue with a further error that since Socrates was a woman, Socrates was a Greek; or one might argue without further error that since Greeks are mortal, women are mortal.

13 A physical reason is invoked to support the crucial "contrary proportion" argument, discussed at length below.

14 *Opere*, vii, 75–76; *Dialogue* (tr. Drake), p. 51.

15 Contrary to the opinion of Koyré, the experiments described in 1638 were quite adequate to verify the law; cf. T. Settle, "An experiment in the history of science," in *Science*, 133 (1961), pp. 19–23.

16 *Opere*, i, 302 " ... hoc solum animadvertentes quod, sicut supra dictum de motu recto, ita etiam in his motibus super planis accidit non servari has proportiones quas posuimus ... " Cf. *Galileo on Motion and on Mechanics*, ed. I.E. Drabkin and S. Drake (Madison, 1960), p. 69.

17 *Opere*, xiii, 348. S. Moscovici has translated this passage as saying that Galileo was unable to adduce a demonstration, perhaps having read *però* as *potere*. Baliani, however, implies if anything that Galileo had a demonstration but did not show it to him, giving only the argument mentioned.

18 Constantijn Huygens sent the youthful production of his son to Mersenne, and it was published in 1649 by Tenneur: it is a derivation of the odd-number rule without geometrical considerations, assuming only the concept of uniform acceleration.

19 M. Clagett, *Science of Mechanics in the Middle Ages* (Madison, 1959), p. 271; the quotation is from Heytesbury.

20 *Two New Sciences* (tr. Crew and De Salvio), all editions p. 173.

21 *Opere*, viii, 208.

22 As previous mentioned, Galileo treated velocities as numbers. Adding the successive speeds as 1, 3, 5, ... would give square numbers, as would adding the successive spaces traversed from rest. There is no reason to think that Galileo carried out such an operation, which would give some kind of undefined "cumulative speeds from rest," but if he did, the proportionality to square numbers would not have disturbed him with regard to the restricted diagram shown above. By his hypothesis, the terminal velocities were as the total distances fallen from rest, so it was not necessary for him to determine them in any other way. Thus the anomalies that strike us could escape him, if he proceeded in the manner described here. I do not wish to conceal Galileo's error, but to show how it may have remained concealed from him in 1604.

23 The figure references are altered here to conform to the previous diagram.

24 Book V, Definition 3. Heath gives, "A ratio is a sort of relation in respect of size between two magnitudes of the same kind," with a very extensive note on "ratio" and "proportion": see *Euclid's Elements* (Dover eds., N.Y., 1956), ii, pp. 116–119.

25 This statement is to be amplified in a later paper. Until Galileo's discovery of the error and its correction, discussed below, it was universally assumed that space and time were proportionally related to one another in all matters of local motion. Until actual acceleration of falling bodies came under study by mathematicians such as Galileo, Beeckman, Descartes and Baliani, the error of that assumption was not recognized. Philosophers related speed to distance fallen, and mathematicians related uniform difform change to time elapsed, without anyone questioning whether the two were really compatible. Even Domingo de Soto, who related free fall to the Merton Rule, did not assert that this contradicted the numerous authors who by that time (1545) had related acceleration in free fall to the distances fallen.

7

Uniform Acceleration, Space, and Time
(Galileo Gleanings XIX)

I

The most reliable source for a reconstruction of Galileo's progress toward a science of motion is the series of undated fragmentary notes on that subject preserved in Codex A of the Galilean manuscripts at Florence. A gathering of such fragments was published by Favaro in the National Edition of Galileo's works, following the *Discorsi*. The more sophisticated fragments are clearly associated with the composition of that work, and show a definite and consistent understanding of acceleration. Eliminating those, it will be found that the earlier notes fall into recognizable groups. First, there are some that refer to "moment of gravity," or to the impetus of a body along a line of descent, and are associated with the discussion of inclined planes in *De motu*. Second, some refer to descent along arcs and chords of circles, associated with Galileo's letter of 29 November 1602 to Guido Ubaldo. These first two groups of notes do not explicitly refer to accelerated motion, and should not be assumed to do so implicity, where such an assumption can be avoided.[1]

A third group of fragments specifically refer to questions of accelerated motion; it is with those that I am here concerned. I shall mention them in what I believe to be the order of their composition.

The earliest fragment of this group survives only in a copy made by Mario Guiducci. It is virtually a Latin version of the 1604 fragment, except that the times-squared ratio is replaced by the equivalent assertion that the times of motion through any two intervals from rest are in the ratios between those inter-

This article is reproduced with the permission of the Council of the British Society for the History of Science. It was first published in *BJHS* 17 (1970), 21–43.

vals and their mean proportional. This copy cannot have been made before 1614, and it has been struck over by two crossed lines.[2]

The second fragment concerning accelerated motion, in my opinion, is numbered folio 179 in Codex A. As originally written it contained the words: *Cum enim assumptum sit, in naturali descensu velocitatis momenta eadem continue augeri secundem rationem elongationis perpendicularis a linea orizontali, in qua fuit lationis initium ...* ,[3] referring apparently to the Latin version of the 1604 fragment. A change in this assumption and a change on the verso of this folio enable us to date the fragment and to pick up the thread of Galileo's thought in 1609, when he projected a systematic treatise on motion.

The proposition being proved was that the times of motion of a given body along the vertical and along an inclined plane of the same height were in the ratio of the length of the plane and its vertical height. This was a correction of the erroneous assumption that the speeds of such a body were proportional to the said length and height previously made in *De motu* about 1590 when Galileo had not yet recognized the essential role of acceleration in natural motion.

The proof proceeds by a very clear statement of one-to-one correspondence between points on lines of different length and between intercepted segments of such lines.

The verso of folio 179 begins with a proof of the corollary that the times of motion (to the same horizontal) along different inclined planes of the same height are as the lengths of those planes.[4]

Next, Galileo undertook to prove that in free fall from rest to two different points, the speeds through the two intervals were as the squares of the distances fallen. This was implied in the 1604 fragment. The diagram is only partly lettered, and the proof is abruptly broken off. Over it are pasted two slips of paper bearing propositions of an entirely different nature, referring to the "moments of gravity" of bodies along the vertical and along inclined planes of the same height.[5]

On 5 June 1609, Galileo wrote to Luca Valerio, seeking his opinion on the validity of introducing two propositions relating effective weight to speed as a basis for a science of motion. Galileo's letter is lost, but Valerio's reply, dated 18 July 1609, enables us to identify the propositions with those pasted over the abandoned demonstrative treatment on folio 179.[6] That treatment may therefore be regarded as having been composed as part of a projected systematic treatment of motion, immediately before Galileo's attention was diverted by the telescope to a series of occupations that interrupted his project for a considerable period of time.

The partial diagram and interrupted proof present strong evidence that Galileo first realized in mid-1609 that there was something wrong with his 1604

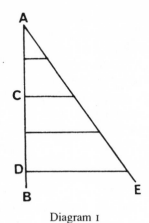

Diagram 1

demonstration, but that he did not immediately abandon the false assumption on which it was based. This is what he had written:

> If two unequal distances are taken along the line of natural descent, the moments of speed with which the body traverses those distances are as the squares of those distances.
>
> Let the line of natural descent be AB, in which from A, the beginning of motion, are taken the two distances AC and AD: I say that the moment of speed with which the body traverses AD, to the moment of speed with which it traverses AC, is in the ratio of squares of the distances AD and AC. Let line AE be placed at any angle with AB ... [7]

Galileo had reasoned about space-proportionality from two previous diagrams, one accompanying the 1604 fragment, and one in its Latin version of 1609. In both diagrams, beginning from an initial point of rest, he had designated by letters the intersections of the line of fall by lines representing velocities acquired. This time, however, he first selected two points on the line of fall, as required by the theorem he sought to prove, and designated them with the letters C and D. Next, starting to develop his triangular-area argument as on previous occasions, he drew the horizontal lines for velocities, including parallels intermediate between A and C, and C and D. In so doing, I believe, he was for the first time struck by the anomalous character of fractional distances. If the distance AC is thought of as a unit, doubled at D and halved at the upper unlettered line, the square law Galileo wished to demonstrate would require the "moments of speed" to fall progressively behind their related distances between C and A, and to run ahead of them from C to D. At C, the space would be identical with the speed. Mathematically, the ratios in which Galileo was interested were unaffected; but physically, the relation of speeds to spaces traversed appeared to change. The paradox had its counterpart in the mean-proportional rule for times, in that no time could be identified for fall from rest to any first point. Having encountered the paradox myself by drawing precisely the same partly-lettered diagram long ago, I think it very probable that Galileo first questioned space-proportionality when he considered the foregoing implications of fractional distances. In any event, he abandoned this attempted deductive method of relating speeds to distances in accelerated motion. But he did not seek another deductive proof; instead, he substituted the propositions on "moments of gravity" as a basis for reasoning about speeds in fall, and sought Valerio's opinion as to the validity of this procedure.

Galileo's abandonment of the area-argument, however, did not mean his abandonment of the space-proportionality assumption entirely at this time. What he did was to alter the words cited above from folio 179 to read: *Cum enim assumptum sit, in naturali descensu velocitatis momenta eadem semper reperiri in punctis aequaliter ab orizonte distantibus iuxta perpendiculares distantias ...* But he left, unaltered, and even underlined, a subsequent passage that begins: *Et quia velocitas semper intenditur pro ratione elongationis a termino A ...* [8]

It is an interesting speculation that by eliminating *continue augeri* in one place, on folio 179r, and greatly altering the supposition used, while he left and emphasized *semper intenditur* in the other place, Galileo had perceived the essential nature of space-proportionality; namely, that it cannot allow continuous growth from rest, but that if we grant the slightest discontinuity at the beginning of motion, the velocity will always increase thereafter – and of course, by hypothesis, it will increase *pro ratione elongationis a termino A*. It is not entirely impossible that Galileo did adopt such a view, and that is what he meant when he wrote, in 1611 (or earlier), *Mobile secundum proportionem distantiae a termino a quo movetur velocitatem acquirens, in instanti movetur.*[9] A discussion of that possibility would lead far afield. Yet it would be easier to understand the survival of the Latin version of the 1604 demonstration to a period as late as 1614, if Galileo in fact did adhere for many years to a belief in space-proportionality with the proviso that actual fall involved a small initial discontinuity; that is, that natural acceleration was not mathematically uniform. It may be significant that there is no mention of uniform acceleration in the fragments, but only of "natural acceleration" or "accelerated motion."

Following Galileo's abandonment in 1609 of his attempted logical derivation of a relation between actual speed in accelerated motion and the instantaneous velocities acquired, his reflections on acceleration appear to have broadened. On folio 163, he derives the double-distance rule for uniform motion following acceleration, and applies it at once to prove the same proposition as before, that the times of motion along inclined planes of equal height are as the lengths of the planes.[10] In this, however, there is no explicit assumption concerning proportionality of velocities to space or to time. There is certainly no assumption of any mean speed, nor is the triangle of the medieval mean-speed theorem completely drawn. Galileo draws only a vertical line meeting a horizontal line at its foot, and argues that if all velocities along it were represented by the horizontal line, then the vertical line would be traversed twice as fast as in natural acceleration from rest, the ratio of growing velocities to constant velocity being as the triangle to the rectangle that could be constructed on those lines. The outline triangle and rectangle are dotted in, without a mean-line.

This proof, and its application to inclined planes of equal height, appear to me to have been put down after folio 179 was written, being simpler in form and requiring less complicated assumptions. Moreover, the double-distance rule would be expected to find a place in the modified Latin version of the 1604 demonstration, if that rule had been already discovered. Thus folio 163 was probably written very shortly after folio 179, while Galileo was troubled about the validity of the lengths-squared rule for speeds over inclined planes. His derivation here of the double-distance rule may very well have been directly suggested by the first diagram on folio 179, in which an equal and parallel line to the base of the triangle was drawn from the initial point of rest.

The same folio (163) includes a new proof of the proportionality of times along vertical and inclined plane of equal height to their lengths, utilizing the mean-proportional relation for times and a proposition concerning equal speed along chords of circles that he had based earlier on considerations of moment, without discussion of acceleration (folio 151r).[11]

Folio 164 seems to be coeval with, and possibly started slightly prior to, folio 163. The form in which it gives a certain paradox (later used in the *Dialogue*) suggests that the double-distance rule had not yet been derived by Galileo.[12] Elsewhere on folio 164 is Galileo's first diagram in which time-measures are graphically represented on the same diagram in which spaces traversed in vertical and inclined fall are shown.[13] This same folio states, in flat contradiction with the 1604 fragment and the abandoned demonstration on folio 179, that speeds of falling bodies are as the square roots (not the squares) of the distances fallen.[14] No proof is offered.

The absence of a demonstration seems surprising, since the notation represents a complete reversal of a view held by Galileo from 1604 to 1609. That the ratio of speeds over two intervals from rest in free fall was as the square of the ratio of distances fallen was involved in the 1604 fragment and in its 1609 Latin counterpart. It was the proof of that specific proposition, by triangular areas, that Galileo was attempting when he detected the difficulty that made him

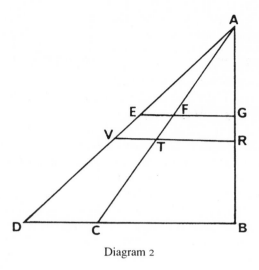

Diagram 2

change to considerations of mechanics and appeal to Valerio. But the sudden appearance of the notation on folio 164v, *Momenta velocitatum cadentis ex sublimi sunt inter se ut radices distantarium peractarum, nempe in subduplicata ratione illarum,* is unsupported, and unattended by a "Eureka" or an accompanying statement of time-proportionality. Why?

To understand this, as well as the absence of any specific reference to time-proportionality anywhere in Galileo's notes, it is good to consider the diagram on folio 164r and its attendant proof.

Here the horizontal lines have nothing to do with terminal velocities; the oblique lines represent different inclined planes. The points G and B represent distances fallen from A, and AR is taken as the mean proportional between AG and AB. The proposition to be proved is that if AB is also taken to represent the total time of fall through the distance AB, then RB represents the time consumed in the fall GB, while TC and VD represent analogous times for the falls from F and E along the inclined planes.

This consideration of accelerated motions separated from any consideration of instantaneous (or terminal) velocities, with its representation of the times of fall through different intervals from rest, the very thing lacking in the 1604 fragment and its 1609 Latin version, leads reasonably enough to the notation in question. With the intervals of fall from rest fixed, and the division of time between them determined, the speeds through the intervals were likewise determined, though not represented and not involved in the demonstration. The notation in question is, in my opinion, to be regarded not as a discovery or a conclusion, but as the writing down of a possibility to be explored. The ratio of speeds, which had become a vexing problem with the failure of the attempted triangle-area proof, ought to be given either by the ratio of distances traversed or that of the times elapsed. Galileo knew that it was not the ratio of distances, as that would not give the odd-number rule and the times-squared law, of which he had long been certain. But it could be the ratio of times, which was the square root of the ratio of distances. Of course it might be some quite different ratio, but this was the simplest and therefore the next to try. And of course, it worked, in the sense of fitting with all his other accumulated propositions. Hence it simply merged into his subsequent work, without any triumphant announcement signalling its first emergence, and without derivation from a principle, because it was not arrived at deductively, but simply by elimination and as a conjecture.

It is highly probable also that Galileo did not see at once that this new approach implied time-proportionality for instantaneous velocities. If he had, it is likely that he would have added a sentence to that effect. More probably, he continued for a time to adhere to space-proportionality, accepting the existence of an instantaneous initial discontinuity in fall from rest. All his known discussions of uniformity and continuity as such belong to a somewhat later date, probably after the edict of 1616. As will be seen, the sharp distinction we make between space-proportionality and time-proportionality simply did not exist in the minds of men at this time. They were not seen as alternatives, disjunctively. Nor was the idea of a discontinuity in accelerated motion repugnant to physicists. Descartes, for example, objected strongly to Galileo's argument for assuming continuity in acceleration from rest as presented in the *Dialogue* and the *Two New Sciences*.

Another memorandum, on folio 79v, summarizes the principal conclusions that Galileo had reached as a result of all the foregoing:

The spaces [covered in] accelerated motion from rest and the spaces in uniform motions following accelerated motions, and made in the same times, maintain the same ratio between them; the latter spaces are doubles of the former. The times, however, and the velocities acquired, have the same ratio between them; this ratio is the square root of the ratio of the said spaces.[15]

This at last puts the acquired velocities directly proportional to the times. The memorandum may be of much later date than the letter from Valerio in mid-1609. It was precisely then that Galileo's attention was diverted from mechanics by the advent of the telescope. For several months he was occupied with the improvement of that instrument, the publication of his astronomical discoveries, and the move from Padua to Florence in 1610. This was followed by discovery of the phases of Venus, an illness, a journey to Rome, the controversy on floating bodies of 1611–12, and the investigation on sunspots in 1612–13. It is difficult to say just when Galileo returned to his researches on motion, but the memorandum in question contains a concept that he had certainly reached by 1615, when he communicated it orally to G.B. Baliani. The memorandum was possibly written down as late as 1616, when his forced abandonment of the Copernican campaign probably diverted him back to the systematic investigation of motion. It has to me the appearance of having been written not as the record of a new development, but as a summary of earlier conclusions such as one often jots down upon resuming a previously uncompleted inquiry.

In the whole process there is no trace of any use of a mean speed, but only of ratios and one-to-one correspondences, interrupted by a fruitless attempt to compare areas under an assumption that concealed a discontinuity. Galileo's mean-proportional conception of times had evolved separately from the inclined-plane proof in *De motu*,[16] through the idea of momentum (folio 151r) and the comparison of chords of circles (folio 147r), quite independently of acceleration. It was first associated with acceleration in the 1609 Latin version of the 1604 fragment, and was dropped again in favour of the times-squared ratio when times were finally related to velocities.

The ultimate discovery of the essential difference between the time-proportionality and space-proportionality emerged as a result of a mass of conclusions that Galileo could not doubt and wished to reconcile, not as a result of his exploring first one alternative and then the other. For they no more appeared as alternatives to him than they had to his predecessors. As he later admitted publicly, he had at first believed them equivalent.

II

The history of science may go astray when the standard translation of a critical passage in a fundamental work misrepresents the author's intention. This is particularly true when translators into two or more languages agree in the misinterpretation. An example is the supposed reference to inverse proportion in the 1604 fragment.[17]

Alexandre Koyré, seeking the true meaning of Newton's *Hypotheses non fingo*, had occasion to remark of its English and French translators: "As the Italian proverb has it, *traduttore-traditore* (translators are traitors) ... They did not limit themselves to translating; they made an "interpretation," and in so doing gave to Newton's assertion a sense that was not Newton's sense."[18]

Koyré was perhaps too severe with the offenders. Real difficulties exist in translation; not only the translator's prepossessions, but what he regards as common knowledge, will enter into the product. A striking illustration is provided by Koyré's own rendition of a passage from Galileo's *Two New Sciences* – a passage of considerable historical significance, and one that has been the subject of dispute from the time of its original publication in 1638 down to the present. Yet Koyré's translation in 1939 was even more defective than the English of 1914 or the German of 1891. It read:

Lorsque la vitesse a la même proportion que les espaces franchis ou à franchir, ces espaces seront franchis en temps égaux. Car si la vitesse avec laquelle le grave franchit l'espace de quatre coudées était double de la vitesse avec laquelle il a franchi les deux premiers, etc.[19]

The English translation, a quarter-century earlier, at least preserved the initial plural noun, in accordance with Galileo's invariable treatment of all such measures as ratios:

If the velocities are in proportion to the spaces traversed, or to be traversed, then those spaces are traversed in equal intervals of time; if, therefore, the velocity with which the falling body traverses a space of eight feet were double that with which it covered the first four feet (just as the one distance is double the other) then the time-intervals required for these passages would be equal. But for one and the same body to fall eight feet and four feet in the same time is possible only in the case of instantaneous [discontinuous] motion; but observation shows us that the motion of a falling body occupies time, and less of it in covering a distance of four feet than of eight feet; therefore it is not true that its velocity increases in proportion to the space.[20]

The German translation, two decades earlier still, was virtually identical with the English:

Wenn die Geschwindigkeiten proportional de Fallstrecken waren, die zurückgelegt worden sind oder zurückgelegt werden soll, so werden solche Strecken in gleichen Zeiten zurückgelegt; wenn also die Geschwindigkeit, mit welcher der Korper vier Ellen Uberwand, das doppelte der Geschwindigkeit sein sollte, mit welcher die zwei ersten Ellen ...[21]

Thus all modern readers of this text, except perhaps some who have used the original Italian, are likely to have received the impression that pervades all recent discussions of it, and many earlier discussions. That impression is that Galileo, in his published argument against proportionality of velocity to space traversed in uniform acceleration, relied on some concept of average speed in free fall, and on a naïve assumption that such average speed would obey the rule applying to uniform motion. Under that impression, some historians have advanced reconstructions of Galileo's thought in which he is supposed to have known and misapplied the Merton Rule.[22] This in turn has strengthened the widespread conviction that the historical inspiration behind Galileo's law of falling bodies was his study of certain medieval writings.

But the original wording shows that whatever the reasoning was that Galileo relied on in this argument, it had nothing to do with a mean-speed comparison, and did not rely on the application of any theorem derived from the analysis of uniform motion. Galileo's own words were:

Quando le velocità hanno la medesima proporzione che gli spazii passati o da passarsi, tali spazii vengon passati in tempi eguali: se dunque le velocità con le quali il cadente passò lo spazio di quattro braccia, furon doppie delle velocità con le quali passò le due prime (si come lo spazio è doppio dello spazio) ... [23]

That is to say:

If the velocities have the same ratio as the spaces passed or to be passed, those spaces come to be passed in equal times: thus if the velocities with which the falling body passed the space of four braccia were doubles of the velocities with which it passed the first two (as the space is double the space) ...

Although the word velocità remains the same in Italian whether singular or plural, the definite articles, verb forms, and relative pronouns here leave no doubt that Galileo meant the plural. The first use of "doubles" is also plural (doppie), unlike the second "double" (doppio). Early English translations by Salusbury (1665) and Weston (1730) correctly gave the plural "velocities," but for doppie reverted to a singular. The anonymous Latin translator of 1699 preserved all the plurals of Galileo's Italian.

It is hardly a mere coincidence that all three early translators were almost perfectly faithful to Galileo's precise wording, while their three modern counterparts agreed almost completely in ignoring it. The modern translators were so well informed about the truths of physics that Galileo's strange syntax did not attract their attention. In the case of Koyré, scientific knowledge was reinforced by a philosophical prepossession of no small importance to the present state of opinion regarding Galileo's thought. The question of what Galileo said was trivial; the important thing was what he must have meant. But Koyré's treason to Galileo was loyalty to a higher cause; it was at worst a feat of legerdemain, of traduttore-tragittore. His fault was that of excessive knowledge, even greater than that of his German and English counterparts. The virtue of the early translators was that of scientific and historical ignorance, a state in which it was best to let the author speak for himself.

The passage is of greater importance to our understanding of Galileo's thought and to the history of science than may appear at a glance. It was discussed in the seventeenth century by Cazrae, Fabri, Gassendi, Mersenne, Fermat, Tenneur and Blondel; in the eighteenth by Riccati, Andres and Montucla; in the nineteenth by Ernst Mach, and in the twentieth by Duhem, Koyré, Cohen and Hall – a list sufficiently impressive without being complete. Only one of all those writers (Tenneur), even came close to reconstructing Galileo's own reasoning, and only one other (Andres), among those writing before the present century, consciously set out to do so. Fermat provided a rigorous proof of the correctness of Galileo's conclusion, while Mach developed a mathematical formula that cloaks its correctness under an apparent refutation. Fermat did not translate the passage, nor did Mach, whose critique was written a decade before the German translation cited above.

But apart from its previous historical interest, the passage can still throw light on medieval and Renaissance views of accelerated motion in free fall as well as on the progress of Galileo's understanding of that topic. This can be achieved only at the cost of a substantial revision of many currently prevailing conceptions. It is hardly to be expected that any such revision will be made without a thorough review of many arguments and documents which can be merely touched on in this paper. But it is time to begin.

III

The disputed passage has customarily been examined not only in mistranslation (or incorrect paraphrase), but also out of context. It lent itself to the latter treatment because it was presented by Galileo as a clear proof in a single long sentence. It is easy to jump to a conclusion about what it is that was to be proved, something that may much better be determined by reading carefully the discussions that precede and follow the passage in question. Here I shall merely summarize that context, but the reader is urged to examine it carefully for himself.

Salviati has begun by reading, from a Latin treatise of Galileo's, the definition of uniform acceleration as that in which equal increments of velocity are added in equal times. There follows a lengthy discussion of another matter, the relevance of which will be pointed out below. Returning to the definition, Sagredo suggests that its fundamental idea will remain unchanged, but will be made clearer, by substituting "equal spaces" for "equal times." To this he adds an assertion that in actual fall, velocity grows with space traversed.

Salviati replies that he once held the same view, and that Galileo himself had formerly subscribed to it, but that he had found both propositions to be false and impossible.

Now, it is universally believed that Salviati was here asserting that the definition of uniform acceleration in terms of equal space-increments was false and impossible, implying in it an internal contradiction. His words, however, do not support such a view. Sagredo's two propositions are: (a) that there is no fundamental difference between relating velocity in uniform acceleration to time and relating it to space, and (b) that in actual fall, speed is in fact proportional to space traversed from rest. Salviati is thus obliged to show Sagredo that those two propositions are false and impossible.

But at this point, Simplicio intervenes to assert his belief that an actual falling body does acquire velocity in proportion to space traversed, and that double velocity is acquired by such a body in fall from a doubled height. Nothing is said by Simplicio about the definition of uniform acceleration, nor does he overtly deny that doubled time of fall would equally produce doubled velocity. Both of Simplicio's assertions are restricted to falling bodies; he asserts first that their speeds are proportional to distances traversed, and second that this is a simple geometric proportionality.

It is to Simplicio, not Sagredo, that Salviati replies with the disputed argument, which he prefaces with the words, "and yet [your two propositions] are as false and impossible as that motion should be completed instantaneously, and here is a very clear proof of it." Thus if we pay attention to the logical structure of Galileo's book, the proof in question relates only to actual falling bodies, and therefore invokes observation as a step. Salviati's answer to Sagredo is by no means completed after that proof. In order to satisfy Sagredo, Salviati still must show that there is a difference in the consequences that flow from time-proportionality, and that those consequences are compatible with the observed phenomena of actual falling bodies. That Galileo is perfectly aware of all this is shown by the fact that those additional steps are carried out, in an orderly manner, in the ensuing pages. That part of the discussion, however, does not concern us here.

Yet it appears to concern all but one of the many persons who have examined Galileo's argument in the single sentence that we are discussing. The usual assumption is that Galileo made use in that sentence of some consequence of the correct definition of uniform acceleration that he had given previously, in which case he was guilty of begging the question. In support of that assumption, various authors have borrowed from the later section one or another of its correct deductions, to use in attempted reconstructions of his thought. Andres used the double-distance rule for uniform motion after free fall from rest.

Cohen and Hall used the mean-speed theorem, or Merton Rule. Cazrae, who started the whole controversy in 1642, accused Galileo of assuming the validity for accelerated motion of a law he had previously developed for uniform motion, though the Latin treatise sharply separated the two sections, and no one of his day was more acutely aware than Galileo of the fallacy of any such assumption. Gassendi failed to support Galileo properly against this charge, and seems not to have noticed that Cazrae had replaced Galileo's "velocities" with a "velocity" in the manner of modern translators. All these (and most later) discussions assume that Galileo was trying to prove that motion in which velocity increased with space traversed was impossible, in the sense of involving a logical contradiction.

If Galileo had thought that instantaneous motion was impossible in that sense, he very probably would have said so. No argument was more convincing than that. But he did not say so; he said that for actual fall (Simplicio's case), space-proportionality was as false and impossible as instantaneous motion, and he appealed to observation as evidence that such motion is not found in falling bodies. No one can prove a logical contradiction in that way. And as a matter of fact, there is evidence that Galileo did not consider either space-proportionality or instantaneous motion to be impossible, either logically or in nature. In 1611, he seems to have been seeking some example of space-proportional acceleration in curvilinear motion. In the *Assayer* he attributed instantaneous motion to light, and though he changed that opinion in the *Two New Sciences*, it was not because he thought that instantaneous motion was logically impossible.

IV

Let us now examine Galileo's argument, correctly translated, in the hope of discovering his line of thought.

If the velocities [passed through] have the same ratio as the spaces passed or to be passed, those spaces come to be passed in equal times: thus if the [instantaneous] velocities with which the falling body passed the space of four *braccia* were doubles of those with which it passed the first two *braccia* (as one distance is double the other), then the times required for these passages [over the spaces named] would be equal.

No diagram accompanies this statement, and none has preceded it. Galileo expects his readers (indeed, his imaginary auditors) to grasp his meaning without a diagram. If he wanted them to conceive of and compare mean speeds, he would have had to introduce and to illustrate that concept. Instead, he called their attention to the varying velocities with which the falling body moved, not to any uniform velocity that might represent them. If the plural "velocities" leaves any doubt on that score, the plural "doubles" removes it. Salviati did not slip inadvertently into the unusual and rather awkward plurals; they were essential to his argument, and he stressed them. If each conceivable velocity passed through in the whole descent is the double of a velocity passed through in the first half of the descent, then there is no way of accounting for a difference in the time required for one descent as against the other. That is all there is to his argument. The first statement does not invoke a rule for comparing uniform motions, as is generally believed; the phrase "or to be passed" can only refer to continuing acceleration. The opening words simply state in general terms what the balance of the passage applies to Simplicio's numerical exemplification.

But how many velocities are meant in each case by the deliberate plural? The answer to that question had been given in the long discussion that intervened between the definition of uniform acceleration and the argument with which we have been concerned. The relevance of that discussion becomes apparent only when the above argument is correctly understood; I, at any rate, regarded it as one of Galileo's habitual digressions, made to keep things interesting, as long as I accepted the general view that Galileo had erred in his "clear proof."

Sagredo and Simplicio had at once objected to Galileo's definition on the grounds that it could not apply to real bodies, for it would require them to pass through an infinite number of speeds in a finite time. Salviati pointed out that this is in fact possible, because the body need not remain in any one velocity for a finite time. He satisfied them that there could thus be infinitely many velocities in any uniformly accelerated motion, however small. That concept was fresh in the minds of Galileo's hearers when he spoke to them of the doubles of all the velocities in the whole motion as compared with those in its first half. No diagram was necessary for them, nor was any diagram appropriate to Galileo's meaning. The conception he desired was inescapable – or so Galileo thought. Removed from its context, it nevertheless seems to have escaped everyone who analysed the passage – except Tenneur.

Replying to Cazrae, Tenneur employed the same diagram that his opponent had produced to argue against Galileo. Omitting Tenneur's references to Cazrae's triangles, his conception of Galileo's argument was this:

If possible, let the heavy body fall through two equal spaces AB and BC so that its speed at C has become double that which it had at B. Certainly, under the hypothesis, there is no point in the line AC at which its speed is not double that at the homologous point in the line AB ... Therefore the speed through all AC was double the speed through all AB, just as the space AC is double the space AB: therefore AC and AB are traversed in equal times.[24]

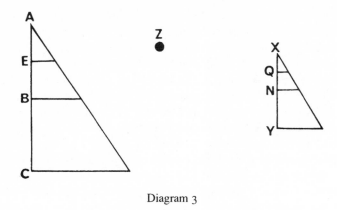

Diagram 3

Tenneur had grasped the essential clue from Galileo's plurals, as shown in the second sentence above. If in the final sentence Tenneur introduced the singular for each over-all speed, instead of comparing the spaces and times for the infinite assemblages directly, as Galileo did, it was not a fault in this case, for Tenneur did not substitute any particular value, or assume any particular rule of "compounding," for it. That he understood Galileo's reasoning exactly is shown by his argument after he had recited Cazrae's position, which amounted merely to asserting that Galileo's conclusion was absurd. "Let Cazrae prove what he likes against it," says Tenneur, "and I concede all that to him"; not the falsity of the consequence, but its implication by the definition is to be proved; and this is the same whether the speeds are separated for two bodies, or joined in one.

Which I prove. Let the same heavy body Z, in two different times, say today in four hours, and tomorrow in three hours, run through the two different spaces AC and XY, which first, AC, is double the second, XY. It cannot be denied that the body Z will run through AB, the first half of the space AC, in equal time with the second space XY, equal to the half-space AB. It is to be proved that, under the Cazrean hypothesis, the spaces AC and XY are traversed in equal times. Therefore divide XY midway at N, just as AC is bisected at B, and bisect AB and XN at E and Q. Then since by hypothesis, speeds are acquired as the spaces, and space AC is double XY, the speed at C is double the speed at Y, and the speed at B is double the speed at N, and the speed at E, [double] the speed at Q, and there is no part in AC, whether a third, or a quarter, or any other ratio whatever to its whole, at the end of which the speed is not double the speed at the end of some homologous part in the space XY; otherwise the motions would not be uniformly accelerated. Therefore the speed through all AC would be double the speed through all XY, just as AC is double XY. Whence AC and XY are run through in equal times, for they include neither any spaces nor any speeds that belong to diverse motions, nor any times that are different.[25] But the spaces AB and XY are traversed in equal time, since these are equal spaces, and it is the same heavy body Z; therefore AC and AB are traversed in equal times, since each is run through in the same time as XY. And do not say that the argument is invalid because the bodies situated at the points B and Y, having arrived in equal times, must have acquired equal degrees of speed. For what that really proves is that the spaces AC and XY cannot be traversed in equal times, a thing that no one doubts and that I can prove by other arguments; and the force of our argument is not thereby destroyed, in which nothing can be denied, and which easily and manifestly shows that these spaces that are in double proportion are traversed in equal times. Therefore Galileo admitted no paralogism at all, and the same was demonstrated by him for overlapping spaces – that two spaces in double proportion are traversed in the same time, on the hypothesis that velocities are acquired as the spaces – just as this was demonstrated here for separate spaces.[26]

Cazrae's diagram was naturally used by Tenneur in the polemic, but Galileo did not need one. Any diagram implying rest under the space-proportionality hypothesis must incorporate a fiction, as he knew only too well, long before 1638. All he needed was an imaginary bisected line and the notion of one-to-one correspondence between two infinite aggregates. That concept had already been developed earlier in the same book.[27] Thus Salviati's reply to Simplicio may be properly paraphrased thus:

"If all the infinite instantaneous velocities occurring in actual accelerated fall from rest over any space, say one of four *braccia*, were the respective doubles of all the infinite velocities occurring over the first half, or two *braccia*, of the same fall, then no difference in the times of fall could be accounted for. But we observed a difference in the times; hence proportionality of speed to space traversed from rest cannot govern the fall of actual bodies." In this there is no appeal to the correct definition of uniform acceleration, or to any of its consequences; neither is any contradiction asserted to exist within the incorrect law for falling bodies, which is merely shown to be in conflict with experience. Least of all is there any illicit use of a rule restricted to uniform motion.

v

Sagredo's opinion that it is a matter of indifference whether time or space be taken as the measure of velocity-increments in uniform acceleration has important historical implications. One might suppose it to have been introduced merely to enliven the dialogue. In that case, however, it would scarcely have been necessary to have Salviati admit that he had once accepted the notion, and still less for Galileo to acknowledge publicly his own former misapprehension. The purpose of diversion would be as well served by having Simplicio offer the incorrect definition as a rival to the correct one, and state that one must be false if the other was true. Instead we have Sagredo, who is never the spokesman for foolish positions, assert that he believes the two to be equivalent. Salviati's admission that he had once held it, to say nothing of Galileo's similar admission in print, suggests that the view was widely held and needed refutation. What is important about Galileo's admission is that it concerns not just the correctness of space-proportionality, but its equivalence to time-proportionality.

Inability to believe that the two could ever have been considered equivalent is so natural to us that we tend to impute one view or the other to Galileo's predecessors. We note that the Merton School writers and Oresme put velocity in uniform acceleration proportional to time, while many Peripatetics and Tartaglia put speed in free fall proportional to space. De Soto gave free fall as a case of uniform acceleration. All these things are true, but they do not imply (as we are likely to think) that any of these views was regarded as contradicting another.

Sagredo's assumption seems to have been universally adopted up to the time of Galileo. It reflects this kind of reasoning: "Space and time are both measures of velocity, so they must be proportional to each other with respect to velocity. Acceleration is merely increase of velocity, and if uniform acceleration is proportional either to spaces traversed or times elapsed, it must thereby be proportional also to the other. Since it is easier to think equal spaces than equal times, it is more sensible to say that uniform acceleration increases in proportion to spaces traversed. But if anyone prefers to relate it to times, there is no reason he should not do so." No one worked out the "proportion" for space, and only the "average" had been worked out for time.

The Merton Rule writers, who were quite specific in using time as the measure of uniform difform motion, were in fact discussing change in general, of which local motion was only a special case. The Merton Rule is really a mean-degree theorem, not just a mean-speed theorem. Looked at in that way, it is clear that the existence of that rule did not automatically call attention to a problem of changing speed any more than to a problem of changing heat, or redness, or any other variable quality. Thus Albert of Saxony could remark that in free fall the speed grew with the growing spaces traversed, and at the same time know that the Merton Rule was valid for all changes with respect to time, without perceiving a contradiction.

Had the contradiction been perceived, one would expect it to have been raised as a question in at least some of the many commentaries on Aristotle's *Physics* written in quaestio-dubitatio-responsio form. At least some writers of such commentaries may be presumed to have been familiar with both views, and to have been fond of disputation; yet they did not pose the question "Whether velocity increases in a falling body in proportion to distance or to time?" The right answer to that question first appeared in 1545, when Domingo de Soto linked free fall to the Merton Rule, but he did not ask the question. In using free fall to exemplify the Merton Rule, it is likely that he also considered this linkage compatible with space-proportionality in free fall, since he did not contradict that popular opinion.

The first person known to have both asked and then answered the question was Galileo, and that was precisely his purpose in introducing the disputed argument into the *Two New Sciences*. The probability that his was indeed the first overt statement that the two views were inconsistent is heightened by the wide debate over it that immediately ensued, particularly in France, as contrasted with the previous silence. This, I think, is the key to what has been called "the enigma of Domingo de Soto," consisting of the double question: "Why was the linking of the Merton Rule and free fall delayed as long as it was?" and "Why did Soto's linkage of the two fail to have any immediate effect?"

As to the question whether Galileo's work was itself inspired belatedly by Soto's, two reasons incline me to think not. First, Soto related uniform acceleration specifically to time, divided it into time intervals, and dealt with it in terms of a single middle point of velocity; none of these concepts appears in Galileo's notes for a very long period of time, and it is unlikely that he read Soto's *Questions on the Physics* at an advanced age. And second, Galileo's thought followed an entirely different line from that of the Merton Rule. The plural "velocities" and a comparison of compounded infinite instantaneous speeds are already to be found explicitly in the 1604 demonstration. Those concepts, rather than any measure of average speed, characterize Galileo's approach to the problems of acceleration.[28]

Similarly, the apparently strange coincidence of Galileo's early mistaken view with that of Beeckman and Descartes loses all its mystery, if we take Sagredo to have been speaking for a commonly held opinion. Beeckman wrote to Descartes, asking for a rule of acceleration based on times; Descartes answered with a rule based on spaces; Beeckman, without even noting the difference, wrote out the correct rule as a consequence of Descartes's reasoning; and long afterwards, Descartes recalled mistakenly that he had reached the same result as Galileo.

Thus, even among the best mathematicians, the conflict of concepts was not recognized before 1638. In one striking instance it was not even recognized then. Mersenne, who had quickly adopted time-proportionality from the 1632 *Dialogue* and discussed it at length in his *Harmonie Universelle* five years later, published in 1639 a French paraphrase of the *Discorsi*. There he wrote, concerning the disputed argument:

> Il refute en suite le pensée de plusieurs, qui disent que la vitesse croist en mesme raison que les espaces, par exemple, que la vitesse acquise en quatre espaces est double de celle qui est acquise en deux, car il s'ensuiveroit de là que le mobile seroit aussi-tost quatre brasses comme deux: neantmoins cecy se peut entendre d'une veritable façon; car pourquoy ne peut-ons pas dire que la vitesse est plus grande à proportion des plus grandes espaces parcourus? Mais il ne faut pas nous esloigner de l'intention de l'Autheur, qui suppose qu'un mesme mobile roulant sur des plans differens, acquiert un esgal degré de vitesse, lors que ces plans ont une mesme hauteur ... etc.[29]

That Mersenne accepted both indifferently at this time is evident from his *Remarque* added at the end of the same section:

> J'ay trouvé les mesmes proportions en laissent tomber des boules de plomb, et de toute autre sorte de matiere, en toutes sortes de hauteurs depuis un pied jusques à cent quarante sept pieds de haut ... [30]

– referring, of course, to Galileo's times-squared proportions.

VI

Although I do not believe that Galileo's reasoning in his rejection of space-proportionality involved any illicit application of consequences deduced from time-proportionality, there is one such attempted reconstruction of his thought that historically deserves attention. It is difficult of access, having been published only once and in an obscure eighteenth-century journal. Not only is it the most plausible attempt at a reconstruction of Galileo's own thought along those lines, but the circumstances of its composition are interesting. Riccati, who had at first supported Galileo's argument as valid, subsequently decided against it. Confusing it with another passage in the same work, however, he then asserted that it was not Galileo's at all, but had been added to later editions by Viviani.[31] Giovanni Andres, who had found it hard to believe that Galileo would have committed a fallacy in a demonstration he himself had described as clear and evident, accepted the idea that it was apocryphal.

But later, coming upon a copy of the already rare edition of 1638, he found that the argument was indeed Galileo's. He then set himself to rediscover Galileo's thought on the basis of propositions enunciated in the same book, much as Galileo had reconstructed the reasoning of Archimedes in his *Bilancetta*.

As set forth by Andres, the reconstruction is open to objection because it invokes a consequence of the correct definition; namely, that a body attains in accelerated fall a velocity which, in a further time equal to that of fall from rest, would carry it in uniform motion through double the distance fallen. Andres does not make it clear how the rule was obtained. In presenting his argument and diagram here, I have embedded them in a pastiche that frees the assumption from dependence on the correct definition. This interchange is supposed to follow Sagredo's praise of Salviati's reply to Simplicio; the argument of Andres is enclosed in quotation marks (p. 228).

SIMP: Perhaps your demonstration is clear and simple to Sig. Sagredo, but I confess that for my part these mathematical subtleties are unconvincing. I should like to hear an argument more closely tied to events we can see before our eyes, if that is possible.

SALV: I shall try my best to serve you, making use of certain observations that have been made by our Author. But first let me ask whether you recall that in a certain discussion of ours many years ago, I remarked that if a body thrown upward passed through the same degree of speed, diminishing, that it acquired increasingly in returning downward, its maximum speed would be exactly that which would carry it, in uniform motion, through twice the space of its descent in another equal period of time.

SIMP: I recall your having said this, and Sagredo's having added to it an illustration of your meaning. You then gave a mathematical proof of it, which I could no more follow than I can your present demonstrations in terms of "velocities" and "doubles" and "instantaneous motions." And I believe that I said then that in physical science there is no occasion to look for the precision of mathematics.[32] But since that time, looking once more into some good Peripatetic writers, I encountered again that same rule that you now mention. Their authority made me the more willing to believe it true, though I could not follow their demonstrations.

SALV: Their demonstrations were also abstract and mathematical, while you have asked for more physical proof. But I can tell you that while you were studying the commentators of Aristotle, our Author devised a certain way of measuring times and speeds, by which means he was able to confirm that in the actual descent of heavy bodies, this rule of double space in equal time holds for the speeds acquired at the end of the free descent. Therefore I ask you to concede to me, not as a thing proved by reason but as an observed fact, that the speeds acquired by falling bodies commencing from rest are sufficient to move them through twice the distance, in uniform motion in the same time, as that through which they previously passed.

SIMP: I shall grant this for argument, without conceding it as necessary or proved, for it seems to agree with that "uniformity" that our Author seeks in natural acceleration. And it would not be proper for me to dispute what he says he has observed with his eyes, when I have made no experiments. Nor do I see how they could be conducted.

SALV: You shall have the experiments in good time, when you see later how this observed physical rule accords well with our Author's definition of uniform accelera-tion. But first let us see whether it accords with your own belief that the velocity of fall increases in proportion to the distance fallen. For this, I draw the following diagram:

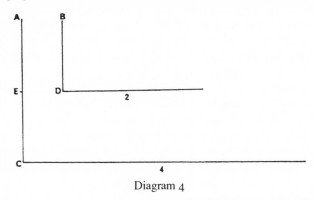

Diagram 4

Let A and B be two equally heavy bodies falling from rest. By your own hypothesis, "A at C has double the velocity of B at D; therefore A has such velocity as to run in uni-form motion four *braccia* in the time that B in uniform motion runs two *braccia*. The time in which B at D runs two *braccia* is the time in which it has run through BD; there-fore the time in which A at C will run four *braccia* in uniform motion is equal to that in which B, accelerated, has run through BD. Now, the time in which A at C runs four *braccia* uniformly is equal to that in which it has run through AC with accelerated motion; therefore the time of motion of A through AC is equal to the time of motion of B through BD. But the time of motion of B through BD is equal to the time in which A has passed through AE; therefore the time of motion of A through AC, which is seen to be the time of B through BD, is equal to that of the motion of A through AE. Therefore EC was passed over without time."[33] And that is what I said.

SIMP: This satisfies me better than your abstract reasoning. But I do not see that there is any difference between increasing velocities with space, as Sagredo suggested and I approved, and increasing them with time, as your Author wishes to do. Perhaps Nature acts in neither way, nor in any other way conformable to abstract mathematics.

SALV: That is what we are now going to investigate, using this time our Author's defi-nition of uniform acceleration. Then I shall tell you of his experiments, by which he has been able to confirm many remarkable things, including the rule which you have cour-teously granted for argument, but only provisionally and if demonstrated by observa-tion. Whether there may be in Nature motions that are indeed instantaneous does not concern us, since I believe you will agree that they can hardly be called "uniformly accelerated."

VII

In conclusion, something may be said about Galileo's physics in the light of Ernst Mach's formula for acceleration under space-proportionality, $S = Ae^{aT}$.[34] As Cohens[35] has already pointed out, if in Mach's formula S is ever zero, it must be always zero; not only is there no acceleration, but there is no velocity, and we have a universe at rest. So in order to have motion at all under Mach's formula, we must stipulate that S is never zero. This makes it difficult (to say the least) to conceive of a scale for measuring distances in any ordinary physical sense. For whether or not any measured distance ever happens to be zero, it should make physical sense to speak of zero distance.

But that is not all. If T is zero, $S = A$, a constant other than zero. Then if average velocity is measured, as it usually is, by S/T, and S is not zero (as in Mach's universe it cannot be), T cannot be zero without implying infinite velocity. But this general observation may be regarded as a quibble, and the possibility of $T = 0$ must be explored more carefully.

By definition, terminal velocity, V, is for Mach $dS/dT = aS$. Since it is hard to imagine any physical meaning for average velocity other than $v = S/T$, we may use that concept to derive from Mach's formula the ratio of terminal velocity V to average velocity v, obtaining $V/v = aT$. Now, in any discussion of accelerated motion, the value $V/v = 1$ must be excluded, for terminal velocity an equal average velocity only for uniform motion or for some kind of instantaneous motion. Therefore we cannot have $aT = 1$, or $T = 1/a$. But if T is ever zero and can increase without limit, it must at some point be equal to $1/a$. At that point, $v = V$, and the preceding net motion must have been uniform or instantaneous. But that will hardly be called accelerated motion.

In sum, Mach's formula is free of physical contradiction only if we read $S = Ae^{aT}$, $S \neq 0$, $T \neq 0$ – a restriction that requires us to give up our usual methods of measuring either of the fundamental conceptual entities (space and time) that enter into the definitions of velocity and acceleration.

The difficulties involved are thus considerably greater than Mach suggested. They also coincide rather well with the difficulties deduced by Galileo from his diagramless consideration of the problem of actual fall, in terms of comparing infinite assemblages of velocities in pairs. It would be a remarkably happy chance if Galileo hit on these difficulties by a clumsy fallacy, after three decades of thought on uniform and accelerated motion, on continuous quantities, and on infinite divisibility. But there is nothing remarkable in his having got the right general idea without being able to solve differential equations. The infinitesimal calculus is an aid to the avoidance of error, but it is not the only available means of analysing change in continuous quantity. Careful thought on geometrical principles came first. It was from Galileo that Cavalieri got his start on "geometry by indivisibles." Galileo got his start, in turn, from Archimedes. It is therefore hazardous to assume that he had to resort to the Merton Rule in order to win through to an understanding of uniformly varying velocities.

VIII POSTSCRIPT

At the time of submission of this (and my preceding) paper for publication, I believed the evidence to show that Galileo was unfamiliar with the medieval mean-speed rule. Discussion of it seems to have gone out of fashion in Italy during the sixteenth century, though not at Paris or in Spain. Subsequent studies lead me to believe that Galileo may have known the mean-speed rule and rejected it as inapplicable to the analysis of unbounded accelerated motion. This possibility must be reserved for future treatment, which may throw light both on some internal relationships of medieval physics and on its history in the sixteenth century.

I also assumed the application of Euclidean proportion to one-to-one correspondence to have been original with Galileo. Since then, I have noted this in an argument of Richard Swineshead: "Then in every instant of the first proportional part it will be moved twice as swiftly as in the corresponding instant of the second proportional part, and similarly for the succeeding, as is evident."[36] Galileo's extension of such proportionality between parts to a proportionality between part and whole in an infinite set gave him the key to his ultimate rejection of space-proportionality. Such a part-whole relationship also had medieval ancestry.[37] If there is a true link between medieval kinematics and the work of Galileo, I believe that it is in this line of thought, rather than the so-called Merton Rule.

NOTES

1 For example, the first diagram on p. 380, *Opere*, viii, suggests that accelerated fall along the diameter is under consideration. But reference to p. 378 shows that the proposition was probably derived without any thought of acceleration. The note accompanying it is a memorandum for the order of proofs in the projected treatise of 1609, discussed below, in which static, dynamic and kinematic considerations were integrated.

2 *Opere*, viii, 383. It is possible that the document is not a copy, but an attempt by Guiducci to resurrect space-proportionality, like Sagredo in the *Discorsi*. But Guiducci copied many other Galilean notes, preserved like this one with Galileo's own papers, so it is probably authentic. The copies were probably made after 1616, when Galileo had to drop astronomy and return to physics.

3 Ibid., 388. *Cum assumptum sit* almost certainly refers to the opening word, *Assumo*, of the document copied by Guiducci.

4 Ibid., 389.

5 Ibid., 380. For the two substituted propositions, see pp. 376–377.

6 *Opere*, x, 248–249. Valerio's reply is confusing because he ignored the phenomenon of acceleration in one of the propositions but not in the other. Valerio also acknowledges the receipt of a theorem, which was probably the Latin version of the 1604 fragment which Valerio called "most elegant, and worthy of you."

7 *Opere*, viii, 380.

8 Ibid., 388.

9 *Opere*, xi, 85.

10 *Opere*, viii, 384–385.

11 Ibid., 378.

12 Ibid., 375: cf. vii, 48–50 and *Dialogue*, tr. S. Drake (Bekeley, 1967), 24–26. The same paradox had been noted by Oresme: see M. Clagett, *Nicole Oresme and the Medieval Geometry of Qualities and Motions* (Madison, 1968), p. 279.

13 Ibid., 381.

14 Ibid., 380.

15 Ibid., 387.

16 *Opere*, i, 297; cf. *Galileo on Motion and on Mechanics*, tr. I.E. Drabkin and S. Drake (Madison, 1960), p. 64.

17 "The 1604 Fragment on Falling Bodies, Galileo Gleanings XVIII," *The British Journal for the History of Science*, iv (1968–69), 340–358.

18 A. Koyré, *Newtonian Studies* (Cambridge, Mass., 1965), p. 36.

19 A. Koyré, *Études Galiléennes* (Paris, 1939), ii, p. 98, n. 2.

20 *Two New Sciences*, tr. H. Crew and A. De Salvio (New York, 1914 and later eds.), p. 168.

21 *Unterredungen und mathematische Demonstrationen*, tr. A. von Oettingen (Leipzig, 1891 and later eds.), vol. 2, p. 16.

22 See, for example, A. Koyré, *Études*, ii, pp. 95–99; I.B. Cohen, "Galileo's Rejection of the Possibility of Velocity Changing Uniformly with Respect to Distance," *Isis*, 47 (1956), pp. 231–235: A.R. Hall, "Galileo's Fallacy," *Isis*, 49 (1958), pp. 342–346; *Discorsi*, ed. A. Carugo and L. Geymonat (Torino, 1958), pp. 776–778.

23 *Opere*, viii, p. 203.

24 J.A. Tenneur, *De motu naturaliter accelerato tractatus ...* (Paris, 1649), p. 8.

25 That is, every doubled space is associated with a doubled velocity.

26 Tenneur, op. cit., pp. 10–11. The diagram eliminates some additional lines used by Cazrae, but not referred to in Tenneur's present argument.

27 *Two New Sciences*, pp. 31 ff.

28 See note 1, above.

29 *Les Nouvelles Pensées de Galilée* (Paris, 1639), p. 184.

30 Ibid., p. 188.

31 Cf. *Two New Sciences*, p. 180 n.

32 Cf. *Dialogue*, tr. S. Drake (Berkeley, 1967), pp. 227–230.

33 *Raccolta di opuscoli scientifici, e letterarii* (Ferrara, 1779), p. 64.

34 E. Mach, *Science of Mechanics* (La Salle, 1942), p. 309.

35 Cohen, op. cit., p. 235.

36 Swineshead, *The Book of Calculations*, tr. Marshall Clagett, *The Science of Mechanics in the Middle Ages* (Madison, 1959), p. 291.

37 Cf. H.A. Wolfson, *Crescas' Critique of Aristotle* (Cambridge, Mass., 1929), pp. 189–191; 346–347. For this reference I am indebted to Rabbi Nachum L. Rubinovitch, whose paper "A 14th-Century Insight on Affinity" will soon appear in *Isis*.

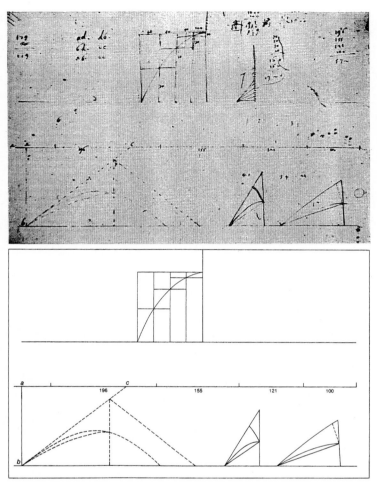

Plate I:
Photograph and transcription of folio II7, in Ms. Galileiano 72, shows Galileo's attempts to establish the restricted principle of inertia he needed. Such a principle, used in conjunction with his times-squared law of free fall, would enable him to prove that the trajectory of an object travelling through a nonresisting medium was indeed a parabola.

The numbers on the horizontal line probably represent the diminishing distances traversed by a ball in successive equal times after it had been given an initial push along a level grooved plane. The little parabolas are preliminary sketches he made of likely trajectories. The steps under the larger parabola at the top represent the height of the ball above the ground at equal horizontal distances along its path. Reproduced by permission of Ministero per i Beni e le Attività Culturali, courtesy of Biblioteca Nazionale Centrale, Florence.

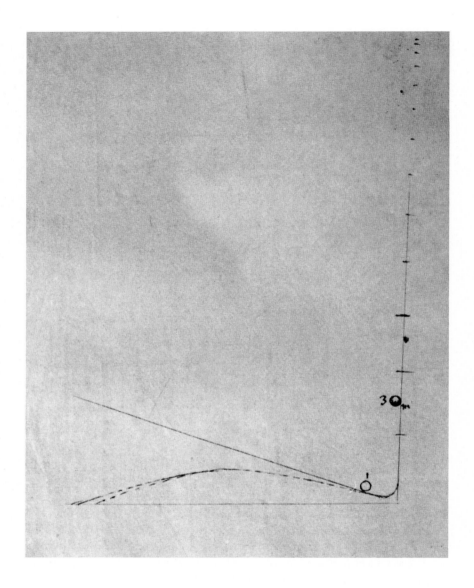

Plate 2:
Folio 175, Ms. Galileiano 72. The ball bounced into the air for a short distance when it rolled directly off the inclined plane onto a flat surface deflecting it horizontally into the air. Galileo was aware of this problem, as is shown by his sketch of the ball's path in the document designated f. 175. In order to reduce the bounce and make the transition smooth he devised a curved deflector. Reproduced by permission of Ministero per i Beni e le Attività Culturali, courtesy of Biblioteca Nazionale Centrale, Florence.

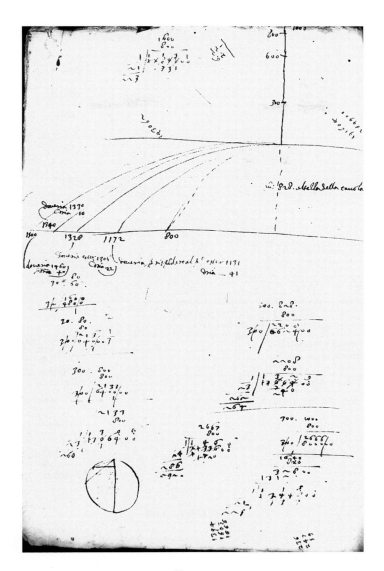

Plate 3:

Folio 116v from Ms. Galileiano 72 records one of the experiments conducted by Galileo by which he discovered the law of the parabolic trajectory in 1608. The diagram at the top shows the data Galileo obtained by measuring the actual distance the ball moved from the edge of the table after being rolled down an inclined plane from specified vertical heights. The numbers at the bottom are his calculations of expected distance that ball should have traveled from different heights. Reproduced by permission of Ministero per i Beni e le Attività Culturali, courtesy of Biblioteca Nazionale Centrale, Florence.

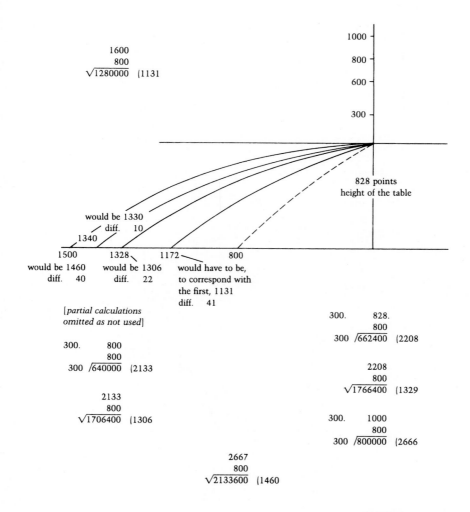

$$
\begin{array}{l}
1600 \\
\underline{800} \\
\sqrt{1280000} \quad (1131
\end{array}
$$

1000 -
800 -
600 -
300 -

828 points
height of the table

would be 1330
diff.   10
1340

1500   1328   1172   800
would be 1460   would be 1306   would have to be,
diff.   40   diff.   22   to correspond with
                                  the first, 1131
                                  diff.   41

*[partial calculations
omitted as not used]*

300.   800
            800
300 /640000 (2133

            2133
            800
√1706400 (1306

300.   828.
            800
300 /662400 (2208

            2208
            800
√1766400 (1329

300.   1000
            800
300 /800000 (2666

            2667
            800
√2133600 (1460

Plate 4:
English transcription of folio 116v, showing Galileo's calculations of horizontal
distances expected under his mean-proportional rule, using shortest drop as basis.
Unused partial calculations are omitted, as are trial divisors in root extraction and related
reminders.

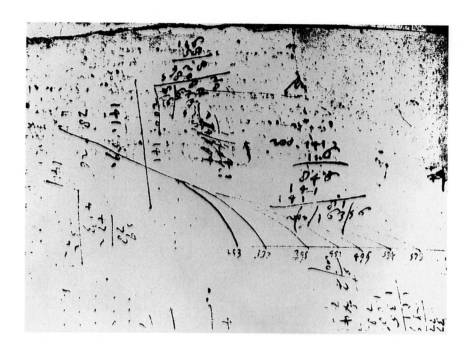

Plate 5:
Folio 114 shows Galileo's measurements of the distances the ball fell from the end of the inclined plane when it was not deflected horizontally. He had devised this experiment to try to resolve the apparent conflict of the data he recorded in f. 116 for the horizontal distances the ball traveled after it fell through the vertical drop of 828 *punti* above the table. Galileo could not have computed the set of seven three-digit numbers shown (253, 337, 395, 451, 495, 534, 573), and so he must have obtained them by careful measurements. Results from experiment duplicated by the authors are in close agreement with Galileo's. Reproduced from Ms. Galileiano 72, by permission of Ministero per i Beni e le Attività Culturali, courtesy of Biblioteca Nazionale Centrale, Florence.

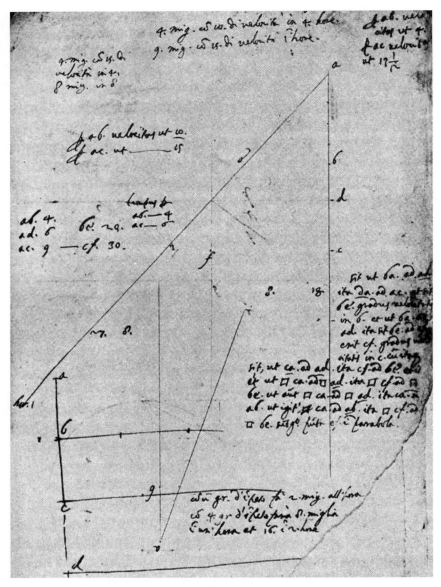

Plate 6:
Folio 152r, Ms. Galileiano 72. Unpublished manuscript records Galileo's first discovery
of the law of free fall, before October, 1604. The crucial diagram is the large one at the
top; the lower diagram and the bottom paragraph refer to calculations for horizontal
motion. Reproduced by permission of Ministero per i Beni e le Attività Culturali,
courtesy of Biblioteca Nazionale Centrale, Florence.

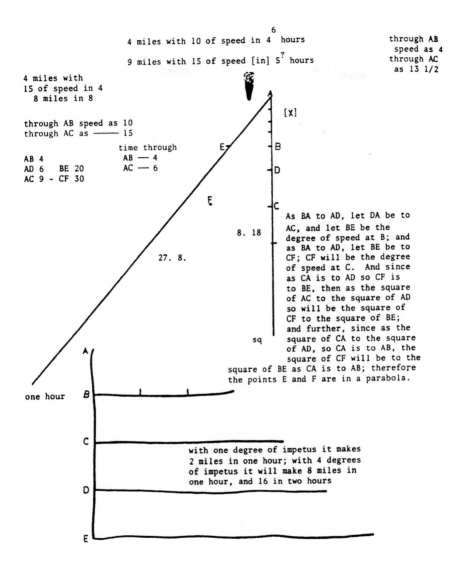

4 miles with 10 of speed in 4 hours

9 miles with 15 of speed [in] 5? hours

through AB
speed as 4
through AC
as 13 1/2

4 miles with
15 of speed in 4
  8 miles in 8

through AB speed as 10
through AC as ――― 15

          time through
AB 4   AB ― 4
AD 6   BE 20   AC ― 6
AC 9 - CF 30

[X]

8. 18

27. 8.

sq

As BA to AD, let DA be to
AC, and let BE be the
degree of speed at B; and
as BA to AD, let BE be to
CF; CF will be the degree
of speed at C. And since
as CA is to AD so CF is
to BE, then as the square
of AC to the square of AD
so will be the square of
CF to the square of BE;
and further, since as the
square of CA to the square
of AD, so CA is to AB, the
square of CF will be to the
square of BE as CA is to AB; therefore
the points E and F are in a parabola.

one hour

with one degree of impetus it makes
2 miles in one hour; with 4 degrees
of impetus it will make 8 miles in
one hour, and 16 in two hours

Plate 7:
English transcription of folio 152r, Ms. Galileiano 72.

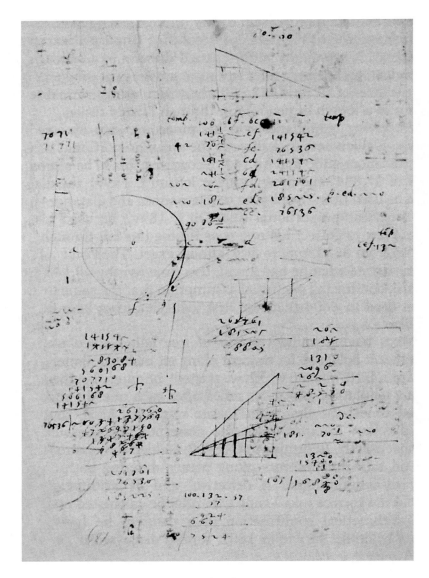

Plate 8:
Galileo's sketch of parabola in triangles on f. 189r from Ms. Galileiano 72; drawn over
calculations relating to an earlier theorem. Numerical confirmation of a theorem derived
in 1602 on incorrect assumptions. Using the rule found for planes differing in length and
slope (Plate 19), Galileo vindicated his theorem, for which he later found a valid proof
based on uniform acceleration. Reproduced by permission of Ministero per i Beni e le
Attività Culturali, courtesy of Biblioteca Nazionale Centrale, Florence.

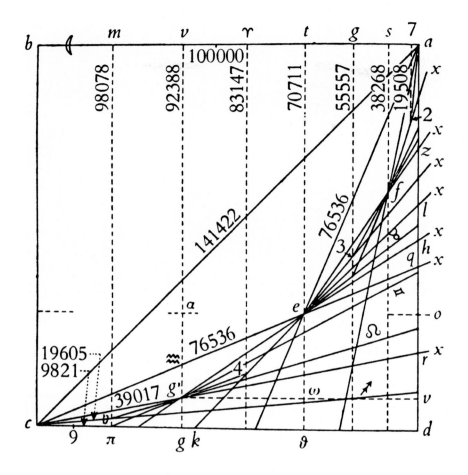

Plate 9:
Folio 166r, Ms. Galileiano 72. Reproduced from A. Carugo and L. Geymonat, ed.,
Galileo Galilei: Discorsi e Dimostrazioni Matematiche Intorno a Due Nuove Scienze
(Torino, Paolo Boringhieri, 1958), p. 544. The notes Galileo wrote across
the diagram or elsewhere on the same leaf are given separately on Plate 13.

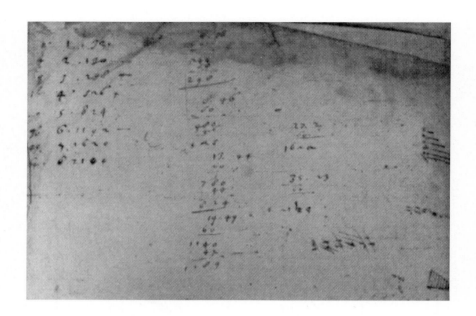
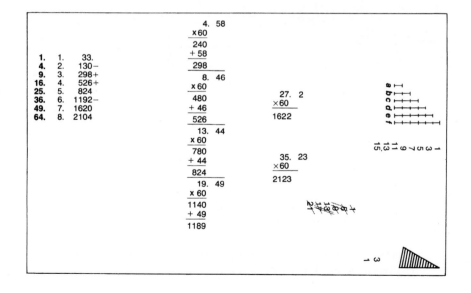

Plates 10/11:

Galileo's results of his 1604 experiment appear on the manuscript page shown in the photograph at top (Plate 10). The page, designated f. 107v, is reproduced in part here by permission of Ministero per i Beni e le Attività Culturali, courtesy of Biblioteca Nazionale Centrale, Florence. The transcription at bottom gives the relevant figures in a modern typeface. The figures listed in the third column at upper left in the reproduction represent very nearly the distances from rest of a ball rolling down an inclined plane at the ends of eight equals times. The figures differ slightly from the products of the first number, 33, by the squares of the numbers from 1 to 8. The correct square-number multipliers (1, 4, 9, 16, 25, 36, 49, 64) were later added in the first column with a different pen and ink. The calculations at centre show how Galileo actually got the numbers he entered in the third column. In each case he multiplied 60 by some integer and then added a number less than 60, probably reflecting the fact that he measured the distances with a short ruler divided into 60 equal parts. The series of alternate odd numbers (1, 5, 9, 13, 17, 21) written. down at lower right (and then cancelled) appears to represent his first guess at a rule for the growth of the figures in the third column. The significance of the other notations is discussed in the text.

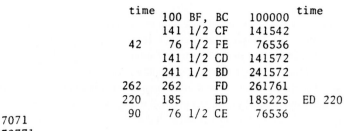

time					time
	100	BF, BC	100000		
	141 1/2	CF	141542		
42	76 1/2	FE	76536		
	141 1/2	CD	141572		
	241 1/2	BD	241572		
262	262	FD	261761		
220	185	ED	185225	ED	220
90	76 1/2	CE	76536		

7071
70771

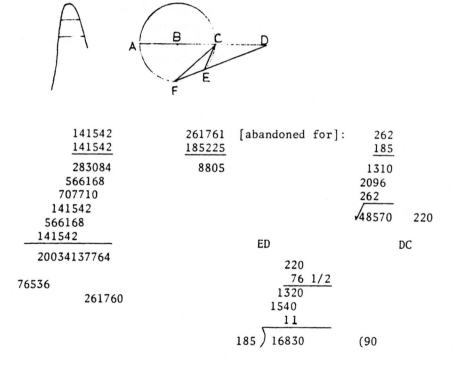

```
    141542          261761   [abandoned for]:     262
    141542          185225                        185
   ─────────        ───────                       ───
    283084            8805                        1310
    566168                                        2096
    707710                                        262
    141542                                            _____
    566168                                          √48570      220
    141542
   ─────────                              ED                    DC
  20034137764
                                              220
   76536                                        76  1/2
                                              ─────────
        261760                                 1320
                                               1540
                                                11
                                              ─────────
                                      185 ) 16830        (90
```

Plate 12:
A partial transcription of folio 189r, Ms. Galileiano 72.

Cum semidiameter sit 100000, quadrantis circumferentia

est $\begin{cases} 157143 \\ 157042 \end{cases}$; seu si semidiameter sit 1000, circumferentia

quadrantis $\begin{cases} \text{plus} & 1570 \\ \text{minus} & 1571 \end{cases}$; si ☐ sit 1000000, ◁ erit 785250.

Tempus quo conficeretur circumferentia quadrantis, si esset recta et ad perpendiculum, 125331.

Tempus per *ac* ad tempus per 2 *aec* est ut 1000 ad 937$^1/_2$ fere; tempus per *ec* ad tempus per 2 *egc*, ut 1000 ad 866$^3/_5$; tempus per 8 *c* ad duas suas, ut 1000 ad 733 $^2/_3$.

ad longa puncta 180; sit tempus casus per ipsam m^1 180, et per ambas *adc* m^1 270.

ac ————————— 254 $^3/_5$; ————————— m^1 254 $^3/_5$.

ae ————————— 138; tempus casus per illam m^1 164.|

ec ————————— 138; tempus casus per eam post *ac* m^1 75, et   *420* per ambas *aec* m^1 239.

af recta — 70 $^1/_2$; tempus etc.   113 $^1/_2$ ⎫ et per ambas
fe ————— 70 $^1/_2$; tempus casus post *af* 48 ⎬ *afe* m^1 161$^1/_2$ ⎫ et per
eg ————— 70 $^1/_2$; tempus ————— 39 ⎬ et per ambas ⎬ 4 *afegc*
gc ————— 70 $^1/_2$; ————— 36 ⎭ *egc* 75 ⎭ 236$^1/_2$

Considera num tempus per *ac* ad tempus per duas *aec* sit ut radix radicis lineæ quæ a centro *b* super *ac* cadit perpendiculariter, ad radicem radicis perpendicularis ex eodem centro super *ae*. Tempus per 2 *egc* ex quiete in *e* est 66326; deberet autem esse 71757,|si casus per *egc* ad tempus per *ec* haberet eandem rationem quam casus per *aec* ad casum per *ac*: movetur ergo citius per *egc* quam per *aec*. Et ex quiete in 8 tempus per duas 8 *c* ad tempus per solam 8 *c* est ut 14378 ad 19598: longe igitur adhuc citius movetur quam per 2 *egc*.

Plate 13:

Folio 166r, Ms. Galileiano 72. Reproduced from A. Carugo and L. Geymonat, ed., *Galileo Galilei: Discorsi e Dimostrazioni Matematiche Intorno a Due Nuove Scienze* (Torino, Paolo Boringhieri, 1958), pp. 542–4. Transcription by A. Favaro, pp. 419–21.

```
kel ──────  82843; ──────  50404
lx, ld ───  58579; ──────  91018
la ───────  41422; tempus casus 64360   per 3 alekc
elx ──────  100000.                        135475
kelx ─────  141422.
kc ───────  41422; ──────  20711


a2 ───────  19604; tempus ──── 44385    72 longa 19508; tempus 44168
2x ───────  20386; ──────  46156        2x ─── 20586; tempus ...
fzx ──────  39990; ──────  64644  per ambas
2f ───────  19604; ──────  18488  a2f 62873

                                          per 4 a2f3e
f3 ───────  19604; ──────  14372             89605
fx ───────  43392; ──────  70144  per ambas
3fx ──────  62996; tempus  84516  f3e 26732
e3 ───────  19604; ──────  12360        linea 39 55552
e3x ──────  91475; ──────  108783       tempus 74536
3x ───────  71871; ──────  96423

4e ♉ ─────  131072; ──────  143743
e ♉ ──────  111468; ──────  132558      4 ♈ 83147
e4 ───────  19604; ──────  11185        tempus 91185
g4 ♊ ─────  195993; ──────  203906  per ambas
4 ♊ ──────  176389; ──────  193439  e4g 21652
g4 ───────  19604; ──────  10467

                                        per 4 e4g8c
8g ♌ ─────  338035; ──────  341316          41473
g ♌ ──────  318431; ──────  331287
8g ───────  19604; ──────  10029    per ambas
c8 ♐ ─────  1019979; ──────  1019979  8 ♍ ...
8 ♐ ──────  1000375; ──────  1010187  g8c 19821  tempus 99030
c8 ───────  19604; ──────  9792

ad longa ─ 100000; tempus casus 100000   media inter ad, te 84090
ac ───────  141422; ──────  141422       tempus per te ── 84090
ae ───────  76536; ──────  91017
eqx ──────  184777; ──────          per ambas aec
xqec tota ─ 261313; ──────  261313  tempus 132593
ec ───────  76536; ──────  41576

af recta ─ 39017; tempus casus 63045   media inter da, fs 61861
fzx ──────  46022; ──────  74408       tempus per sf ── 61861
efzx tota ─ 85039; tempus ──── 101129  per ambas
ef recta ─ 39017; tempus ──── 26721   afe 89766

                                        per ambas
ehx ──────  127228; tempus ──── 151300  4 131319;
gehx ─────  166245; ──────  172957      tempus per
eg ───────  39017; ──────  21657           vg 96118

grx ──────  472242; ──────  491363  per ambas
cgrx ─────  511259; ──────  511259  egc 41553
cg ───────  39017; ──────  19896
```

Plate 14:
Folio 183r, Ms. Galileiano 72. Reproduced from A. Carugo and L. Geymonat, ed.,
Galileo Galilei: Discorsi e Dimostrazioni Matematiche Intorno a Due Nuove Scienze
(Torino, Paolo Boringhieri, 1958), pp. 545–6. Transcription by A. Favaro, pp. 421–2.

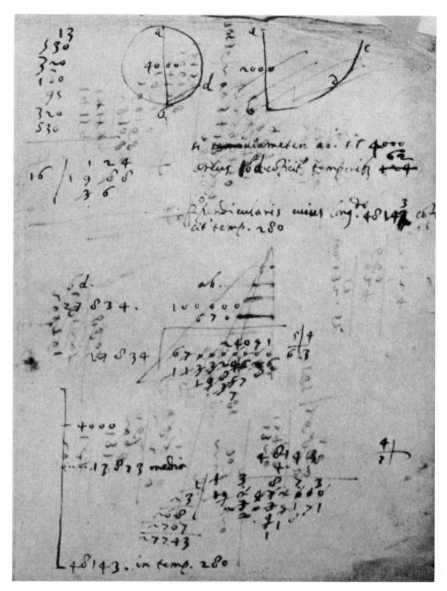

Plate 15:

Folio 189v1, Ms. Galileiano 72. The triangular diagram at centre is not a part of 189v1, but shows through from f. 189r. At centre, calculation of distance fallen in 280 *tempi*. At top, and not required for this calculation, are timings of pendulum of length 1,740 *punti* through a large and a small arc. Reproduced by permission of Ministero per i Beni e le Attività Culturali, courtesy of Biblioteca Nazionale Centrale, Florence.

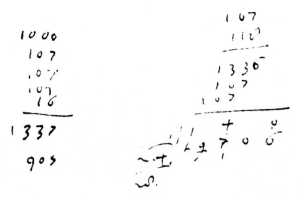

Plate 16:

Folio 154v, Ms. Galileiano 72. First recorded timings of fall, and a later mean-proportional pendulum calculation. Reproduced by permission of Ministero per i Beni e le Attività Culturali, courtesy of Biblioteca Nazionale Centrale, Florence.

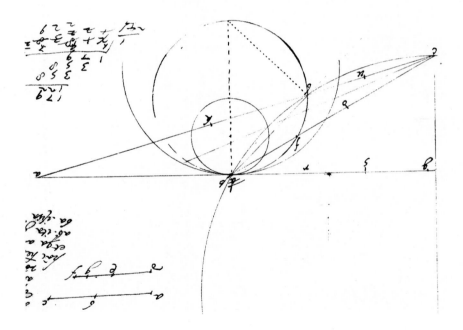

Plate 17:

Part of folio 185r, Ms. Galileiano 72. The part is restored to its original position to show Galileo's geometric model of pendulum and fall, and his accompanying pendulum sketch. Other lines and letters were added later for a different purpose. Reproduced by permission of Ministero per i Beni e le Attività Culturali, courtesy of Biblioteca Nazionale Centrale, Florence.

Plate 18:
Folio 151v, Ms. Galileiano 72. Recorded measurement of 1,590 punti for the pendulum
timing flow of 903 grains of water, with geometric models for pendulum and fall.
Reproduced by permission of Ministero per i Beni e le Attività Culturali, courtesy of
Biblioteca Nazionale Centrale, Florence.

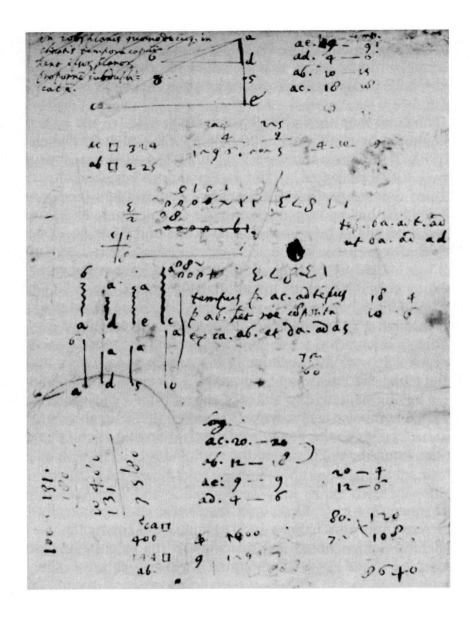

Plate 19:

Folio 189v2, Ms. Galileiano 72. The first entry was made with the page inverted and before f. 115v (Plate 20). This corrected the first timing of fall through 4,000 *punti*. The other entries show how Galileo found the rule for descents along planes which differ both in length and in slope. Reproduced by permission of Ministero per i Beni e le Attività Culturali, courtesy of Biblioteca Nazionale Centrale, Florence.

Plate 20:
Folio 90bv, Ms. Galileiano 72. Surviving fragment of pendulum calculations made shortly before Galileo's discovery of the law of fall. Folio 115v4. Calculation after discovery of fall law, and after first entry on f. 189v2 (Plate 19). Time for pendulum of length 27,834 *punti* was taken at 235 *tempi*. Folio 115v6. Summary of adjusted findings; see text. Reproduced by permission of Ministero per i Beni e le Attività Culturali, courtesy of Biblioteca Nazionale Centrale, Florence.

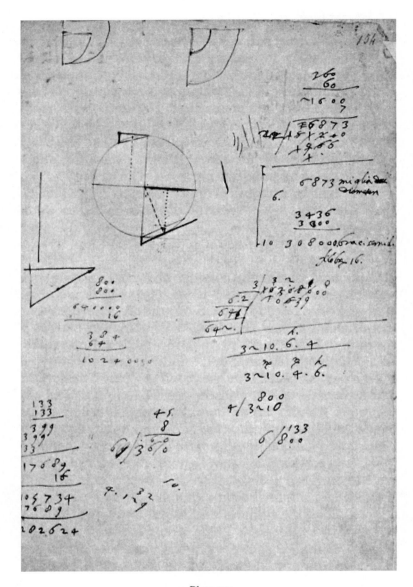

Plate 21:
Folio 154r, Ms. Galileiano 72. Note *filo br. 16* at right centre, first entry on the page.
After confirming the general pendulum law, Galileo here calculated for a pendulum as
long as Earth's radius. Reproduced by permission of Ministero per i Beni e le Attività
Culturali, courtesy of Biblioteca Nazionale Centrale, Florence.

8

The Uniform Motion Equivalent to a Uniformly Accelerated Motion from Rest (Galileo Gleanings XX)

I

Theorem I on accelerated motion in Galileo's *Two New Sciences* is generally regarded now as his version of the medieval mean-degree theorem, or Merton rule.[1] This view is acceptable, but in quite a different sense from that in which it is usually meant: it is equivalent in result but not in concept or method of proof. The matter is of prime importance to deciding Galileo's relationship to medieval physics, to an understanding of the basis that underlay his analysis of free fall in his most mature work, and to the mathematical ideas of his disciple Buonaventura Cavalieri. It is therefore deserving of careful study. The analysis set forth herein has not, so far as I know, been put in the field before. It is in effect a continuation of my recent reconstruction of Galileo's steps in arriving at his rejection of space-proportionality of velocities in free fall.[2]

I have previously noted that the word *mean* in the Crew and De Salvio translation of Theorem I is a completely unjustified insertion.[3] Where Galileo said *motu aequabili ... cuius velocitatis gradus subduplus sit ad summum et ultimum gradum velocitatis*, "in uniform motion ... whose degree of speed is as one-half to the highest and last degree of speed," they have: "a uniform speed whose value is the mean of the highest speed and the speed just before acceleration began." This error of translation has doubtless affected the ideas of many English and American historians of science. Alexandre Koyré, in his influential *Études Galiléennes*, rendered it thus: *un mouvement uniforme et dont la vitesse est le degré moyen entre les degrés maximum et minimum dudit mouvement uniformement accéléré.*[4] Besides incorporating the word *moyen*, this departs still

Reprinted from *Isis* 63 (1972), 28–38, by permission of the University of Chicago Press. Copyright 1972 by the History of Science Society, Inc.

further from Galileo's text, and has doubtless influenced many French and Italian historians.

Half the final speed is conceptually quite a different matter from the mean speed in accelerated motion, even though the values coincide for motion starting from rest. The medieval concept of mean speed was that of a single value chosen from the infinite set of instantaneous speeds which could be taken to represent the motion as a whole. Galileo never made use of that concept in any of his works known to me, and I cannot recall that he ever used the word *mean* or any of its equivalents, though he may have done so apart from his discussions of accelerated motion. What his concept was, and how it developed by stages into Theorem I, is the chief subject of this paper.

Marshall Clagett has given a literal translation of the theorem with which we may begin.[5] I have added a few of the Latin words in brackets to facilitate the ensuing discussion; those in parentheses are his.

The time in which a certain space is traversed by a moving body uniformly accelerated from rest is equal to the time in which the same space would be traversed by the same body travelling with a uniform speed (*motu aequabili*) whose degree of velocity (*velocitatis gradus*) is one-half [*subduplus*] of the maximum, final (*summum et ultimum*) degree of velocity of the original uniformly accelerated motion.

Let there be represented (*repraesentetur*) by extension (*extensio*) AB the time in which the space CD is traversed by a moving body accelerated from rest at point C. The maximum and last degree of velocity in the instants of time AB we let be represented by EB, constructed [*utcunque constitutam*] on AB. And, with AE drawn, all the lines drawn from the individual points on the line AB and parallel to BE will represent the increases [*crescentes*] of the degree of velocity after instant A. Then I bisect BE at F. Parallelogram AGFB will be formed of the parallel lines FG, BA and AG, BF. It will be equal to the triangle AEB, its side GF dividing AE into equal parts at I.

For if [*quodsi*] the parallel lines of triangle AEB are extended up to IG, the aggregate of all the parallel lines contained in the quadrilateral will be equal to the aggregate of those contained in triangle AEB, for those in triangle IEF are equal [*pares sunt*] to those contained in triangle GIA, while those which are in the trapezium AIFB are common. Since each and every [*singulis et omnibus*] instant of time corresponds to each and every [*singula et omnia*][6] point on line AB, and since the parallel lines drawn from these points and included in triangle AEB represent the growing [*crescentes*] degrees of the increasing [*adauctae*] velocity, while the parallel lines contained within the rectangle represent in the same way just as many [*totidem*] degrees of non-increasing but uniform velocity, it appears that there are just as many [*totidem*] moments of velocity (*momenta velocitatis*) consumed [*absumptae*][7] in the accelerated motion represented by the growing parallel lines of triangle AEB as in the uniform motion represented by the parallel lines of

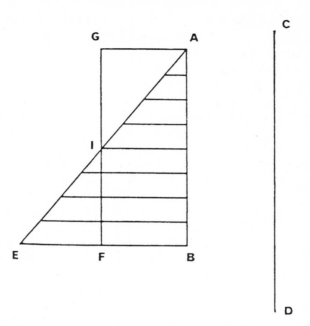

GB. For [*quod enim*] the deficiency of velocity moments in the first half of the accelerated motion – the deficient moments being represented by the parallel lines of triangle
AGI [*deficiunt enim momenta per parallelas trianguli AGI repraesentata*] – is compen-
sated for [*reficitur*] by the moments represented by the parall lines of triangle *IEF*.

It is obvious therefore that equal spaces will be traversed in the same time [*aequalia
futura esse spatia tempore eodem*] by two moving bodies, one of which is moved with a
motion uniformly accelerated from rest, while the other is moved with a uniform motion
having a moment half of the moment of the maximum velocity of the accelerated
motion. Q.E.D.

A reader acquainted with current histories of science may feel dissatisfied
with this proof even in its correct translation. It might easily seem to him that
historians are right who say that Galileo has merely established in two different
ways that the areas of the triangle and the rectangle are equal, using first an
obvious geometric property (mutual bisection at *I*), and then reinforcing this
with a discussion of the velocities represented by parallel lines drawn in the
large triangle, the two smaller triangles, and the rectangle. Then at the end he
repeats the physical relation of space, time, and speed that was stated in the theorem, as if somehow that relation had been proved by the relation of areas seen
as sums of lines.[8]

If that were indeed all there was to Galileo's proof, I should be the first to condemn it, as I have criticized on similar grounds the celebrated demonstration by Jordanus of equilibrium on inclined planes.[9] But there is much more than that to it. I shall speak presently of the reasons for which historians generally regard it in the above way and associate it with the Merton rule, in agreement with various vernacular translations. The reason that the proof may appear repetitious and inconclusive even in Clagett's literal translation lies in the phrase "For if ..." which precedes the diagram above. These words suggest that Galileo intends to support further something that he has already proved sufficiently for any geometer by pointing out the bisection of *AE* by *BF* at *I*; that is, the equality of areas. But Galileo's word *quodsi* should, in my opinion, be read as "Now if ... ," as introducing a new thought.[10] Having already established the equality of areas, he now asserts that the aggregates of parallel lines contained within them are thereby matched (*pares sunt*, not *aequales*, though I have let Clagett's choice of words here stand). These aggregates of parallel lines, however, are not to be thought of as sums, despite the use of that word in the German translation,[11] that of Crew and De Salvio, and the new French translation.[12] Aggregates here are sets, assemblages, or collections; Koyré rightly supplied the word *ensemble*[13] and Carugo-Geymonat the Italian *insieme*.[14] The concept of comparing matched sets is made perfectly clear by Galileo in what follows.

The mathematical part of the demonstration concluded, Galileo turns to the physical part. Each instant of time and all instants of time are first related to each point and all points on line *AB*. "Each and every" is a valid translation, but I think that this phrase is now generally used redundantly, for emphasis, as "each one and every one, individually," whereas Galileo's Latin says "each single one and all taken collectively." He does not identify the latter with an area. Next, the same number of speeds is shown to exist in the uniform motion as in the accelerated motion. Finally (and here I would read *quod enim* as "Now indeed," and the phrase in brackets as "and in fact they do fall short by ...") he matches the paired lines by defect and excess, using an implied recurrence formula for the individual and collective lines in the smaller triangles. The theorem can then be, and is, restated as proved.

The thrust of Galileo's argument may be reasonably paraphrased in this way:

Now, you may think that in uniform motion there is just one single velocity, while in accelerated motion there are many velocities; in fact, an infinitude of them. But I say that in both cases there is an infinitude of velocities, the only difference being that in uniform motion they are all equal, whereas in accelerated motion they are all different. I proved this by representing instants of time by points on a line, and velocities at these points by lines through them, so that for every instant there is one and only one velocity in uniform motion, and one and only one velocity in accelerated motion. And just as we have paired off these velocities, we pair off these excesses and defects with respect to one another; and indeed the velocities in uniform motion represented by lines in triangle *AGI* are lacking from the accelerated motion, but they are exactly made up by those in *IEF*. Hence the spaces covered in the same time *AB* are the same for both motions, the velocities singly and collectively being equivalent; so these spaces are both represented by *CD*, which was drawn to represent the space covered in accelerated motion. Q.E.D.

II

In neither the mathematical nor the physical part of the proof is there any reference to areas or surfaces as such. Everything is handled by reference to the collections of lines included respectively in the triangle and the rectangle, and the purpose served by those figures is that of limiting the lengths of lines; that is, it is the boundaries of the figures rather than their areas that concern Galileo in his proof. In *Two New Sciences* no physical meaning is suggested for the areas of the figures.

The case is different in the *Dialogue*, in a passage which probably represents a slightly earlier stage of Galileo's thought, discussed below. But the development of his thought was not toward a kind of integration, deriving areas from parallel lines, as most historians believe; it was away from such an idea, toward a technique of dealing with infinite sets that is found not only in this isolated theorem, but in the First Day of the same book and in the Third Day for rejection of space-proportionality. Since I have discussed this development elsewhere,[15] I shall not repeat it here, but its emergence in several places in the *Two New Sciences* should no longer be ignored by historians, particularly historians of mathematics. Of its relation to the geometry of Cavalieri I shall speak later.

The impression prevailing among historians that Galileo was trying to sum his parallel lines rather than to keep them separated and orderly for comparison in sets is, I believe, largely the result of errors in vernacular translations of the proof of Theorem I, quite apart from errors in the translation of its statement already mentioned. The translation of *aggregatum* as "sum" is one such error, found in English, German, and French. It is absent from the translations of Salusbury in 1665, Weston in 1730, and from Koyré's translation, as well as that of Carugo and Geymonat. Nevertheless, Koyré linked Galileo's proof with medieval area proofs, as Clagett also seems to do in his brief commentary, while Carugo-Geymonat follows Koyré in all essentials. Thus it appears that ideas of historical relationships have been more influential in the interpretation of this theorem than has a close attention to Galileo's words, since the same conclusions are reached from correct and incorrect translations. This seems to me a risky business, unless we are prepared also to reinterpret medieval writers with every new interpretation of Galileos texts, and that seems to me not to be writing history at all. It puts historians in a class with astrological prophets, of whom Galileo remarked that it is easy to see the correctness of their predictions, but only after the events have occurred.

The other principal error in the Crew and De Salvio translation of this proof was that it completely ignored the two occurrences of the word *totidem* – "just as many as." This phrase, essential to Galileo's proof by correspondence, was given by Salusbury and Weston, who had no theories about what Galileo ought to mean and hence let him speak for himself. It was given in the German the first time as *ebensoviel*, but left out the second time, nullifying its force in the physical argument. Koyré gave it both times as *autant*, but ignored it in his commentary. The more recent French translation gives it only once, and where it should appear again we find *somme*, the word *totidem* perhaps having been misread here as *totum*. Carugo and Geymonat of course give the correct Italian, *altrettanti*, both times, just as Clagett gives the correct English. But again the role of this word is equally ignored in all the commentaries. It is similarly ignored by all who assume that Galileo had in mind summing the areas from lines.

As remarked earlier, Galileo did have such an idea when he wrote the passage on uniform acceleration and free fall in the *Dialogue*, probably some time between 1624 and 1630. But there, where he did assign a physical meaning to the areas, it was not at all the meaning of "space traversed" as Koyré believed was intended in Theorem I. Koyré says:

Nothing is more curious than the figures with which Galileo accompanies his demonstration.[16] He seems to be conscious how his manner of representing the space traversed, the trajectory of movement – that is, a *line*, by a *surface*, is hardly natural, and how easily this mode of representation might lead us to geometrization a priori, an error that he himself had committed. It would also have to be possible to represent the trajectory by a line. But Galileo did not know how to do this. Thus he was limited to drawing a line alongside his diagram and without any relation whatever to it.[17]

Thus Koyré believed that Galileo was not serious about the line *CD* and really meant the areas to represent spaces traversed, no matter what he said. But Galileo was perfectly serious, as shown by his abandonment of any physical meaning whatever for the areas. The meaning he had previously assigned to them in the *Dialogue* was that of "total speed," a troublesome concept that had its origin as far back as his attempted demonstration of the times-squared law in 1604.

In the *Dialogue* Galileo was concerned only with the rule that states that a body continuing in uniform motion after acceleration from rest will travel twice the distance in the same time that it underwent acceleration. This of course follows directly from the theorem we have been discussing, but for his purposes in the *Dialogue* he offered instead a justification for the rule directly. This was done by considering a triangle for accelerated motion and comparing it with a rectangle on the same base. He then overtly equated the totality of lines with the area, as he is said to have done again, but did not do, in the *Two New Sciences*. Here is the passage, with the distinctive parts italicized:

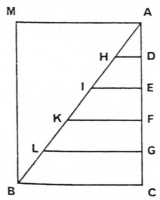

To represent the infinitude of degrees of speed that come before the degree *DH*, there must be understood to be an infinitude of lines ... drawn through the infinity of points of the line *DA* and parallel to *DH*. *This infinitude of lines is ultimately represented here by the surface of the triangle AHD*. Thus we may understand that whatever space is traversed by the moving body with a motion that begins from rest and continues uniformly accelerating, it has consumed[18] and made use of an infinitude of degrees of increasing speed corresponding to the infinitude of lines ... starting from the point *A* ... .

Now let us complete the parallelogram *AMBC* and extend to its side *BM* not only the parallels marked in this triangle, but the infinitude from all points on the side *AC*. And just as *BC* was the maximum of the infinitude [of lines] in the triangle, representing the highest degree of speed acquired ... in accelerated motion, while *the whole surface of the triangle was the sum total of all the speeds* with which such distance was traversed in the time *AC*, so the parallelogram becomes the *total and aggregate* of just as many degrees of speed but with each of them equal to the maximum, *BC*. This *total of speeds* is double that of the *total of increasing speeds* in the triangle, just as the parallelogram is double the triangle. And therefore if the falling body makes use of the accelerated degrees of speed conforming to the triangle *ABC* and has passed over a certain space in a certain time, it is indeed *reasonable and probable* that by making use of the uniform velocities corresponding to the parallelogram it would pass with uniform motion during the same time through double the space which it passed with the accelerated motion.[19]

Whether or not this is "reasonable and probable" depends largely on whether we wish to accept an intuitive and undefined equivalence between a sum of infinitely many finite line segments and the area of a plane figure. We balk at the idea of a sum if we think of it as necessarily equivalent to laying infinitely many line segments end to end, in which case the result could not be half or double another such construction. The same objection need not apply to lines of no breadth, seen as "stacked up," so to speak. Galileo obviously had this in mind, and in the *Dialogue* he did not scruple about its intuitive aspect, though he was careful to call it only reasonable and not proved.

For a direct proof of the double-distance rule, or any other comparison of nonequivalent uniform and accelerated motion I believe that either some concept of mean speed or some concept of total, effective, or compound speed had to be introduced overtly or by implication. For the identification of exactly equivalent uniform and accelerated motions, on the other hand, such a concept can be avoided. In the *Two New Sciences* the double-distance rule appears only in a discussion following the scholium to Problem IX, where it is derived independently by "returning to the same line of argument which was employed in the proof of the first proposition"; that is, by the parallels in a triangle and rectangle on the same base. The language employed in this derivation is rather like that of the *Dialogue*, however, for Galileo speaks of the parallels *ex quibus, cum infinitae sint, ... exurget superficies ipsa trianguli*, and *ex talibus gradibus conflabitur aggregatum consimile parallelogramo*. Unusual terms like *exurget* ("is squeezed out"?) and *conflabitur* ("is melted together"?) make the sense a little vague, but however these passages are translated, they more resemble the treatment of areas in the *Dialogue* than the use of one-to-one correspondence in Theorem I. (A reference in this scholium to *CD* should read *CT*, as in the 1638 edition, and applies to the preceding diagram, as noted by Weston in 1730.)

Since the existence of equivalent motions had been established in Theorem I, it would have been easy to derive the double-distance rule from it directly, without going back to the area concept. This is a further reason for supposing the above passage to have been written before Theorem I reached its final state.

In any event, to achieve the analysis of free fall without employing either the mean-speed or total-speed concept as such was a great advance, since either concept raises in one way or another the thorny question of the existence or nonexistence of a first instant of motion. This question had plagued Galileo's early researches in the form of a mistakenly assumed equivalence of space-proportionality and time-proportionality. Although he did not discuss it specifically, it can be retraced through certain crucial notes and diagrams.[20] The substitution of half the terminal velocity avoided the problem entirely, on which opposite sides seem to have been taken by various mean-speed mathematicians. One ingenious medieval solution to it deserves attention as illustrating the difference of approach.

First it may be mentioned that the usual medieval analysis of accelerated motion proceeded by considering deceleration first; there were Aristotlian reasons for this, and it led to some interesting conclusions about convergent series. One standard procedure was to find the mean of a given motion by bisection and then to bisect successively toward zero degree, or rest. Diagrams in the 1494 edition of Heytesbury's *Regule solvendi sophismata*, reproduced by Clagett,[21] illustrate this. Elsewhere in the same book we find the statement:

Since any degree of velocity whatever differs by a finite amount from zero velocity ... which is rest, therefore any body may be uniformly accelerated from rest to any assigned degree of velocity, and likewise it may be decelerated uniformly from any assigned velocity to rest.[22]

There could scarcely be a better illustration of the problems of infinity, the discrete, and the continuous. Why does Heytesbury not feel that the very fact adduced as a premise operates against the idea of a *uniform* acceleration from *rest*, instead of in favor of it? Granted the tiniest discontinuity in passing from zero velocity to some finite initial speed, the trouble vanishes; but how is he to get around that initial nonuniformity, the appearance of some finite velocity after nothing? Just making it smaller will not help in principle, so long as any discontinuity exists at all, and this seems to be assured by his premise. If he had said that there are degrees of speed that differ only by an infinitesimal amount from zero, we wold not feel this difficulty, but he seems to have closed this road in his premise. He appears to be confronted with the problem of applying the mathematics of the discrete to a problem of the physically continuous. How was this done?

It was done quite simply. Heytesbury postulated that each increment in uniform acceleration corresponds to its *mean*.[23] Now, by repeated bisections toward zero degree, new means are always created between each degree and zero; thus one is assured that there will be no "first instant of motion" or "first velocity after zero" that would involve a discontinuity. This manoeuvre amounts to an introduction of the concepts of limit and of infinitesimals that seemed to us to have been cast out in the premise. For Heytesbury is saying in this way that velocities in uniform acceleration may be found that are smaller than any previously assigned quantity, however small. It was only his word *finite* that perplexed us into thinking that he required some quantum jump from zero to a finite velocity.

Heytesbury offered a proof that the mean degree in uniform acceleration is exactly one-half the terminal degree of speed; Galileo's Theorem I circumvents any reference to mean degree or "total speed." Heytesbury's proof depends on the application of continued proportion to the division of time; Galileo's, on the equipartition of time, so far as his diagram is concerned, and on the consideration of all instants in his proof by correspondence. Heytesbury sums the power series $(1/2)^n$ for $n > 2$ and shows this to equal $1/2$; Galileo sums no mathematical or physical elements of any kind. Heytesbury concludes his demonstration in this way: "This is not only true of the latitude of velocity of motion commencing from zero degree, but it could be proved and argued in just the same way in the case of latitudes of heat, cold, light, and other such qualities."[24] Galileo restricted his conclusion to local motion. It seems to me implausible that these wide conceptual differences arose from Galileo's study of medieval physics; rather, it appears that he either neglected it or thoroughly opposed its approach.

Medieval writers seem to have considered uniform acceleration as essentially replaceable by a rapid succession of uniform motions all differing in value. This was not Galileo's concept. Uniform acceleration is conceptually related, however, to Buridan's accumulating increments of impetus in explanation of acceleration. That concept is found among many of Galileo's contemporaries: Benedetti, Beeckman, Descartes, and Baliani. But it is not so much as mentioned by Galileo, who seems to lump it with all other fantasies relating to the cause of acceleration. It is precisely in causal investigations of nature that one encounters the perplexities of a last instant of rest, a first instant of motion, and the occurrence of discontinuity or uniform change in nature.

These puzzles are of the kind that most medieval physicists delighted to explore, as do modern physicists in their way. Out of them have come all sorts of ingenious mathematics and useful physics. My point is simply that Galileo did not delight in this kind of puzzle as the content of physics but eliminated it wherever he could from his physical demonstrations. I think that this is seen in the evolution of Theorem I from 1604 through the *Dialogue* to the *Two New Sciences*. In place of examining the question of the existence of a first instant of motion, we find him saying only (and proving as early as 1590) that there can be no velocity in acceleration that endures for any time whatever, and treating rest as "the infinite degree of slowness." Rest and motion are not disparate things in Galileo's physics; bodies are indifferent to both. Speculations on the relation of the discrete and the continuous, on finite and infinite, are relegated to the discursive parts of his last book and excluded from the mathematical demonstrations, which he separated in its very title. This is not the medieval pattern, in which one was part and parcel of the other.

Heytesbury's mathematics is admirable. That it has nothing to do with Galileo's treatment of uniform acceleration does not mean that Galileo was ignorant of the limit concept in mathematics or that he did not apply it elsewhere in his physics. For example, in his youthful *De motu* and *Mechanics* he demonstrated mathematically that a frictionless body on a perfect plane could be moved, in principle, by a force smaller than any previously assigned force. [25] He took the mathematical concept from Archimedes, and it is unlikely that he ever forgot or discarded this. The reasons why he did not apply it in the analysis of acceleration, as Heytesbury did, are worthy of study. Similarly, Swineshead had the idea of analysis by one-to-one correspondence between line segments of different length and made a certain use of it in deriving the 3:1 ratio for spaces traversed in the first and second half of motion (the order is that of deceleration), but he did not use it to reject space-proportionality or to derive the times-squared law, as Galileo did.[26] These are things that seem to me to be more important to the history of science than some identical results derived from different conceptual frameworks and analyses, by which we are presently beguiled by historians.

Galileo concentrated his attention on macroscopic phenomena of no great size, so far as his physics was concerned. We are concerned with subatomic particles and distant galaxies; medieval writers were concerned with the knowledge of change in general through its causes in ultimate principles. Galileo scorned such researches; I think he would have included them with what he called "little flowers of rhetoric that are followed by no fruit at all." That it would be wrong for us to say this today does not mean that it was not right for Galileo to say it in his day. Medieval physics had reached the point of no return long before his time, but it would not remain there forever. Curtis Wilson and William Wallace are among those who have clearly shown the relevance of medieval science to modern science. But despite a wealth of efforts, no one has shown its relevance to Galilean science. It is important always to remember that Galileo's physics was neither medieval nor modern, and especially that it had no followers. Physics after the death of Galileo was Cartesian, not Galilean, and then it was Newtonian; the ideas of cause, essence, and principle that Galileo had excluded from physics were back in 1644 – to stay.

Galileo's choice of a physically real and measurable velocity – one-half the terminal speed – in place of a theoretical construct, the instantaneous middle speed, as the basis for Theorem I is not without significance for mathematical physics. If the medieval concept of a mean speed was known to him, as it may very well have been, it seems to me more likely that he rejected it as unsuitable to his concept of physics than that he worked from it and merely sought ways of avoiding the word *mean*. I remember my own qualms when in my first course in physics *velocity* was defined as a ratio of space to time, $v = s/t$, and immediately afterward I was asked to accept the idea of instantaneous velocity, that is, to put $t = 0$, for I already knew that division by zero was an illicit operation, by means of which one could prove that $2 = 1$. Later I was supplied with the proper mathematics and the correct definitions. But that mathematics did not exist in 1638, and Galileo would have laughed at the idea of definition by a ratio of quantities different in kind. Likewise, I think, he would have distrusted, as a representative of all velocities in an accelerated motion, anything which could not be observed at all. Such was the middle speed. But the terminal speed can be maintained in uniform motion long enough to be observed and measured, and this allowed him to move from pure mathematics to the physically real. That one-half the terminal speed is numerically equal to the mean speed is a very superficial reason for thinking of Theorem I as equivalent to the medieval mean-speed theorem.

III

The relation of Galileo's method of proving Theorem I to Cavalieri's "method of indivisibles" will be touched on briefly in conclusion, a more thorough examination being put off for the future. Cavalieri's *Geometria indivisibilibus continuorum nova quadam ratione promota* of 1635 is often referred to in English as his *Geometry of Indivisibiles* (and in other modern languages by equivalent genitives). It is one thing to refer correctly to his "method of indivisibles," as he himself does on many occasions. But the idea that Cavalieri had in mind some kind of special "geometry of indivisibles," different from ordinary geometry, is quite absurd. His title means "geometry advanced in a certain new method [or perhaps "by a certain new ratio"] by indivisibles of continua." The method consists essentially in the comparison of aggregates (*congeries*) of lines or planes. Cavalieri's indivisibles are not those discussed by Bradwardine, as is often said; Bradwardine rightly rejected the idea of a line indivisible longitudinally, while Cavalieri's lines were correctly taken as indivisible in the sense of having no breadth or thickness. These he treated in aggregates of indefinite number but within bounded figures. His argument did not assume that areas were composed of these lines, though his contemporaries as well as ours attributed this idea to him. He summarized his position by remarking that a continuum is either made up of lines or of lines and something else. If the former, his ratios of aggregates were as valid as ratios of areas. If the latter, then he said it was fair to assume a uniform distribution of the "something else" that went into the continuum, in which case it would not affect his ratios of aggregates.[27]

Cavalieri and Galileo were in correspondence during the years in which the *Dialogue* and *Two New Sciences* were being composed. Cavalieri extended the use of one-to-one correspondence from the cases of equivalence, to which Galileo confined himself, to cases of ratio other than equality, along the lines foreshadowed by Swineshead in the passage cited previously. It may be that Galileo distrusted this extension, or it may be only that he saw no way to apply it directly to the diagram for the double-distance rule. The appearance of that rule in the scholium mentioned previously, and the form of proof given for it, incline me to believe that Galileo preferred not to extend the reasoning beyond cases of equivalence. If so, it was very likely because he believed that the idea of uniform distribution of whatever it is that divides the discrete from the continuous, however acceptable as a mathematical postulate, could not safely be carried over into physics. The discussions of light in the *Assayer* and of the constitution of matter in the First Day of the *Two New Sciences* appear to me to confirm this notion. A great deal remains to be done, however, before anything but tentative conclusions can be put forth about Galileo's conception of the relation of pure mathematics and physical reality.

NOTES

1 The Merton rule states that a uniform motion at the velocity occurring at the midpoint in time is equivalent to a uniformly accelerated motion from rest to a definite *terminus ad quem*.

2 "Galileo's 1604 Fragment on Falling Bodies," *British Journal for the History of Science*, 1967, *4*: 340–358; "Uniform Acceleration, Space, and Time," *Br. J. Hist. Sci.*, 1968, *5*: 21–43.

3 Stillman Drake, *Galileo Studies* (Ann Arbor: Univ. of Michigan Press, 1970), p. 222.

4 Alexandre Koyré, *Études Galiléenes* (Paris: Hermann, 1939), p. 142.

5 Marshall Clagett, *The Science of Mechanics in the Middle Ages* (Madison: Univ. of Wisconsin Press, 1959), pp. 409–410.

6 The order of phrases is here reversed from Clagett's translation to preserve the case endings.

7 The departure from Clagett's translation is made because the same idea of consumption or spending is present in the *Dialogue* passage (see below).

8 The final statement is in terms of equal spaces, whereas the theorem was given in terms of equal times. Kenneth Manning, in a seminar paper presented to Professor I.B. Cohen, disputed the validity of Galileo's proof on those grounds, and in oral discussion most members of the seminar regarded the conclusion as proved, but not the

original theorem. Substitution of the conclusion for the theorem would not have changed Galileo's procedure in any way; the lines *AB* and *CD* would be drawn first, and the comparison of motions would be made on the time-line by reason of the previous definition of uniform acceleration. But Galileo regarded space, time, and velocity as so related that any two being fixed, the third is determined; hence having shown the equivalence of velocities, he regarded it as a matter of indifference whether the times or the spaces where taken as equal in order to establish the equality of the other variable. A similar assumption lay behind his argument in rejecting space-proportionality of speeds in uniform acceleration; cf. Drake, *Galileo Studies*, p. 235. Modern criteria of rigor in proof may demand overt statement of such an assumption, but its omission does not appear to me to be a substantial defect. Since one motion here is uniform and the other is shown to be its equivalent, Galileo's final theorem on uniform motion, preceding this proof, probably appeared to him sufficient for the purpose.

9 Stillman Drake, "Mathematics, Astronomy and Physics in the Work of Galileo," *Art, Science, and History in the Renaissance*, ed. Charles Southward Singleton (Baltimore: Johns Hopkins Press, 1968), p. 312.

10 *Cassell's New Latin Dictionary* (New York: Funk & Wagnall's, 1959), p. 498: "*quod* (3): introducing a fresh sentence, *and, but, now*, etc., esp. foll. by *si. ...*"

11 Galileo, *Unterredungen und mathematische Demonstrationen von Galileo Galilei*, trans. Arthyr von Oettingen (Leipzig: W. Engelmann, 1891), *Dritter und Vierter Tag*, p. 21.

12 Galileo, *Discours et démonstrations mathématiques concernant deux sciences nouvelles*, trans. Maurice Clavelin (Paris: A. Colin, 1970), p. 140.

13 *Études Galiléenes*, p. 142.

14 Galileo, *Discorsi e dimostrazioni matematiche, intorno à due nouve scienze*, ed. Adriano Carugo and Ludovico Geymonat (Turin: P. Boringhieri, 1958), p. 193.

15 See "Uniform Acceleration, Space, and Time," loc. cit.

16 Actually Koyré presents an incomplete diagram, lacking the parallel lines on which Galileo's reasoning depends. This seems to be an attempt to stress the supposed integration of the areas, since he later comments surprisingly: "*le mouvement n'est plus subdivise en fragments* [i.e., as in the *Dialogue*] *mais pris, en quelque sort, en bloc*" (*Études Galiléenes*, p. 143).

17 Ibid., p. 141, note.

18 See n. 7 above. Galileo did not consider the body as having instantaneous velocities, but rather as consuming them.

19 Galileo, *Dialogue*, trans. Stillman Drake (Berkeley: Univ. of California Press, 1953 and later eds.), p. 229.

20 See "Uniform Acceleration, Space, and Time," esp. p. 27.

21 Clagett, *Mechanics in the Middle Ages*, facing p. 273.

22 Ibid., p. 237.

23 Ibid., p. 272.

24 Cf. ibid., pp. 271–272.

25 Galileo Galilei, *On Motion, and On Mechanics; Comprising De motu (ca. 1590)*, trans. with intro. and notes by I.E. Drabkin, and *Le meccaniche (ca. 1600)*, trans. with intro. and notes by Stillman Drake (Madison: Univ. of Wisconsin Press, 1960), pp. 66–67, 170–171, 172.

26 *Mechanics in the Middle Ages*, p. 291, and "Uniform Acceleration, Space, and Time," esp. pp. 34–36.

27 Buonaventura Cavalieri, *Geometria degli indivisibli*, ed. Lucio Lombardo-Radice (Turin: Unione Typografico, Editrice Torinese, 1966), p. 205 (scholium to Bk. II, Prop. 1). The incorrect genitive in this recent Italian title shows how the error persists despite the fact that Koyré had called attention to it in 1954. See Koyré, *Études d'histoire de la pensée scientifique* (Paris: Hermann, 1966), p. 298, n. 3.

9

Galileo's Discovery of the Law of Free Fall

The modern era in physics opened with the publication of Galileo's *Discourses on Two New Sciences* in 1638. It reported basic discoveries he had made 30 years earlier. In describing his book to a friend in January, 1639, the old and blind author dictated these words (here, of course, translated from the Italian): "I assume nothing but the definition of that motion of which I wish to treat and whose properties I then demonstrate ... . I declare that I wish to examine the essentials of the motion of a body that leaves from rest and goes with speed always increasing ... uniformly with the growth of time ... . I prove the spaces passed by such a body to be in the squared ratio of the times ... . I argue from supposition about motion defined in that manner, and hence even though the consequences might not correspond to the events of natural motion of falling heavy bodies, it would little matter to me ... . But in this, I may say, I have been lucky; for the motion of heavy bodies, and the properties thereof, correspond point by point to the properties demonstrated by me."

Not even Galileo's severest critics attribute his discovery of the law of free fall to sheer luck; hence it may seem odd that I, one of his most fervent admirers, should do so now. I believe Galileo meant his remark quite literally. Evidence for this belief exists in an early manuscript of his that has never before been published in full. That document unfolds a fascinating story of scientific discovery through a combination of error, good luck, persistence and mathematical ingenuity.

Historians of science have searched the writings of earlier men for the possible origins of Galileo's analysis of accelerated motion because no documentary

Reprinted from *Scientific American* 228 no. 5 (May, 1973), 84–92. Reprinted with permission.

evidence in his own hand had been found. Nothing surviving from classical antiquity offers a plausible source. In the 14th century, however, there were some very interesting developments in the application of mathematics to physical questions. In particular William Heytesbury and Richard Swineshead of Merton College, Oxford, and Nicole Oresme of Paris analyzed accelerated motion. The roots of medieval investigations lay in a theological problem – the increase of charity in a man – and in its philosophical implications, which spread into the general problem of rate of change. The results that were achieved are impressive, and it seems strange that their authors never thought to apply them to the problem of the changing speed of a freely falling object.

Medieval English writers adopted an arithmetical approach, from which they developed the mean-degree or mean-speed theorem, often called the Merton rule. Under this rule the speed at the middle instant was taken as being representative. Uniform motion at this mean speed over a fixed time was declared to be equivalent to uniformly accelerated motion from rest over that same time. It followed that in any uniformly accelerated motion from rest, one-fourth of the total distance was traversed in the first half of the time. This fact yielded the ratio 3:1 for the distances traversed in the later and earlier halves. Oresme proved the rule geometrically, and in another manuscript he extended the ratio to the progression 1, 3, 5, 7 and so on for equal times. Not even Oresme, however, connected uniform acceleration with free fall, nor did any medieval writer announce that the distances covered are proportional to the squares of the times, a fact that is deducible from the above progression. The prevailing view among historians of science was recently summarized by Edward Grant in his book *Physical Science in the Middle Ages:*

"Oresme's geometric proof and numerous arithmetic proofs of the mean-speed theorem were widely diffused in Europe during the fourteenth and fifteenth centuries, and were [then] especially popular in Italy. Through printed editions of the late fifteenth and early sixteenth centuries, it is quite likely that Galileo became reasonably familiar with them. He made the mean-speed theorem the first proposition of the Third Day of his *Discourses on Two New Sciences*, where it served as the foundation of the new science of motion."

Puzzles nevertheless remained. I again cite Grant's words: "The Mertonians arrived at a precise definition of uniform acceleration as a motion in which an equal increment of velocity is acquired in any whatever equal intervals of time, however large or small." Yet "Galileo, as late as 1604, mistakenly assumed that velocity is directly proportional to distance rather than to time, as he later came to realize." If we were to assume that the medieval writings were Galileo's source, it would be hard to explain why he accepted and extended the earlier results while rejecting the only definition on which they were based. Similarly, if he came on the medieval writings later in life, why did he still make no use of the Merton rule in his proof of the proposition mentioned, either in his notes or in his book?

I once suggested two possible alternative sources for Galileo's times-squared law. The first was that he might have discovered that spaces traversed in equal times follow the odd-number rule by roughly measuring the distance traveled by an object rolling down a gently inclined plane, using the first distance as his unit. I thought he might have made this discovery incidentally in testing an earlier (and mistaken) belief of his that acceleration is only a temporary event at the very beginning of motion.

Alternatively, it seemed to me that Galileo might have arrived at the odd-number rule by pure reasoning, as Christiaan Huygens did many years later. For example, suppose that acceleration adds an equal increment of distance in each equal time. Then in the sequence of numbers representing spaces the ratio of the first number to the second number must be the same as the ratio of the first two numbers to the second two numbers, which in turn must be the same as the ratio of the first three numbers to the second three numbers, and so on. Why must these equalities be preserved? Because we have arbitrarily chosen to use a certain unit of time, and we might have used its double or its triple instead.

Since the number representing each distance must be uniformly larger than the preceding one, the numbers will have to be in arithmetic progression. Does such a progression exist? It is certainly not the progression of consecutive integers. The ratio of the first two integers (1 and 2) is 1:2. The ratio of the first two integers added together (1 + 2) to the second two added together (3 + 4) is 3:7, which is clearly not the same as 1:2.

The progression of the odd integers alone, however, does follow the rule we have set up. First, each number is uniformly two greater than the preceding one (1 + 2 = 3, 3 + 2 = 5, 5 + 2 = 7, and so on). Second, the sum of the first two numbers (1 +3) has the ratio of 1:3 to the sum of the next two numbers (5 + 7), and 1 + 3 + 5 has the same ratio (1:3) to 7 + 9 + 11, and so on. Moreover, the progression of odd integers is the only arithmetic progression that meets these conditions, as Galileo pointed out in 1615 to the same friend to whom he was writing in 1639. (There are plenty of other number sequences, such as 1, 7, 19, 37, 61, 91 ... , that meet the ratio test above, but in no such sequence is each number uniformly greater than the preceding one.)

Alas! My two earlier suggestions must now be rejected along with the notion that Galileo got his idea for the law of free fall from medieval writings. The document I shall present shows no more trace of experimental evidence or of arithmetical reasoning than it does of the Merton mean-speed rule. Moreover, that document cannot be dated after October, 1604, when Galileo wrote to his friend Paolo Sarpi in Venice clearly stating the times-squared law and also

GALILEO'S CALCULATIONS	EXPLANATION
(1) 4 miles with 10 degrees of speed in 4 hours	Galileo first assumed that one more degree of speed is consumed in each mile of distance traversed: the speed is of one degree in the first mile, two in the second, three in the third and four in the fourth. That gave him 4 miles with 10 degrees (1 + 2 + 3 + 4) of speed consumed. The time elapsed was arbitrarily put at one hour for each different degree of speed.
(2) 9 miles with 15 degrees of speed in $5^?$ hours	$5^?$ means that the time was to be examined. Although Statement 2 was intended to follow from Statement 1, and the next increment of speed to be added was indeed five degrees, the number of miles to be added was one and not five. Had his intention been to add a greater increment of speed for each hour, he would have written "$9^?$ miles" instead of "$5^?$ hours," and Statement 1 would still have contradicted Statement 2. (An unsuccessful attempt to get a ratio of overall speeds appears in the top right corner of f. 152r. In it Galileo tried multiplying the ratios of times and distances together.)
(1a) 4 miles with 10 degrees of speed in 4^6 hours (3) 4 miles with 15 degrees of speed in 4 hours (3a) 8 miles in 8 hours at 15 degrees of speed	To get a ratio of distances and times Galileo substituted 6 hours for the 4 hours in Statement 1. With this change, in order to travel at 15 degrees of speed. Galileo noted the implication that if "15 degrees" is an overall rate, then in order to cover twice the distance (8 miles) it would take twice as long (8 hours), which would contradict Statement 2.
(4)	In an effort to resolve the contradiction Galileo drew a vertical line and first lettered it A, B and C. He divided the distance AB into four units; AC is nine units. D he added so that the distance AD would be the mean proportional between them at six units ($\sqrt{4} \times \sqrt{9} = 2 \times 3 = 6$). He let this represent the time taken in traversing AC and let AB represent the time taken in falling from A to B.
(5) through AB speed 10 through AC as . . . 15 distance / speed acquired / time through AB 4 / / AB — 4 / BE 20 / AD 6 / / AC 9 / CF 30 / AC — 6	Using the mean proportional AD, Galileo tabulated new times for the two original distances. A new working hypothesis had emerged in which the original time through AB at 10 degrees of speed was again 4 hours, but the time through AC at 15 degrees of speed had become 6 hours, the mean proportional of the distances from the rest. It happens that if two objects fall from rest through distances that have the ratio 4 : 9, their respective average speeds do have the ratio 10 : 15. In other words, like the times (4 hours and 6 hours), the average speeds have the ratio 2 : 3. The entries for BE and CF, made later, represent the acquired speeds at the end of the falls AB and AC and are exact doubles of the speeds previously assigned through those distances.
(6) As BA to AD, let DA be to AC, and let BE be the degree of speed at B; and as BA to AD, let BE be to CF; CF will be the degree of speed at C. And since as CA is to AD, so CF is to BE, then as the square of AC to the square of AD, so will be the square of CF to the square of BE; and further, since as the square of CA to the square of AD, so CA is to AB, the square of CF will be to the square of BE as CA is to AB. Therefore the points E and F are in a parabola passing through A	Galileo next drew the line AE and placed point E so that BE would represent the speed acquired at B. He wrote out his conclusion, assuming that the ratio of the acquired speeds would be the same as the ratio of the overall speeds through falls from rest. He expected that other termini of horizontal lines representing acquired speeds would fall on line AE, just as the terminus of BE did. When he calculated the placement of point F by the ratio he had developed in his conclusion (BA is to AD as BE is to CF), he found that F lay not on line AE but on a parabola through points A and E. The values BE 20 and CF 30 later added to his tabulation in Statement 5 would increase the horizontal scale of his diagram by about five to one as compared with the vertical scale.

GALILEO'S CALCULATIONS (*left column*) for the law of free fall are explained in detail (*right column*). Portions of Galileo's calculations in lighter type do not appear on *f. 152r* but are inserted for clarity. The term "degrees of speed" is an arbitrary one and is used in much the same way that doctors refer to degrees of burn injury. Galileo's conclusion is analyzed in the chart on page 255.

worked out a curious proof of it. In the letter he remarked that he had known the law for some time but had lacked any indubitable principle from which to prove it. Now, he believed, he had found a very plausible principle, which was borne out by his observations of pile drivers, namely that the speed of an object is proportional to the distance it has fallen. In the ordinary meaning of the words this statement is simply false.

The proof Galileo gave Sarpi has been a headache to historians of science for a long time. In his later *Discourses on Two New Sciences* Galileo correctly stated that the speed of a falling object is proportional to the time of fall. We could forgive him for having started with the wrong assumption and having found the right one before he had published his account. That does not, however, seem to be what happened. Apparently he had reached the assumption earlier and had then written out for Sarpi a demonstration based on the wrong one.

Even that is not the worst of the puzzle. Galileo candidly admitted in his last book that he had long believed it made no difference which assumption was chosen. Now, if Galileo ever actually believed such seeming nonsense, there should be among his notes at least one instance where he made some obvious mistake traceable to the wrong choice. Yet no such mistake is found, even in notes that can be positively dated to the long period during which he adhered to the proof composed for Sarpi. (Even 10 years later he had a version of that proof copied for use in a book on motion that he intended to publish.) Galileo was either pulling the reader's leg in 1638 by confessing to a pretended error, or we are missing some essential point in what he said in 1638 and have been misinterpreting his earlier demonstration concerning the pile drivers, although both seem so clear as to be incapable of being misunderstood.

The answer to all these puzzles now shows that Galileo was no less candid in 1638 than he had been ingenious in 1604. What we have been missing is at last revealed by a document designated *f. 152r* (Plate 6 with transcription in Plate 7) in Volume 72 of the Galilean manuscripts preserved at the National Central Library in Florence. I do not think anyone could have guessed the real answer. Furthermore, if anyone had guessed it, his conjecture would have been laughed at in the absence of a document in its support.

How could such a document have escaped notice for so long? All Galileo's notes were edited and published around the turn of this century. The distinguished editor, Antonio Favaro, omitted only those sheets (and portions of sheets) that contain nothing but diagrams and calculations whose meaning was uncertain. There are many such sheets; Galileo was a paper saver, and until the time of his death he kept many calculations he had made 40 years earlier. So when Favaro published *f. 152r*, he retained only two coherent sentences at the right center and bottom center of the sheet, together with a much modified version of the diagrams. These extracts meant little unless they could be dated. Favaro despaired of putting the 160 sheets of Volume 72 in the order of their composition after they had been chaotically bound together long ago.

Through the generosity of the John Simon Guggenheim Memorial Foundation and the University of Toronto, I was enabled to spend the first three months of 1972 studying the manuscripts in Florence. My purpose was to attempt a chronological arrangement in order to annotate a new English translation of the *Discourses on Two New Sciences*. Like Galileo, I was lucky. It turned out that the watermarks in the paper he had used provided an essential clue. He had lived at Padua until the middle of 1610 and then had moved back to Florence. It was therefore reasonable to expect that there would have been a change in his source of paper. An inspection of his dated correspondence revealed that there was no duplication of watermarks between the letters written from the two cities. Watermarks on the undated sheets thus separate the earlier ones from the later ones.

It also happens that 40 of the 160 sheets in Volume 72 were copies in the handwriting of two of Galileo's pupil-assistants in Florence. Every one of the 40 sheets shows the same watermark, one that also appears on Galileo's letters between 1615 and 1618. It is evident that the copies were made at his house under his direction in the writing of a book on motion. Most of the originals survive, and their watermarks confirm the Paduan origin of the theorems copied, whereas the copies supply some early theorems of which the originals are lost.

A probable order of Galileo's notes thus began to emerge. The watermark evidence sorted out Galileo's early work. Additional clues were provided by his handwriting. Finding such clues presented a harder problem to the skilled editor of 50 years ago than to the relative amateur of today. He could have compared samples of handwriting only by leafing back and forth in a bound volume, whereas I could work with a Xerox copy made from microfilm. I could therefore not only place sheets side by side but also cut them up for the closer comparison of individual entries.

At that stage Galileo's pages of calculations began to fit in order, enabling me to recognize the origin of the undistinguished-looking *f. 152r*. It is pretty certain that the document is Galileo's first attempt at the mathematics of acceleration. His train of thought has now been traced out, and it is summarized in the chart on page 251.

The first two lines on the sheet, written neatly across the top and centered, certainly have nothing to do with any actual experiment. The units – miles, hours and "degrees of speed" – are quite arbitrary. Acceleration in free fall was what basically interested Galileo, but his first step was to seek a general rule of proportionality for uniform growth of distances, times and speeds. In his first working hypothesis he assumed for the sake of argument that one degree of speed was gained for each unit of distance fallen. Thus he assumed that four miles were traversed with one degree of speed consumed in the first unit of distance, two in the second, three in the third and four in the fourth. This gave him four miles with 10 degrees ($1 + 2 + 3 + 4$) of speed consumed. He arbitrarily put one hour as the time elapsed for each different degree of speed. Consistent times for each speed (or distance) could not be, and did not need to be, determined in the initial working hypothesis.

Galileo next wrote: "9 miles with 15 of speed in 5$^{?}$ hours." The question mark is Galileo's, not mine; it indicates that this time to be examined, as we would today write "x hours." I mention the point because in the original manuscript the number looks more like a 1 than a 5, as 5's written in those days often do. This second statement was doubtless meant to be a part of the same working hypothesis as the first, Galileo's purpose being to have two different examples in order to compare ratios. In actuality the two statements are not even consistent. If the increment of speed to be added next was five degrees, bringing the total to 15, then the number of miles to be added was one, not five. I believe the ambiguity of the rule "One more degree of speed for each additional mile" tricked Galileo into adding five miles for the new five degrees of speed. The slip was careless, but it was not fatal; far from it. As James Joyce remarked, a man of genius makes no mistakes; his errors are portals of discovery.

Now that Galileo had two distances, two speeds and two times to work with, he proceeded to apply the Euclidean theory of proportion, which was the only device he trusted for the application of mathematics to physics. His first step was to reduce both hypothetical motions to the same speed in order to compare ratios of distances and times. Accordingly he wrote 6 above the 4 in the phrase "4 hours" of this first statement. In this way he redefined "10 [degrees] of speed" so that four miles at that speed would take six hours rather than four. In order to cover the four miles in four hours, then, a body would have to travel at 15 degrees of speed. Galileo noted this fact to the left of his original statements and continued with its implication: If the meaning of "15 degrees of speed" is "going 4 miles in 4 hours," then in order to cover eight miles at 15 degrees of speed the same body would take eight hours. (This would be true of any motion, however irregular, if the meaning of "speed" is fixed by total time and total distance.) That, however, immediately contradicted his earlier hypothesis that nine miles were traversed at 15 degrees of speed in only five hours.

Here it will be useful to recall that all Galileo was trying to do at this stage was to find a mathematically consistent rule for the use of the phrase "degrees

GALILEO'S CONCLUSION IN EUCLIDEAN RATIOS	GALILEO'S CONCLUSION REPHRASED IN MODERN TERMINOLOGY
	New definitions: s_1 denotes distance AB s_2 denotes distance AC $\sqrt{s_1 s_2}$ denotes AD, the mean proportional of AB, AC t_1 denotes the time AB for an object to fall through AB t_2 denotes the time AD for an object to fall through AC v_1 denotes the velocity (speed) BE acquired by an object at the end of the fall s_1 v_2 denotes the velocity (speed) CF acquired by an object at the end of the fall s_2
By construction of the diagram. $$AB : AD :: AD : AC.$$	$s_1/\sqrt{s_1 s_2} = \sqrt{s_1 s_2}/s_2$ by the definition of a mean proportional.
Let BE be the degree of speed at B. Then CF will be the degree of speed at C if we let (7a) $AB : AD :: BE : CF.$ When each of two ratios is equal to a third, they are equal to each other, so that $$AD : AC :: BE : CF.$$	Let v_1 be the velocity of an object at the end of s_1. Then v_2 is the velocity at the end of s_2 if we let (7b) $s_1/\sqrt{s_1 s_2} = v_1/v_2$. * From the above statements it can be said that $$\sqrt{s_1 s_2}/s_2 = v_1/v_2 .$$
It follows that if both sides are squared, $$(AD)^2 : (AC)^2 :: (BE)^2 : (CF)^2.$$	It follows that if both sides are squared, $$(\sqrt{s_1 s_2})^2/s_2{}^2 = v_1{}^2/v_2{}^2 .$$
Furthermore, since AD is the mean proportional of AB and AC and $$(AD)^2 : (AC)^2 :: AB : AC,$$ and since ratios equal to the same ratio are equal, then (8a) $(BE)^2 : (CF)^2 :: AB : AC.$	Furthermore, since $(\sqrt{s_1 s_2})^2/s_2{}^2 = s_1 s_2/s_2{}^2 = s_1/s_2$, then this statement with the previous one gives (8b) $v_1{}^2/v_2{}^2 = s_1/s_2 .$
Therefore the points E and F are on a parabola passing through A.	Although Galileo did not expressly mention time in his conclusion, he had tabulated the times t_1 and t_2 (AB and AD) through the falls s_1 and s_2 (AB and AC) in such a way that they exhibited the ratio $$t_1/t_2 = s_1/\sqrt{s_1 s_2} ,$$ and that is precisely how he habitually spoke of times and calculated with them. Thus it is not an exaggeration to say that a direct implication of Statement 7b is $$t_1/t_2 = v_1/v_2 .$$ This, together with Statement 8b, at once implies (9) $s_1/s_2 = t_1{}^2/t_2{}^2 .$ In other words, the ratio of the distances traversed by two falling objects is equal to the ratio of the squares of the times from rest. Therefore if the velocities, which are proportional to the times, are plotted against the distances on Galileo's original diagram, the result would be a parabola, as he said. (A parabola results whenever one variable is proportional to the square of the other.)

GALILEO'S CONCLUSION is expressed in his own terms of Euclidean ratios (*left column*) and is rephrased in modern notation (*right column*). Euclidean ratios, such as "$AB : AD :: AD : AC$," are read in the manner "AB is to AD as AD is to AC." The order of the quantities in some of his ratios has been changed for consistency when this does not interfere with his train of thought.

of speed." He was not yet concerned with attaching any physical meaning to the phrase, a task that would seem pretty pointless to him until its use in ratios was first made possible. In order for the phrase to be useful in ratios, any given over-all speed must carry the moving body through proportional distances in proportional times.

Faced with an apparent contradiction from a working hypothesis that he (mistakenly) thought he had expressed consistently in his first two statements, Galileo did not go back to see whether or not he had made an error. If he had discovered his error and corrected it, he would have ended up with the consistent but useless formula "4 miles at 10 degrees of speed in 4 hours; 5 miles at 15 degrees of speed in 5 hours." Such a formula would have equalized the ratios of distances and of times under acceleration, contradicting good sense and the basic idea of acceleration itself, and it would have told him nothing at all about ratios of speeds. In any case he did not check back. Instead he looked for the source of the trouble by drawing a vertical line and lettering it A, B and C to represent distances from rest. He indicated four units from A to B. The distance from B to C is supposed to be five units, making AC equal to nine units, as is shown in his tabulation.

Quite by accident, in my opinion, the two distances in Galileo's working hypothesis were both square numbers: 4 and 9. If Galileo had had any previous inkling of a rule involving squares and square roots when he wrote his first two statements, he would surely have put in the numbers 2 and 3, either for the speeds or for the times. I have already noted three ways in which he might have suspected that a square-root law existed, and there may be still other ways. His choice of numbers here, however, seems to me to exclude all of them. I believe that he first discovered the squaring relation in precisely this search for consistent ratios, and that he confirmed it afterward by experimental test.

It is the fact that 4 and 9 both happen to be square numbers that accounts for Galileo's addition of point D between B and C in his diagram. The distance AD is equal to six units. So far Galileo's problem had been one of conflicting ratios, and this was a difficulty that could be forever eliminated by introducing continued proportion. To the mind of any mathematician of the time, continued proportion was immediately suggested by the squares of two integers. Between any two such numbers there is always an integral mean proportional that is the product of the two square roots. The ratio of the lesser square to the mean proportional is then equal to the ratio of the mean proportional to the greater square.

In Galileo's case the mean proportional is equal to the product of the square root of 4 and the square root of 9, that is, the product of 2 and 3, which is 6. The ratio 4:6 is equal to the ratio 6:9, that is 2:3. Galileo entered point D on his vertical line six units distant from *A* just because it created a continued proportion. It then occurred to him to use the mean proportional to solve the puzzle of the time ratios. If the distance *AB* represented the time in his first statement (4 hours), then *AD* (6) represented the time in the second statement (originally 5? hours). From the mean proportional a new working hypothesis had emerged: The original time through the shorter distance *AB* at 10 degrees of speed was again four hours, but now the time through the longer distance *AC* at 15 degrees of speed had become six hours, the mean proportional of the distances from rest.

From that day forward Galileo had in his hands the principal analytical device that he applied in all his reasoning about free fall. There is no logic to the foregoing procedure except for the logic of discovery. Galileo perceived that this arrangement of numbers would preserve a consistency of ratios, and that was the necessary first move. Whether or not the rule would also agree with observable facts he left until later; in this, as he later remarked, he had been lucky.

I have said that he was just lucky in having two square numbers to start with. If that is so, can anything else be said of the ratio 10:15 chosen for the speeds assigned to those distances at the outset of the inquiry? When two objects fall from rest through distances that have the ratio 4:9, their respective average speeds do have the ratio 10:15. In other words, the speeds, like the times, are in the ratio 2:3. If, however, Galileo had started with any other two numbers, even two squares other than 4 and 9, then neither the ratio 10:15 nor the ratio of numbers obtained by adding up "degrees of speed consumed" would have agreed with the ratio of the smaller number to the mean proportional between the two numbers selected.

Two such lucky coincidences may strain the reader's credulity. All I can say is that I have not succeeded in finding any better – or indeed any other – reconstruction that will account reasonably well for all the entries on *f. 152r*. Moreover, coincidences among very small numbers are not as improbable as they may seem.

Another line from point *A* stretches off to the left at an angle from the original vertical line. Galileo probably intended it to represent the speeds acquired in acceleration. These speeds at the end of the falls *AB* and *AC* happen to be exact doubles of the speeds properly assigned *through* those distances. Galileo seems to have not yet been aware of that relation when he placed point *E* on his slanted line so that the distance *BE* would represent the speed acquired at *B* and wrote out his long conclusion. He did assume, however, that the ratio of acquired speeds would be the same as the ratio of his "overall" speeds through falls from rest. He apparently drew the slanted line in the expectation that other termini of horizontal lines representing acquired speeds would fall on it just as the termi-

COMPARISON OF SARPI PROOF WITH PROOF IN f. $152r$

Galileo arranged matters so that the points representing speeds on his diagram would fall on a straight line rather than on a parabola. All he had to do was replace his original ratio $v_1{}^2/v_2{}^2 = s_1/s_2$ with the ratio $V_1/V_2 = s_1/s_2$, where $V_1 = v_1{}^2$ and $V_2 = v_2{}^2$. In his original proof of the law of free fall for Paolo Sarpi, Galileo often applied the phrase "degrees of speed" to speed acquired at a point but not to speed through a distance. He considered speed through a distance in the Sarpi proof as being proportional to V^2 (instead of proportional to v, as he had done in f. $152r$). Thus he gave the phrase two senses: In f. $152r$ "speed through a distance" was an overall speed, a kind of average treated as being proportional to terminal speed. It is denoted w below. In the Sarpi proof "speed through a distance," which is denoted W, represents Galileo's provisional idea of a "total speed," for which we have no word or concept.

Using these convenient denotations (as well as the previous ones s, t and v), Galileo's two treatments can be displayed along with their modern counterpart:

f. $152r$	Modern usage	Proof for Sarpi
$v_1/v_2 = w_1/w_2$	$v_1/v_2 = t_1/t_2$	$V_1/V_2 = s_1/s_2$
$w_1/w_2 = t_1/t_2$	$s_1/s_2 = v_1 t_1/v_2 t_2$	$W_1/W_2 = V_1{}^2/V_2{}^2$
$(8b)$ $s_1/s_2 = v_1{}^2/v_2{}^2$	$s_1/s_2 = v_1{}^2/v_2{}^2$	W "in contrary proportionality to t," that is $\sqrt{W_1}/\sqrt{W_2} = t_1{}^2/t_2{}^2$
(9) $s_1/s_2 = t_1{}^2/t_2{}^2$	$s_1/s_2 = t_1{}^2/t_2{}^2$	$s_1/s_2 = t_1{}^2/t_2{}^2$

Galileo's choice of $V_1/V_2 = s_1/s_2$ in the Sarpi proof does not mean that he had to reject $v_1{}^2/v_2{}^2 = s_1/s_2$ (Statement $8b$). It did mean, however, that velocitá, defined for Sarpi as being proportional to distances traversed, could not be the same physical entity as the "speed" that had been implicitly defined as being proportional to time (the times-squared law) in f. $152r$. The basic relations that can be derived from Galileo's two treatments are:

For f. $152r$: $v_1/v_2 = t_1/t_2$ $v_1{}^2/v_2{}^2 = s_1/s_2$ $s_1/s_2 = t_1{}^2/t_2{}^2$

For Sarpi: $V_1/V_2 = t_1{}^2/t_2{}^2$ $V_1/V_2 = s_1/s_2$ $s_1/s_2 = t_1{}^2/t_2{}^2$

Galileo sought a proof for only the third and last relation, which remained unchanged by his new definition of "velocity." He believed that he could derive this law as well from his new definition $V_1/V_2 = s_1/s_2$ as from the old one $v_1/v_2 = t_1/t_2$, and he did. Later he found an experimental reason for adopting only the latter (older) definition of "velocities" in free fall.

"Contrary proportionality" in the proof for Sarpi relates the root of one variable to the square of the other. Since Galileo concluded from $W_1/W_2 = s_1{}^2/s_2{}^2$ that $s_1/s_2 = t_1{}^2/t_2{}^2$ by invoking "contrary proportionality," then his phrase meant that $\sqrt{W_1}/\sqrt{W_2} = t_1{}^2/t_2{}^2$. This relation is not derivable from f. $152r$ alone, since there $w_1/w_2 = v_1/v_2 = t_1/t_2 = \sqrt{s_1}/\sqrt{s_2}$. Yet the relation does arise from f. $152r$ and the proof for Sarpi taken together, that is, from the simultaneous assumption that $W_1/W_2 = V_1{}^2/V_2{}^2$ and $w_1/w_2 = v_1/v_2$. Hence it was only if Galileo had already perceived the fact that speeds in free fall are proportional to times $(v_1/v_2 = t_1/t_2)$, and retained that knowledge when he was writing the proof for Sarpi, that he could validly invoke "contrary proportionality." Only on this same basis can we understand the fact that in all his notes and books Galileo never made use of the assumption that our ordinary velocity is proportional to distance $(v_1/v_2 = s_1/s_2)$, as he has often been accused of doing.

PROOF OF THE LAW OF FREE FALL that Galileo wrote in a letter to his friend Paolo Sarpi can be analyzed in modern terms. The quantities s_1, s_2, v_1, v_2, t_1 and t_2 retain the same meaning that they have in the chart on page 255. To Galileo the relation $v_1/v_2 = \sqrt{s_1}/\sqrt{s_2})$ seemed anything but certain. He was dissatisfied that the velocities of falling bodies, when plotted against the distances they had fallen, followed a parabola instead of a straight line. The Sarpi proof redefined "velocity" by taking its square in order to make the velocities fall along a straight line and to justify Galileo's times-squared law of free fall $(s_1/s_2 = t_1{}^2/t_2{}^2)$. The proof does not, as has been supposed up until now, begin by assuming that the velocities of falling bodies are proportional to the distances through which they have fallen.

nus of *BE* did. Here a surprise was in store for him. When he calculated the length of *CF* (the speed acquired through *AC*) according to the ratio he had developed in his conclusion, he found that point *F* lay not on the slanted line *AE* but on a parabola through points *A* and *E*.

Galileo's conclusion on *f. 152r* is in a way the starting point of the modern era in physics. In it he correctly related the acquired speeds to the mean proportional of distances from rest, and he obtained explicitly or implicitly all the essential rules governing acceleration in free fall. Galileo's conclusion is analyzed in his terms of Euclidean ratios of proportion and in 20th-century terminology in the chart on page 255.

Galileo's conclusion on *f. 152r* did not mention time as such. Here I wish to point out once more that the only acceleration he ever applied was the acceleration of free fall. Thus whereas we generally think of acceleration as a function of two variables, distance and time, Galileo's acceleration was completely determined by either one. A rule for speeds in terms of distances, assuming one universal constant acceleration, left no freedom of choice of times. The relations Galileo expressed were thus complete and correct; *f. 152r* implied all the significant relations of distance, time and speed that exist in free fall.

Although Galileo did not expressly mention time in his conclusion, he had explicitly tabulated the times through the two distances from rest using a mean-proportional relation between those distances, and that is precisely how Galileo habitually spoke of times and calculated with them in the rest of his notes and published books. Hence I do not think it is exaggerating to say that a direct implication of *f. 152r* is that the ratio of the speeds at two points in fall from rest is also the ratio of the elapsed times of fall. In modern terminology we write this ratio as $v_1/v_2 = t_1/t_2$, where v_1 and v_2 are velocities at two points and t_1 and t_2 are the respective times that have elapsed from rest. This ratio at once implies that the ratio of the distances is equal to the ratio of the times squared.

That Galileo himself saw the implication is indicated by the fact that the times-squared law was the very proposition he offered to prove for Sarpi, saying he had known it for some time but had lacked any indubitable principle from which to demonstrate it. The obvious conclusion to be drawn is that in 1604 Galileo did not consider a simple statement equating ratios of the times and speeds under acceleration to be of such a nature as to be acceptable as an "indubitable principle."

On the contrary, in writing for Sarpi he took as his principle the seemingly contradictory assumption that speeds acquired are directly proportional to distances fallen from rest. Now, if we take the word "speed" here in the modern sense, which it had had in *f. 152r* and was to have later in *Discourses on Two New Sciences*, then the principle assumed for Sarpi was not only false but also inconsistent with Galileo's own conclusion in *f. 152r*, where it is not the speeds but the squares of the speeds that are proportional to the distances. Thus it must seem at this point that we have no choice but to say that Galileo first knew the right answer and then turned his back on it in favor of the wrong one, and that only years later did he return to the position that had been implied in his very first try at the mathematics of acceleration. That is even worse than what all historians of science have been saying up to the present.

I said "It must seem at this point" because things are not going to turn out that way at all. We have already seen what a very chancy business Galileo's investigations on *f. 152r* had been. His conclusion entirely lacked objective evidence, nor is there any reason that we should think Galileo remained unaware of the flimsy character of his first stab at the perplexing problem. Even if we now know, in a kind of absolute way, that his first result was the right one, there was nothing sacred about it to Galileo, who had arrived at it merely as an exercise in proportionality. He would have had every right to turn his back on it for something better, if indeed he had done so, which he did not. All that the result on *f. 152r* represented to Galileo was an internally consistent meaning for "speed," and for all he knew it might just be one among many.

The result that the ratio of the distances was proportional to the ratio of the velocities squared through those distances ($s_1/s_2 = v_1^2/v_2^2$) was anything but certain. There was as yet no way to measure a speed directly, and the square of a speed was hard to even make physical sense of. Even if the result was rewritten so that the ratio of the speeds was proportional to the ratio of the square roots of the distances ($v_1/v_2 = \sqrt{s_1}/\sqrt{s_2}$), it did not even look probable. How would a speed go about adjusting itself to a quantity represented by a distance through which it had already passed?

Galileo's other relation, that the distances fallen were proportional to the squares of the times through those distances ($s_1/s_2 = t_1^2/t_2^2$), may also have been puzzling, but it differed from the others in one respect: it could be physically verified. Galileo proceeded to verify it, probably in the way I formerly thought he might have discovered it, and in his letter to Sarpi he mentioned the odd-number rule that holds for successive spaces in equal times. Hence when he wrote to Sarpi, Galileo was far more certain of the truth of the times-squared law than he was of the validity of the derivation that had led him to it.

The most surprising implication of the documents, however, is this: Galileo never did adopt distance-proportionality for speeds in acceleration except as a temporary working hypothesis, and it led him at once to time-proportionality. We have been wrong in supposing that he rejected time-proportionality in favor of distance-proportionality in his proof for Sarpi late in 1604, in spite of its wording. Moreover, what he told the public in 1638 was the literal historical truth: For a long time he had thought it was a matter of indifference whether speeds in free fall were defined as proportional to the elapsed times or were related to the distances traversed. He merely neglected to add the statement, perhaps supposing that it was as obvious to others as it was to him: "Provided that the rest of your mathematics remains unaltered by your choice of assumption." It now remains for us to see how Galileo himself managed this, and to consider what it tells us about what he thought was the physical meaning of the word "speed."

We can start by noting what seems at first to be the most sophisticated thing in *f. 152r:* Galileo's reference to the fact that under the normal definition, points representing speeds would fall on a parabola when they are plotted against distances of fall. If we had no idea of the date of this entry and only knew about Galileo's later achievements respecting the parabolic trajectory, we might imagine that this remark greatly pleased Galileo. Although his work on the trajectory was still four or five years in the future when he wrote his conclusion in *f. 152r,* he already knew a great deal about parabolas. His first paper in 1587 dealt with the centers of gravity of paraboloids, and he applied for a chair of mathematics on the strength of it. The parabola as such did not dismay him. In Galileo's lifelong view, however, nature always acts in the simplest way. Since the simplest rule would put the point F on the straight line AE, I fancy that Galileo wrote the last words of *f. 152r* not with the joy of discovery but with something like disgust, and that he regarded them as casting serious doubt on the reliability of this chancy speculation about acceleration in free fall.

It was not hard, however, to rearrange matters so that the points representing speeds acquired by a falling body would fall on a straight line. All Galileo had to do was replace his mean-proportional treatment of speeds in his conclusion with a linear treatment, requiring only a change in his definition of "speed." The substitution would simply make $s_1/s_2 = V_1/V_2$, where the new "velocity" V_1 represents $v_1{}^2$ and V_2 represents $v_2{}^2$. The relation between velocity and distance would then become linear instead of parabolic. Careful examination shows that Galileo did exactly this when he composed the demonstration for Sarpi. He also found a physical justification for the new definition of velocity.

One thing that historians of science have all overlooked is that no definition of "speeds" in acceleration had ever been clearly made in terms of ratios of distances and of times. Archimedes had done it only for the case of uniform motion. Galileo took it from there, seeking a ratio definition for the ever changing speeds under acceleration. The Merton rule that represented overall velocities by their mean speeds threw no light on the problem of these ratios but ingeniously circumvented it. When Galileo tackled it, no physical measure for each and every speed in acceleration had ever been assigned. He therefore felt free to define the measure of speed in any way he pleased as long as experience bore him out, or at least did not contradict him. I might add that it was a long time, probably more than 20 years, before Galileo realized what the medieval writers had always assumed, namely that there does exist a uniform motion equivalent to any uniformly accelerated motion from rest. No trace of this realization is to be found in his manuscript notes. It first appeared as Theorem I in Book II of the Third Day in his *Discourses on Two New Sciences*, a theorem that, with all due respect to Professor Grant, does not employ any mean speed to represent accelerated motion in free fall.

In his letter to Sarpi, Galileo observed that pile drivers strike twice as hard when the weight falls twice as far. That remark shows that what Galileo meant by his word *velocità* in the Sarpi proof could not be the same as what he had meant by *gradus velocitatis* in *f. 152r*. In *f. 152r* the speeds would be in the ratio of $\sqrt{2}:1$ at the end of falls in the ratio of $2:1$. In the conclusion of *f. 152r* Galileo essentially stated that the velocities were proportional to the square roots of the distances from rest ($v_1/v_2 = \sqrt{s_1}/\sqrt{s_2}$), so that speeds in the ratio of $\sqrt{2}:1$ would be found at the end of falls whose lengths were in the ratio of $2:1$. In the letter to Sarpi, however, Galileo appealed to observations of pile drivers. The effect of a pile driver, involving kinetic energy, is governed not by speed but by its square. Hence the effect does not support a distance-proportionality assumption in the sense that we have always imputed to Galileo's words. What it does support is the mean-proportional relation between distances and speeds derived in *f. 152r* and altered to linear form by Galileo in the Sarpi proof by a simple redefinition of "velocity." In modern terms the Sarpi proof is equivalent to the proof of *f. 152r* with the substitution of V_1 for v_1^2 and V_2 for v_2^2. Galileo did this, I believe, in order to place the acquired *velocità* along a straight line rather than on a parabola. With the transformation Galileo's earlier ratio $s_1/s_2 = v_1^2/v_2^2$ becomes $s_1/s_2 = V_1/V_2$. See proof on p. 258.

Galileo's choice of the ratio $s_1/s_2 = V_1/V_2$ when he wrote to Sarpi did not mean, as historians have all previously supposed, that he had to reject his original relation $s_1/s_2 = v_1^2/v_2^2$. It did mean, however, that the *velocità* defined for Sarpi as being proportional to distance could no longer be the same physical entity as the "speed" implicitly defined as proportional to time in *f. 152r*. The times-squared law, however, remained unchanged by Galileo's new definition of *velocità*. He believed that he could derive the times-squared law as well from $s_1/s_2 = V_1/V_2$ as he could from $v_1/v_2 = t_1/t_2$. In fact he could, and he did. Later he found an experimental reason for adopting only the second (older) relation and defining "velocity" as we do.

It is easy enough to say at this point: "But there must have been some criterion of choice at every stage, since in fact the speeds in free fall do not increase according to distances fallen but do increase according to elapsed times." This statement, however, assumes a physical definition of "speed." What we know as "speeds" do indeed increase in that way. In Galileo's proof for Sarpi, however, he chose to use something else, based on his observations of pile drivers, which is the square of our notion of speed. Let us call Galileo's choice *velocità*. We could not very well argue that *velocità* fails to increase according to the distance fallen, basing our argument on the fact that *velocità* in Italian means, after all, the same thing as "speed" in English. It would be necessary to change the usual methods of measuring time and distance to make *velocità* behave like our speed. Galileo made no alteration in his relations of time and distance, however, so that his *velocità* behaved differently. The worst we can say of his definition is that we prefer another one: the same one that he himself later adopted. Galileo reasoned that the effect of a pile driver could change only when the falling weight acquired a different speed. In effect he decided, for a time, to think of "*velocità*" as "whatever it is that changes the striking power of a body falling from different heights." This quantity he could measure, and it does behave like our v^2. Later he found a way of directly observing speed in our sense of the word, but he had no such way in 1604.

In his proof for Sarpi, Galileo invoked a "contrary proportionality" between speeds and times, a curious relation that equates the ratio of the roots of one variable to the ratio of the squares of the other. With this tool he concluded that the square root of what he calls the "total speed" was proportional to the time squared. Did Galileo have any ground for asserting such a relation? He certainly presented none in the proof for Sarpi, since time was not mentioned there until he made his appeal to contrary proportionality. Nor can such a relation be derived from *f. 152r* alone.

Yet the relation does arise from the two documents together, and it could only arise because Galileo assumed that what we call "speed" remained proportional to time even when his new *velocità* was made proportional to distance. It was only because he assumed this when he wrote the proof for Sarpi that he could validly invoke "contrary proportionality" as he did. And that he did assume it is borne out by the solid fact that throughout the 160 sheets of notes on motion, written over a period of 30 years, there is not one single instance in which Galileo ever made use of the assumption that speeds in the ordinary sense are proportional to distances in free fall.

This, then, is the new picture: Galileo obtained the result that speeds in free fall are proportional to times elapsed from rest in the course of his very first try at the mathematics of accelerated motion, probably in the middle of 1604. He never abandoned that conception, although for a time he altered his definition of *velocità* in the interests of elegance and the simplicity of natural phenomena, supported by reasoning about an observed phenomenon of kinetic energy. Ultimately a classical experiment, still unpublished (*f. 116*), induced him to reject the alternative definition.

We do not have, and we do not need, any special name for the physical quantity *velocità* of the Sarpi proof. If we did give it a name, that word (say "punch") would occur frequently in our discussions of energy, and it would seem to us that this newly named physical entity should enter such discussions directly, instead of as the square of something else. We might think of a falling body's velocity as "how fast it will go horizontally if deflected," and its "punch" as "how hard it will hit vertically." Whether there really are two physical entities (velocity and "punch") or whether there is only one entity (velocity) that enters into some calculations as itself (v) and into others as its square (v^2 or V) would be hard to decide.

This question was bound to crop up in some form after Galileo, convinced that nature had forced his hand, suppressed his alternative definition. In one form the question arose late in the 17th century, when Leibniz christened the neglect of v^2 ("punch") as "the memorable error of Descartes." After decades of heated argument the entire problem was recognized to be a merely semantic controversy.

10

Galileo Gleanings – XXIII
Velocity and Eudoxian Proportion Theory

I

Two principal charges have been brought against Galileo's argument, in the Third Day of the *Discorsi*, that speeds in free fall are not proportional to distances from rest. The oldest was that Galileo there misapplied to accelerated motion a proposition valid only for uniform motions. First brought in 1643 by Pierre Cazré, and not properly refuted by Gassendi in their long controversy over free fall, this charge has been revived again in connection with a proposed reinterpretation of Galileo's ingenious but ill-fated demonstration.[1]

A quite different objection to the same argument, more recent in origin and now widely popular, is that Galileo employed a consequence of the correct definition of uniform acceleration in refutation of a proposed alternative definition. In replying to that contention, which had gained ground through a common mistranslation of Galileo's words into English, French, German and Russian during modern times, I did not deal with the older charge except to mention it as of historical interest.[2] Now that it has reappeared in answer to my analysis of Galileo's proof, it is incumbent on me to lay it finally to rest,

It is easy to show that Galileo is unlikely to have had in mind literally the proposition that he is alleged to have intended in his argument. That proposition is the second part of Theorem II in the brief Latin treatise entitled *On uniform motion* which preceded the argument in question. This states that "if the distances are as the speeds [through them], the times will be equal." But in Galileo's contested demonstration two distances are specified, of which one is double the other, while the conclusion he reached implied to him that speeds of

Reprinted from *Physis* 15 (1973), 49–64, by permission.

a single body through both those distances must be infinitely great, and in that sense "equal." ("For one and the same body to pass the four braccia and the two in the same time cannot occur except in instantaneous motion.") Galileo's most intractable foe would hardly accuse him of consciously trying to deduce non-proportionality from assumed proportionality in a brief statement which he himself described as "a very clear proof."

In view of this, the reader may be doubtful that I am correctly representing the position of Dr. Finocchiaro, who is a very strict logician and who indeed wants to make Galileo out as too logically and grammatically precise to have reasoned along the lines of my interpretation of his argument. Yet I believe that what I have just said is not a misrepresentation, and that the implication is not evident to Dr. Finocchiaro only because of his neglect to distinguish consistently between velocity through a distance in accelerated in motion, and velocity acquired at any point in such a motion.

Theorem II in the treatise *On uniform motion* deals only with the speeds of a body *through* specified distances saying that (I) if the times are equal, the distances are as the speeds, and (2) if the distances are as the speeds, the times are equal. Since in Galileo's disputed argument, only the velocities *acquired* at given points in the motion are under discussion, this theorem is not immediately applicable to it. It would become applicable only if some further assumption were made which related acquired velocities to overall speeds through the distances. No such assumption was made overtly by Galileo, a fact which has induced many modern writers to impute to him a covert assumption of what is known as the mean-speed theorem, or Merton Rule. Since that rule is implied by the correct definition of uniform acceleration, such writers have brought against him the other charge, mentioned above, which was dealt with in my previous paper and does not concern us here.

As I have said, Theorem II in the treatise *On uniform motion* is not directly applicable to Galileo's demonstration; but its inapplicability does not depend (as Dr. Finocchiaro asserts) on any restriction to uniform motions. The theorem is, in fact, quite unlimited in its application to uniform, accelerated, or even irregular motions, and Galileo was perfectly well aware of the fact. It is not applicable directly to "instantaneous velocities." But before proceeding, it is advisable to establish those points beyond doubt.

II

The little Latin treatise which Galileo entitled *On uniform motion* set forth six theorems, of which four contain restrictions of applicability and two do not; the restrictions, where they exist, are italicized below:

Theorem I.   If a moveable *carried uniformly at the same speed* traverses two spaces, the times of movements will be to one another as the spaces run through.

Theorem II.   If a moveable runs through two spaces in equal times, these spaces will be to one another as the speeds. And if the spaces are as the speeds, the times will be equal.

Theorem III.   At unequal speeds through the same space, the times of movements are inversely as the speeds.

Theorem IV.   If two moveables *are carried in uniform motion* but at unequal speeds, the spaces traversed by them in unequal times will have to each other a ratio compounded from the ratio of the speeds and the ratio of the times.

Theorem V.   If two moveables *are carried in uniform motion* but at unequal speeds, and unequal distances are traversed, the ratio of the times will be compounded from the ratio of the speeds and the ratio of the times.

Theorem VI.   If two moveables *are carried in uniform motion*, the ratio of their speeds will be compounded from the ratio of the spaces traversed and the inverse ratio of the times.

It is evident that Galileo did not omit the restriction from Theorems II and III by any oversight; he considered these two to be true without restriction of any kind. And so they are; but Galileo might have been mistaken in his confidence about this. In order to know whether or not he was justified, it is necessary to examine his proofs; and for this, we must first look at the four axioms in the treatise "On uniform motion."

Axiom I.   The space traversed in a longer time *in the same uniform motion* is greater than the space traversed in a shorter time.

Axiom II.   The time in which more space is traversed *in the same uniform motion* is longer than the time in which less space is traversed.

Axiom III.   The space traversed at greater speed in the same time is greater than the space traversed at less speed.

Axiom IV.   The speed at which more space is traversed in the same time is greater than the speed at which less space is traversed.

Here the restriction to uniform motion occurs only in the first two axioms. The latter two are in effect no more than formal statements of the rules Aristotle had given for the use of the term "greater speed," which Aristotle (and Galileo) intended to be of perfect generality for all motions whatever, regular or irregular.[3]

Now, if Galileo proved Theorem II without recourse to any axiom restricted to uniform motion, then that theorem will be of general applicability wherever it is relevant, and will not be applicable only to uniform motions, as maintained by my adversary. It remains now to examine Galileo's proof. This proof depends on the Eudoxian definition of "same ratio," of which more will be said below.

Let AB and BC be two distances traversed in equal times, distance AB [being traversed] with speed DE, and distance BC with speed EF. I say that distance AB is to distance BC as speed DE is to speed EF.

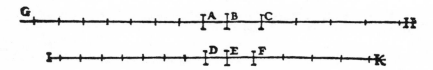

As before [i.e., in Theorem I], equimultiples are taken on both sides according to any multiples; that is, BG and EI are equimultiples of AB and DE, as likewise BH and KE are [equimultiples] of BC and EF. It will be concluded as above that the multiples BG and EI agree in falling short of, equalling, or exceeding the equimultiples BH and EK [respectively]. The proposition is therefore manifest.

For the modern reader unfamiliar with Euclidean proofs, this leaves something to be desired. Galileo's readers were accustomed to filling in the missing steps from Theorem I, indicated by his words "as above." Taking m and n as arbitrary multipliers, the argument is as follows:

$$BG = m\text{AB}, \quad EI = m\text{DE}; \quad BH = n\text{BC}, \quad EK = n\text{EF}$$

Now, if BG is equal to BH, then EI must equal EK, by the definition of "equal speed" given by Aristotle and universally accepted; that is, in equal times, equal distances are traversed at "equal speeds."

But if BG is greater than BH, then EI is greater than EK, by Axiom IV above (the speed at which greater space is traversed in the same time is the greater speed).

And if BG is less than BH, then EI is less than EK, by Axiom IV and the definition of "equal speed."

Hence, according as BG $\gtreqless$ BH, EI $\gtreqless$ EK. But BG and EI, BH and EK, are respectively equimultiples of AB and DE, BC and EF. Therefore, by Euclid V, Def. 5, it follows that AB:BC :: DE:EF.

This completes the proof of the first part of Theorem II. The second part is proved in the same way, using Axiom III in place of Axiom IV. And since those are the unrestricted axioms, Theorem II is also unrestricted to uniform motion, precisely as Galileo stated it.

It may be added in passing that Theorem III is then proved by Galileo without recourse to any previous proposition except Theorem II, and is therefore likewise of general applicability, as Galileo worded it. The same cannot be said of any of the other theorems, and Galileo has duly introduced into their statements the restriction to uniform motions. I believe that this should remove from any reader's mind the suspicion that Galileo himself might not have known exactly the range of applicability of each of his theorems.

III

Pierre Cazré in the mid-seventeenth century, and Dr. Finocchiaro in his recent analysis, assumed that the opening clause in Galileo's contested argument against space-proportionality was simply a statement of the second part of Theorem II, above. This was also assumed by Pierre Duhem and by Rafaello Caverni a half-century ago, but with a very curious difference. Whereas Cazré and Finocchiaro took this to invalidate the demonstration, Duhem and Caverni thought that it made it conclusive. This ought to give the reader some idea of the complexities of the affair which I have undertaken to unravel.

To begin with, no one is to be blamed for assuming at first glance that the opening clause of Galileo's argument is just the second part of Theorem II restated in different words. The first clause of the argument reads: "If the velocities have the same ratio as the spaces passed or to be passed, those spaces come to be passed in equal times."[4] Theorem II concludes by saying that: "If the spaces are as the speeds, the times will be equal." Apart from a reversal of order of distances and velocities, which is inconsequential in asserting this proportionality, there seems to be no essential difference between the two statements. But as soon as an analysis of the contested argument is attempted, the most profound difference should be detected at once; for the velocities referred to in the argument are those acquired at any point in free fall under the hypothesis that these are proportional to the distances fallen, whereas the speeds referred to in Theorem II are the overall speeds (otherwise undefined) of bodies passing through given distances. Theorem II says nothing about speed possessed by the body at the end of the motion; this could be the same as at every other point (uniform motion), or it could be the highest speed (accelerated motion), or it could be the least speed (decelerated motion), or it could be neither, and different from all others, as in some irregular motions. In every case the overall speed would be the same, provided only that the distances and times of the total motion remained unchanged.

But the whole point of Galileo's contested argument turns on the velocities *acquired in fall*. It was only the rule for acquired velocities that had been under discussion. Galileo undertook to show that if these were made proportional to the distances fallen, as proposed by his interlocutors, it would follow that all distances would be fallen in the same time, which is seen not to be the case. It could be of no help to him in this demonstration to appeal to a rule that speeds *through* various distances produced the desired identity of time, because no one had conceded that *those* speeds were proportional to the instantaneous velocities, which were in turn assumed proportional to the distances fallen. It happens that in actual free fall, speeds through distances from rest are proportional to the acquired velocities, but Galileo was here obliged to discuss a fanciful kind of free fall in which, as he proceeded to show, they are not. In this fanciful law of free fall, all the speeds *through* distances from rest are equal, being infinite, and hence are not proportional to those distances in any sense whatever.

It is true, then, that by the nature of the argument itself and by the character of Theorem II, Galileo could not start out from that theorem. But it is not true, as my adversary asserts, and as Cazré had asserted before him with withering sarcasm against Galileo,[5] that Galileo could not legitimately appeal to Theorem II by reason of any restriction of that theorem to uniform motions.

It is my position that all the disputes over the interpretation of Galileo's argument hinge on one point: does any use of the word "velocity" in it apply to "speed through a distance from rest," and if so, which use (or uses)? I contend that throughout his argument, Galileo's word "velocity" applies to acquired speed at an instant, and never to speed through a distance. If I am right about this, then the opening clause cannot invoke or represent Theorem II, which theorem has no bearing whatever on any instantaneous velocity, but exclusively concerns overall speed through a distance. Accordingly I read the opening clause in this way:

If the [acquired] velocities [at each and every point] have the same ratio as the space passed or to be passed, those spaces come to be passed in equal times ... .

Dr. Finocchiaro does not declare explicitly how he would gloss the same passage, but in his paraphrase of the argument as a whole, he puts it in the form "whenever velocities are as the spaces, such spaces are passed in equal times," and he then identifies this as "a theorem Galileo has already proved," referring the reader to *Opere* VIII, p. 193, which contains Theorem II; and he goes on to remark that this is "alas, one about non-accelerated motion (uniform velocities)." Hence it is evident that he would not gloss the passage as I do, but considers it to refer to velocities *through* two spaces passed.

The effect of this difference is that whereas I regard the opening clause as the statement of a new proposition, never before considered, which is to be proved by what follows, Dr. Finocchiaro regards the same clause as a theorem already proved, but not applicable in a proof that "the velocity of a falling body is not proportional to the space passed." Here again there is no gloss, and his use of the singular, "velocity" leaves it uncertain whether an overall speed or a terminal velocity is meant.

I note that in the discussion that led up to Galileo's proof, only instantaneous velocities were ever in question, and never any finite speeds through distances in accelerated motion. It seems to me highly unlikely that Galileo suddenly switched the meaning of the word "velocity" in this way without any warning or explanation. I note also that the immediately ensuing clause permits no doubt about what is intended there, since it uses the plural "velocities" for a passage through a single distance, and these can only be instantaneous velocities:

... thus if the velocities with which the falling body passed the space of four braccia were doubles of the velocities with which it passed the first two (as the space is double the space) ... .

Finally, the only other occurrence of the word "velocity" in the argument comes at the end, where it again certainly refers to velocity at each instant, inasmuch as the hypothetical rule to be rejected by this demonstration referred only to the growth of instantaneous velocities with distance from rest:

... therefore it is not true that its velocity increases in proportion to space.

Accordingly, I paraphrase the gist of Galileo's argument in this way:

If all the infinitely many instantaneous velocities occurring in actual accelerated fall from rest over any space, say one of four braccia, were the respective doubles of all the infinitely many instantaneous velocities occurring over the first half of the same fall, then no difference in the times of fall through two braccia and four braccia could be accounted for.

Dr. Finocchiaro objects that he sees no comprehensible connection between the "if" and the "then" clauses above. As often happens, such connections are self-evident to one person and entirely non-existent to another. I recall my own utter astonishment, as a youth, when my father and I both heard for the first time the puzzle, "If there are more men in London than there are hairs on any man's head, does it follow that there are two men in London with the same number of hairs on their heads?" I saw no connection whatever, whereas my father replied, after a moment's reflection, "Of course it follows." The necessary connection was not clear to me until he said, "Well, start counting them," and I saw that after I had put aside all the men with *different* numbers of hairs, there would still be men left over. Once I saw this, I was sure that anyone else of whom I asked the question would see the answer at once; but in fact, many have been baffled at first, just as I was.

Now let me ask Dr. Finocchiaro: "If a body moves by a simple law that gives it infinitely many different instantaneous speeds, and these are all the speeds it can have in traversing a certain distance from rest in a given time, then how far will it go in the same time, from rest, at the simple doubles of those speeds?" I am sure he will reply, "Twice as far," though he may add "but I don't see how one would go about doubling each and every speed without leaving gaps in between, so to speak." And perhaps it is this sort of sensed difficulty, not yet thought out, that makes him feel that there is something missing in Galileo's argument, or rather in my paraphrase of it. That the sensed difficulty in this case is valid only for finite aggregates, and disappears in infinite aggregates, is seen in a flash only by people who, like Galileo, have spent a lot of time contemplating the nature of continuous magnitude, and who, having done so, sometimes fail to mention things that have become obvious to them.

Given my interpretation of Galileo's argument (which, incidentally, was offered by one of his younger contemporaries, J.A. Le Tenneur, and hence did not require mathematical insight then out of reach), it is easy to see why Galileo singled it out for special praise by Sagredo – praise which Dr. Finocchiaro regards as so excessive that it may represent a warning by Galileo that something is wrong with his proof. My view is precisely the opposite; like the eighteenth-century historian, Giovanni Andres, I believe that Galileo's unusual praise means that we should look in this demonstration for something that captivated his fancy, whether or not we endorse the proof. Nothing of the sort can be found by attempting to interpret this demonstration in terms of elementary logical blunders, such as the assumption of things not yet proved, or the misapplication of previous propositions. Some novel concept, on the other hand, is to be expected in a proof that Galileo singled out by special praise.

Just as my interpretation of the argument struck my adversary in this matter as omitting a necessary logical step, so his reinterpretation strikes me. The difference is that the former is compatible with the recognition of a new mathematical method now acknowledged as one of great power and beauty, whereas the latter is compatible only with a logical obtuseness or carelessness not characteristic of Galileo even in his old age. However the argument is analyzed, it is necessary to provide a bridge between instantaneous velocities determined by a rule, and a statement about times through two given distances. To offer the Merton Rule (or any single value taken as representative or an infinite set) as such a bridge is not only to imply that Galileo thought he could refute one definition by the consequences of another definition; it is also to invoke a concept never found elsewhere in any of his published books or private notes. On the other hand, reasoning by one-to-one correspondence between infinite aggregates is found frequently in both. It was thus that he discovered the double-distance rule while still at Padua, probably before 1608; it was thus that he reasoned about this rule and about the times-squared law in the *Dialogue*; and it was thus that he proved Theorem I on accelerated motion in the *Discorsi*.

The logical gap that Dr. Finocchiaro himself identifies in his own reinterpretation of Galileo's argument (on p. 141 of his paper) resides in a supposed limitation of Theorem II *On uniform motion*, which limitation has been shown above not to exist. But even with that limitation removed, a gap remains for me in Dr. Finocchiaro's summary of the argument. This gap may be seen by glossing each occurrence of the word "velocity" in his summary with either the word "instantaneous" or the word "overall." However this is done, the former word must appear somewhere, because it is essential to the rule of velocities under debate; and the latter must also occur, at least in the step which is supposed to involve Theorem II. But one can get from instantaneous to overall velocities only by (a) assuming a representative value (such as the mean speed) or (b) reasoning about infinite aggregates. Neither of these is to be found in Dr. Finocchiaro's reconstruction.

IV

It is not yet generally realized that the Eudoxian theory of proportion lies at the basis of most of Galileo's applications of mathematics to physics and was also the source of most of the contemporary opposition to his laws of motion, which was considerably greater than is reflected in current histories of science. That theory was not exploited in the Middle Ages because of the defective definitional apparatus of the fifth book of Euclid's *Elements* as it passed into Latin from an Arabic text, and because of an erroneous commentary on the crucial definition, inserted by Campanus of Novara. Medieval writers accordingly built a beautiful theory of proportion, arithmetical in character, on Euclid's seventh book, going far beyond any ancient development of the pre-Eudoxian arithmetical theory. Indeed, at the hands of Nicole Oresme, this theory very nearly achieved the arithmetization of the continuum that was in fact only made rigorous less than a century ago, and was made so only by the explicit use of the Eudoxian definition of "same ratio" by Richard Dedekind, Karl Weierstrass, and others.

An arithmetical theory of proportion was highly satisfactory for the development and extension of Aristotelian physics in the Middle Ages, because Aristotle himself believed all motion to be successive, rather than continuous, with the sole exception of motion induced directly by unmoved movers. The theory of impetus gave an explanation of acceleration in free fall which was essentially cinematographic in character, occurring in vanishingly small discrete quanta of velocity added in successive "physical instants" which were sharply distinguished from mathematical points in time. These concepts were raised in opposition to Galileo's law of free fall, most comprehensively by Honorato Fabri, but also in part by others.[6] Even Descartes, in letters to Mersenne, opposed Galileo's idea that a body in reaching any speed in free fall from rest must have first passed through every possible lesser speed.

The idea of continuous motion in the fully modern sense of the continuum, espoused by Galileo, was incapable of development without the Eudoxian theory of proportion of magnitude in general. This theory seems to have been first made completely clear in modern times only in 1543, when Niccolò Tartaglia published his acute specific critiques of the commentary of Campanus. The correct text of Euclid, Book V, had been translated into Latin from a Greek manuscript in 1505 by Bartolomeo Zamberti, and was often reprinted; but Zamberti had not explicitly corrected the traditional interpretation, and his text was usually reprinted together with that of Campanus during the sixteenth century. Hence it is no wonder if Galileo was the first to study seriously the mathematical concept of continuous motion, or at any rate the first to vanquish some of its most formidable paradoxes. Outside Italy, Targalia's Euclid was probably not read, as it certainly was not read in the universities, having been published in Italian (and a very colloquial and barbed Italian at that).

It is therefore of prime importance, in attempting to trace (or to divine) Galileo's thought on motion, that the historian first master the fifth book of Euclid. Galileo lectured on that book at Padua, and very late in life he realized the necessity of making it easier to understand by the general public for whom he wrote, so he dictated a little dialogue on it when he was blind and near death. Girolamo Saccheri, celebrated for his adumbration of non-Euclidean geometry in 1733, considered the definition of "same ratio" to be the only rival of the parallel postulate among the logical obscurities in Euclid, and devoted half his famous book to it. Most subsequent mathematicians have regarded this definition as among the supreme achievements of mathematics in all time.

But the traditional arithmetical theory of proportion had not by any means been put aside in Galileo's day, and it quite naturally gave rise to a kind of quantum-physics of motion as an alternative to Galileo's continuity-physics. This is found in Baliani, Fabri, and Beeckman, among others, as a serious rival physics that is neglected by historians because it soon became indistinguishable from continuity-physics through the marriage of algebra with geometry in the work of Descartes. Because that marriage was finally legitimized late in the nineteenth century, we tend to forget its common-law status at first. It is interesting that Le Tenneur, who in 1649 was the only known man to understand correctly Galileo's argument from one-to-one correspondence, had written a book in 1640 against Stevin's *L'Arithmetique* in protest against the consideration of unity as a number. Seen in modern light, his arguments, though elementary, amounted to questioning whether algebraic numbers exhausted the continuum. He upheld the classic Greek distinction between the continuous and the discrete, with its emphasis on the concept on incommensurability. Others swept that distinction under the rug; but it was still there, and had its effects on physics no less than on mathematics. The rivalry between early wave theories and particle theories of light was no less related to it than was that between infinitesimals and indivisibles as routes to the calculus.

Because of the same distinctions, as reflected in two different theories of proportion, I think it may be misleading to assign the roots of Galileo's ideas on motion to medieval speculations, which themselves are as deserving of admiration but were different from his, both in purpose and in mathematical basis. Though the medieval theory of proportion was far from extinct in Galileo's time, it did not pervade his work. His models were Euclid and Archimedes, as he never tired of saying; but they were the fifth book of Euclid rather than the seventh, and the authentic works of Archimedes rather than attributed treatises. Theorem I of the treatise *On uniform motion* was taken directly from Archimedes *On spiral lines*, which in turn depended solely on Euclid V, Definition 5. But thus far no trace of Eudoxian proportion theory in the Middle Ages has come to my attention.

It is curious that Aristotle himself did not formulate Galileo's theorems concerning uniform motion, or at least Theorems II and III of his little treatise.[7] Aristotle seems to have known about Eudoxus's new theory of proportion, according to T.L. Heath, and those theorems are rigorously demonstrable from the definition of equal and greater velocity.[8] But I do not believe that they are rigorously deducible from Aristotle's definitions *except* by Eudoxian proportion theory. Their absence not only in Aristotle, but in the works of acute medieval physicists, suggests that those writers either had a very different purpose from Galileo's, or that they remained ignorant of Eudoxian proportion theory – or perhaps both.

v

Dr. Finocchiaro objects to my reliance on the whole context of Galileo's contested argument for clues as to its proper interpretation, preferring a strict logical and grammatical analysis of that argument in isolation. In that connection he questions a matter of punctuation and word-order in my translation. If the dispute were over a principal theorem essential to Galileo's book, I should probably agree with the strict logical approach. But we are concerned rather with a spirited interchange between the interlocutors, and here Galileo's style is as important as his words. I shall try to explain why I say this, leaving the specific points of translation to the end.

Anyone who labors through Galileo's more complicated Euclidean proofs in the *Discorsi* will learn what he regarded as necessary for rigorous demonstration. From algebraic paraphrases of those proofs, with which most of us are content, we will never discern Galileo's techniques in the weaving of an argument. We do know, however, that he greatly admired Archimedes and tried to emulate him. Now, a complicated proof by Archimedes can be quite maddening. He will start with a lemma of which the purpose is far from evident, ramble on with a proof of something else, put in another lemma or two, and just as we despair of his ever getting to the point, he suddenly springs it on us unawares, like the skillful writer of a detective story who knows when we have finally got all the scattered clues in hand and then will not allow us time to put them together before he has revealed the solution. It is this pattern that is seen in the contested argument about space-proportionality when its whole context is viewed, and in my opinion that fabric was skillfully woven by Galileo with malice aforethought. Such an analysis then helps us to understand why the whole argument is put in a single long sentence, stripped even of one step that Dr. Finocchiaro regards as needing not only to be present, but to be further explained.

If Galileo's work on motion belonged squarely in the mainstream of the history of physics (and if there had been only one such mainstream, rather than two, as I believe), then the context properly applicable to the understanding of a single proposition would be that of the writings of Galileo's contemporaries. As I look at matters, the context of his own writings is much more fruitful in appraising Galileo's thought and his logical consistency. Let us take, for instance, the faults of his treatise "On uniform motion." It omits the definition of "equal speeds" and it includes two axioms and two theorems that are not restricted to uniform motions, despite the title. Are these indications of logical deficiency, or faults of style? The context of Galileo's writings as a whole indicates the latter; in conjunction with other evidence, it also affords a plausible source of the faults.

The opening of the Third Day is exceptional among all Days of the *Dialogue* and the *Discorsi* in its lack of any preliminary conversation to open the debate. The treatise begins without even an indication that Salviati is reading aloud, as is later made clear. The conclusion of the Second Day similarly lacks a conversational conclusion of the usual sort. Now, there exists a letter to Galileo from the Italian representative of the publishers, telling him that they have divided the material received at this point.[9] I surmise that Galileo had intended to send additional copy, closing the Second Day and opening the Third Day. Near the end of the book, a certain resumption and duplication of ideas that had started near the end of the Second Day suggests that a part of the proposed material had been written but not sent, and was inserted elsewhere by Galileo after the publisher's division had been made.

As to the intended beginning of the Third Day, I conjecture that Galileo may originally have intended to include a treatise "On motion in general," but found that it would contain only a definition, two axioms, and two theorems; moreover, separating this would leave the section on uniform motion very short. To make that of reasonable length, he may have transferred to it the axioms and theorems on motion in general, putting them second in place because of the title. The definition of "equal speed" would then have appropriately gone into the conversational opening that was not sent to the printer (if it was ever actually written) in time to be used. Nothing could have been more appropriate to open the Third Day than a dialogue virtually identical with one which had already been used in the now forbidden previous book:

SALV.   ... Tell me, Simplicio, when you think of one body as being faster than another, what concept do you form in your mind?

SIMP.   I imagine one to pass over a greater space than the other in the same time, or to travel an equal space in less time.

SALV.   Very good. Now as to bodies of equal speed, what is your idea of them?

SIMP.   I conceive them to pass equal spaces in equal times ...

SAGR.   Let us add something, however, and call the speeds equal when the spaces passed over are in the same ratio as the times in which they are passed. That will be a more general definition.

SALV.   So it will, because it includes equal spaces passed in equal times, and also those which are unequal but are passed in times proportional to them.[10]

This general definition is precisely the one made possible by Eudoxian proportion theory combined with Aristotle's definitions. In the *Dialogue*, it was used in resolving the seeming paradoxes of equal speed entailed by accelerated motion, and this would have made an excellent introduction to the Latin treatises of the *Discorsi*. But I hasten to add that all this is merely a conjectural explanation of seeming logical defects, based on certain stylistic curiosities at the beginning of the Third Day.

VI

As to my punctuation of the opening clause of the contested argument, this was certainly deliberate on my part in the English version, and inadvertent in my citing of the original, where I put an unjustified colon in place of the textual semicolon. In translating, I never hesitate to use the punctuation that seems to me best suited to convey the sense of the original. The original edition of 1638, as well as the second of 1655, reads:

> *Quando le velocità hanno la medesima proporzione che gli spazii passati o da passarsi, tali spazii vengon passati in tempi eguali; se dunque le velocità con le quali il cadente passò lo spazio di quattro braccia ... .*[11]

My English version then read: "If the velocities have the same ratio as the spaces passed or to be passed, those spaces come to be passed in equal times: thus if the velocities with which the falling body passed the space of four *braccia* ... ."

Dr. Finocchiaro prefers to read the conjunction of clauses:

> ... in equal times; if therefore the velocities ...

I certainly have no objection to this, merely preferring to make what I believe to be the thought as simple and clear as possible in translation. I do not think that "if therefore" means "if on that account," because the hypothesis about a doubling of velocities does not depend on the initial clause, but on the previous discussion. Even if I am wrong about that, it seems to me that "thus if" leaves the reader free, if he wishes, to interpret the meaning as "if on that account," whereas "if therefore" tends a little more to coerce him. But apart from what I think, it seems to me that the Latin version published in 1699–1700 reflects a similar view, for it reads:

> *... talia spatia aequalibus decurruntur temporibus: si itaque velocitates ... .*[12]

Now, there is a very interesting possibility about this Latin translation which may give it great weight; for it was published at Leyden without any indication whatever of the name of the translator, jointly with a new edition of the 1635 Latin translation of the *Dialogue* which was originally published by the Elzevirs. That same firm printed the *Discorsi* at Leyden in 1638, and Galileo was then negotiating with them for an edition of all his scientific works translated into Latin, including a Latin version of the *Discorsi*. But in 1640, Galileo received a letter from Elia Diodati saying that "The Elzevirs wrote to me that they were putting off for some time the printing of the Latin translation of your works until they have sold a greater number of those already printed," meaning the *Letter to Christina* and the *Discorsi* as well as the *Dialogue*. The Latin edition of Galileo's works was never printed; but the translation later printed at Leyden may very well have been one made under Galileo's direction that had long lain in that city unpublished.

Both the early English translators used the colon. The first of these, Thomas Salusbury, wrote in 1665 and could not have seen the colon of the Latin translation; he wrote:

... passed in equal Times: if therefore the Velocities ...

Thomas Weston's translation was published in 1730, and he gave:

... passed thro' in equal Times: If therefore the Velocities ...

Weston made frequent use of semicolons, but never followed them by capitalizing the next word. His use of colon and capital suggests a separation of thoughts more strongly than does Salusbury's punctuation, or mine. The Latin and early English translations are of much greater authority than modern versions, since they correctly gave Galileo's plural, "velocities," whereas modern translators have probably been influenced by their own mistranslation into a singular, as if overall velocity or a mean speed were meant.

NOTES

1 [Pierre Cazré] *Physica demonstratio qua ... accelerationis motus ... determinatur* (Paris, 1645), pp. 8–9; [Pierre Gassend] *De proportione qua gravia decidentia accelerantur ...* (Paris, 1646), pp. 14–19, 110–134; M.A. Finocchiaro, "Vires acquirit eundo," *Physis* xiv, 2 (1972), pp. 125–145.
2 S. Drake, "Uniform acceleration, space, and time," *BJHS* V, 1 (1970), pp. 28–36; *Galileo Studies* (Ann Arbor, 1970), pp. 229–237.

3 Aristotle, *Physica*, esp. 232 a-b, 249 a-b; for example: "Two things are moving at the same rate if they take equal time to accomplish a certain amount of motion"; "Two things are of equal speed when they move over equal stretches or intervals in a given time"; "To cover the same distance ... the quicker cannot take either an equal or a longer time"; "The swifter moveable would cover an equal distance in less time."

4 See § VI, below, for the original Italian.

5 Cazré, loc. cit.: *Et tamen mirum quantum Galilaeus de hac (ut putat) subtili, clara, evidenti, ac mathematica demonstratione sibi applaudat, quam integrâ paginâ mirificis laudibus exaggerat. Sed illud multò adhuc mirabilius, quòd Lynceus Philosophus ac Mathematicus, Lynceorúmque princeps, in tam aperta luce caecutiat; et vir eius nomini facilè delundatur. Ut enim prima illius paralogismi assumption, in motu uniformi, ac perpetuò sibi aequali vera et necessaria sit; in motu tamen accelerato minimè necessaria sit ... Assumptio igitur Galilaei falsa est, et tota eius ratiocinatio merus paralogismus.* Dr. Finocchiaro is, on the contrary, very charitable to Galileo, holding the misapplication of a mathematical theorem to be merely an unsuccessful manoeuvre. Had Galileo been guilty, and then praised his defective argument, Cazré's view would be mine.

6 Fabri regarded the hypotenuse of Galileo's famed triangle as truly a step-function – a denticulated line as he called it – in which the steps were too small to be observed, but were necessarily physically real.

7 Sometimes he came close to doing so: "For since the distinction of quicker and slower may apply to motions occupying any period of time and in an equal time the quicker passes over a greater length, it may happen that it will pass over a length twice, or one-and-a-half times, as great as that passed by the slower: for their respective velocities may stand to one another in that proportion." (*Physica* 233b. 19 ff).

8 See note 3, above.

9 *Opere*, XVII, p. 187: "... They have commenced the Third Day with the treatise *De motu locali*, and say they do not find Fifth Day, at least before the Appendix. Moreover, they want to know if you have sent all the copy, asking to be notified as soon as possible ... ." (26 September 1637).

10 *Opere*, VII, p. 48; *Dialogue* (tr. Drake, Berkeley, 1953), p. 24.

11 *Opere*, VIII, p. 203.

12 Galileo, *Discursus et Demonstrationes Mathematicae* (Leyden, 1699), p. 148.

11

Galileo's Work on Free Fall in 1604

I

On 16 October 1604, Galileo wrote to Paolo Sarpi that if granted the assumption that speeds in fall increase as do the distances from rest, he could prove the distances to be as the squares of the times. A purported demonstration, written out at that time, is preserved as f. 128 in volume 72 of the Galilean manuscripts at the Central National Library in Florence.[1]

Much has been written on the question why Galileo initially chose to put speeds proportional to distances, rather than to elapsed times as he did in his published works. No truly satisfactory answer to that question has been given.[2] To dispose of the question it is necessary to consider Galileo's unpublished notes.

At present it is widely believed that Galileo arrived at his times-squared law by some kind of reasoning from the medieval mean-speed rule. If so, the choice he made in 1604 is still more puzzling, since all mean-speed writings related increase of speed to time. The problem is aggravated by the absence of mean-speed reasoning from all Galileo's writings. No other logical basis for deduction of the times-squared law has won acceptance, though at least one ingenious alternative was offered.[3] On the other hand, few historians have granted the possibility that Galileo may have discovered his law by even rough experiments, and none declared him to have found it by precise observations.[4] Thus the question why he at first chose distances, rather than times, remained effectively unanswered.

It often happens that when the answer is found to a question that has long been debated, it turns out that the trouble had been that the wrong question was being asked. Such is the case in the present instance. The puzzle has existed

Reprinted from *Physis* 16 (1974), 309–22, by permission.

because our alternatives of distance and time were not the alternatives between which Galileo was compelled to decide in 1604. Or, more accurately, he did not then recognize them as the alternatives, as he did later, and as they were presented in the *Two New Sciences*. The question that actually confronted Galileo late in 1604 was whether to put speeds in free fall proportional to distances, or to the square roots of distances. We may say that that is the same thing mathematically as choosing between distances and times, because distances in fall are as the squares of the times from rest. Furthermore, we might say that Galileo himself was well aware of the latter fact in 1604, since that is precisely what he tried to prove. All this is true, but it does not follow that Galileo saw his choice as one between distances and times when he first sought something to represent a measure of increasing "speeds." If we wish to know how matters looked to Galileo in 1604, we must study his working papers. How matters looked to his contemporaries and his precursors, or how they ought to have looked to him, affords no more than a basis for speculation and probable reasoning about Galileo's procedure. On many matters we have no better basis for reconstructing his thought, but in this case we have evidence from his own hand that I consider incontrovertible.

Of the three most essential documents bearing on the question, ff. 107*v* (Plate 10 with transcription in Plate 11) and 189*r* (Plate 8) were not published in the *Opere* of Galileo edited by Antonio Favaro, while f. 152*r* (Plate 6 with transcription in Plate 7) was published only in part and with a mistaken diagram. The last named was reproduced in facsimile and analyzed in 1973.[5] It threw new light on the 1604 purported demonstration, but taken alone it suggested that Galileo arrived at the times-squared law solely by a mathematical device (the mean proportional of two distances) adopted simply to obviate a conflict of ratios arising from an initial false hypothesis. Since the times-squared law is very easy to test once it is known, by utilizing the odd-number rule for successive distances implied by it, Galileo's evident confidence in the truth of those two relationships when writing out the 1604 demonstration was explained as arising from *subsequent* observational tests, between the time of completing f. 152*r* and October 16, 1604. Galileo's reason for putting increasing speeds proportional to distances was evident from his own words in the demonstration, since he there appealed to our experience of machines that act by striking. In effect, he had taken as his criterion for "speed" what we call v^2, and not v. It was noted that the conclusion of f. 152*r* was phrased in terms of speeds rather than times (squares of speeds proportional to distances from rest), but since a tabulation on the same page correctly showed two times as a distance and a mean proportional between two distances, it seemed that the times-squared law was already implicit, to be made explicit in the demonstration intended for Paolo Sarpi.

Thus it appeared that f. 152r alone sufficed to account for Galileo's knowledge of the times-squared law, without any prior observational corroboration; that a simple test using the odd-number rule accounted for his confidence in it, and his remark that he had been asserting it for some time; and that a plausible but mistaken definition of *velocità* accounted for the form of the 1604 demonstration. In short, the choice he had made still seemed to have been between proportionality of speeds to times (as implied on f. 152r) or to distances, suggested to him by percussion effects.

These conclusions arose in the process of ordering all Galileo's notes on motion chronologically, a task still in progress. The difficulties in it arise from the necessity of finding a single chronological order for *all* the notes, and not just for an illuminating selection. A single page that cannot be placed without creating logical or psychological contradictions puts in doubt any ordering of all the rest. The criteria of "contradiction" are not entirely objective and may permit some leeway. Thus a man might forget that he had already found a certain result, and derive it again many years later; but this becomes psychologically less probable as the mass of theorems swells and their interdependence becomes more evident. Changes in handwriting, vocabulary, and watermarks of papers used have to be considered, as well as logical priority and psychological compatibility. One must assume that each entry among the notes had some purpose that makes sense for the state of Galileo's knowledge at the time it was made. Such considerations impose quite stringent conditions on the ordering, and though they cannot all be described in detail with profit to the reader in each presentation of some crucial documents, the reader is assured that many of the objections that may be raised to what ensues here, without considering all the notes, would only be replaced by graver objections if different orderings were tried.

II

On f. 107v there is a column of figures that conforms almost exactly to the distances that would be run on an inclined plane at the ends of eight equal time intervals from rest. To the left of these are the numbers from 1 to 8. In the extreme left-hand margin are the squares of the first eight integers, entered in lighter ink and barely legible in reproduction; these numbers are slightly larger, and differ as to the form of the 6. They were probably added on another occasion. Had a need for them been anticipated when the page was begun, they would more naturally appear after the integers than before them, particularly if the intention had been to compare them directly with the data in the third column. In any event, the latter figures are not exact multiples of the squares by the first datum, 33, though they are remarkably close to being so.

How the figures in the third column were actually obtained is evident from a series of calculations on f. 107, in each of which Galileo multiplied 60 by some integer, and then added a number less than 60. This is the procedure by which distances would be computed if measured by applying a short ruler divided into 60 equal small spaces. On f. 166r a line described as "180 *punti* long" shows that the *punto* of his ruler was 29/30 mm, and this is corroborated by several diagrams in other notes measured by the same ruler.

The data listed on f. 107v were probably obtained by rolling a ball on a plane tilted 60 *punti* in 2000 (about 1.7°) and determining its positions at intervals of 0.55 seconds, probably by musical beats but perhaps by pendulum observations or the sound of regular drops of water into a resonant basin. The timing was probably done by adjusting gut frets tied around the grooved plane, so that the ball made a sound after passing over each fret. In any event, the data in column three are evidently empirical, since they are not exact multiples of squares of times.

Now, it is in my judgment impossible to place f. 107v after Galileo's discovery of the times-squared rule. The obvious way to test the rule, once it is known, is to adjust the frets (or other marks of distances passed) to the rule, and then to test the equality of times by ear or pendulum or the like. Such an experiment would render the calculations shown on f. 107v quite unnecessary, and it is unlikely that any record of it would be made at all, since it would have been only for Galileo's own reassurance. Moreover, on f. 107v there is a list of the alternate odd numbers, 1–5–9–13–17–21, neatly cancelled; the only reasonable explanation of this is that it was his first guess at a rule that turned out to be 1–4–9–16 ...

In the right hand margin of f. 107v we find the series of odd numbers from unity, uncancelled; there are exactly eight of these (all visible on the original), and they are the gnomon of the first eight square numbers. These were put down by Galileo to represent the successive speeds; for since the times were equal, the successive speeds must be as the successive distances. To the left, Galileo drew a series of uniformly increasing vertical lines, as Archimedes represented an arithmetical progression. He then recorded the numbers 3/1, accompanied by a symbol of uncertain meaning (repetition? approximation?), probably when it struck him that in the series of odd numbers, the ratio of the first term to the second holds also for all equal groupings; that is, 1:3 :: 1+3 : 5+7 :: 1+3+5 : 7+9+11, and so on. Hence the choice of unit is arbitrary. Finally Galileo drew the familiar triangular diagram in which increasing speeds were represented by lines parallel to the base and the "overall speed" from rest to any such line was represented by the corresponding triangle. In effect, Galileo's triangle enclosed the Archimedean diagram for arithmetical progression. A basis was thus laid for later puzzles, since the triangle must come to a point, while the lines representing "speeds" remain lines, however short. Galileo did not confront the paradox until later,[6] on f. 179v, and did not dispose of it until still later, when he abandoned the idea of successive speeds and treated speeds in free fall as literally continuous in the mathematical sense.

Chronologically, f. 107v comes between the Italian commencement of f. 152r with a false hypothesis concerning the relations of distances, overall speeds, and times, and its completion in Latin not very much later with a correct tabulation of numerical examples and an argument that put the increasing speeds as the square roots of the distances fallen from rest. Galileo added the remark that in that case the speeds lie along a parabola with distances taken along its axis. He had deduced his result by assuming speeds attained to be proportional to the overall speeds from rest.

III

It was precisely here that Galileo first confronted the necessity of choice as to the proportionality of speeds in fall. As he then saw it, that choice was not between distances and times, for times as such had not entered into Galileo's conclusion as written on f. 152r. To him, the choice was between distances, as implied by the triangular diagram already drawn on f. 107v, and the square roots of distances, forced on him by the assumption used on f. 152r. Between these two alternatives, Galileo would probably in any event have chosen the former, for several reasons. First, it was simpler, and Nature operates in the simplest way. Second, the triangle was more amenable to use for exploring ratios than was the parabola, and ratios dominated Galileo's physics. Finally, the evidence for the triangle of speeds rested on its (supposed) derivation from an irrefutable relation borne out by careful test, whereas the parabola of speeds depended for its support only on an assumption – one that happened to be correct, but that Galileo had at the time no means of testing. For without the double-distance rule, not yet known to Galileo, there was no way to compare the speeds acquired at different places during fall by any kind of direct measurement.

Whether or not Galileo would have made the same choice anyway (speeds proportional to distances rather than to their square roots), he did make the choice on the evidence of the pile driver, as will be seen in the 1604 demonstration. A pile driver works with double effect when the weight is dropped from double distance, and it was reasonable to believe that since the weight remains the same, only the "speed" can account for this, whence the "speed" is proportional to the distance of fall. We say that the square of the speed also changes, and we know that impact effects depend on v^2, but Galileo did not know this and he made quite a different use of the square of the acquired speeds in his 1604 demonstration.

It was while Galileo was reflecting on this choice, just before he wrote to Sarpi and composed the demonstration that has so puzzled historians and biographers of Galileo, that he drew a little sketch on f. 189r. That document is of very great interest, though only the sketch made on it, after all the other work had been done, is of interest to us here. On f. 189v, which was partly published by Favaro, Galileo had found a rule for times on planes differing in slope and length, simply by a kind of numerical trial and error; and he had also related the time consumed in the swing of a given pendulum to the distance fallen vertically in a different time, though in this he made an error that he subsequently corrected after mastering the times-squared law. On f. 189r he had verified numerically his earlier theorem, sent to Guidobaldo del Monte in 1602, that less time is consumed in descent along two conjugate chords to the bottom of a vertical circle than in descent along the single chord connecting the extreme points.[7] Originally he had established this result under the mistaken assumptions of *De motu* that speeds along planes may be treated as constant and as depending only on slopes, being inversely as the lengths for planes of equal height. Now, making the times for such planes directly as the lengths, and applying the mean-proportional rule, he verified the previous theorem. Two of his early attempts to prove it while taking acceleration into account (ff. 148r, 153r, 156–157) include calculations in the same hand as f. 107v, supporting the chronological proximity of these documents.

The sketch in question on f. 189r was added with the sheet turned at right angles to the previous work, of which one line drawn for taking a sum appears to be a continuation of the parabola in the sketch, though it is unrelated to it. The sketch shows the semiparabola mentioned on f. 152r, with its axis, and with several perpendiculars from axis to curve, originally seen as speeds by f. 152r. A straight line was then drawn from the vertex, making an acute angle with the axis, and the parallels were extended out to it. Finally a line making a larger angle was similarly drawn and the parallels were extended to it. Repeated drawing caused ink to go through the paper, obliterating a number on f. 189v, proving this sketch to have been last to be made.

Galileo drew this sketch while reflecting on the choice of parabola or triangle to represent "speeds." The final straight line probably represented the particular triangle that would result from squaring the abscissae of the parabola. Let us suppose that Galileo next ruminated as follows:

I could say that the squares of speeds are proportional to the distances fallen from rest, justifying this through my assumption on f. 152r that speeds have the same ratio as overall speeds from rest. But then this triangle could not be used to represent the overall speeds, because ratios of triangles would not be as the ratios of speeds, or their squares, but as the squares of squares of speeds, far from that assumption. Yet surely if the increasing speeds are as lines, the total speeds are as sums of lines; that is, as areas. Well, suppose instead that separate speeds are as the distances from rest, as our experience of pile drivers seems to show. Then the overall speeds are as triangles, being sums of changing speeds. Of course in that case the other lines, in the parabola, were not speeds at all, as they were on the other assumption. In fact they are the exact opposite of speeds, so to speak. For using the triangle, you square the distances to get the overall speeds, whereas to get those other lines you must take the square roots of those same distances. But wait; aren't the times the exact opposite of the speeds? For it is the same thing to increase speed or diminish time. Well, then, what you get from the square roots of the distances is the times, not the speeds. And sure enough, here in my tabulation on f. 152r, the times are 4 and 6 for the distances 4 and 9, so the times are to one another as one distance is to the mean proportional of both distances. So there is no problem at all, because that is just what I wanted to prove – that the distances have the squared ratio of the times. Moreover, the proof can be made very simple; I shan't need this parabola at all, but merely the well-known contrariety of speeds and times in motion.

This may not have been Galileo's actual line of thought, but it conforms very closely to the 1604 demonstration. It is also agreeable with his unpublished notes. There was nothing explicit on f. 152r to direct Galileo's attention to the other possibility – of identifying speeds with times. Although the ratio 3:2 was implied for speeds on the one hand and for times on the other, the actual numbers 3:2 were not present. The speeds were 15:10, as originally derived by summing integers, while the times had been derived as 6:4 by employing the mean-proportional relation. To link speeds directly with times would contradict good sense in the above framework, as it would also conflict with the idea of speeds as little additive increments. It was in that way that everyone had thought of them, because no speed could exist without some motion and hence some time. That was also Galileo's view up to 1604, and for some time afterward, as evidenced by many notes, and particularly by ff. 172 and 179.[8]

IV

In translating Galileo's 1604 demonstration, I have preserved his word *velocità* wherever instantaneous speed at a place is meant, and have used "speed" where speed through a distance is implied. That Galileo thought of the two as different is evident from his demonstration, in which the latter is treated as the square of the former. In bracketed glosses, *velocità* is designated by *V* and "overall speed" by W; S stands for distance, t for time, and v for speed in modern terms, in which Galileo's *velocità* would be v^2. The glosses will enable the reader to follow the internal consistency of Galileo's argument, however physically unreal its referents, while also interpreting it in modern terms.

I suppose (and perhaps I shall be able to demonstrate this) that the naturally falling body goes continually increasing its *velocità* according as the distance increases from the point from which it parted [$V \propto S \propto v^2$]. ... This principle appears to me very natural, and one that responds to all the experiences we see in instruments and machines that work by striking, where the percussent works so much the greater effect the greater the height from which it falls. And this principle assumed, I shall demonstrate the rest.

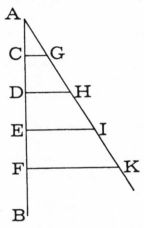

Draw line AK at any angle with AF, and through points C, D, E, and F draw the parallels CG, DH, EI, and FK. Since lines FK, EI, DH, and CD are to one another as FA, EA, DA, and CA, therefore the *velocità* at points F, E, D, and C are as lines FK, EI, DH, and CG. So the degrees of *velocità* go continually increasing at all points of line AF according to the increase of the parallels drawn from all those same points. Moreover, since the speed with which the body has come from A to D is compounded of all the degrees of *velocità* it had at all the points of line AD, and the speed with which it has passed over line AC is compounded of all the degrees of *velocità* that it has had at all points of line AC, therefore the speed with which it has passed line AD has that ratio to the speed with which it has passed line AC which all the parallel lines drawn from all the points of line AD over to AH have to all the parallels drawn from all the points of line AC over to AG; that is, the [ratio of the] square of AD to the square of AC. Therefore the speed with which it has passed line AD has to the speed with which it has passed line AC the square of the ratio that DA has to AC [$W \propto S^2 \propto V^2 \propto v^4$].

And since speed has to speed "contrary proportionality" to that which time has to time – because it is the same to increase speed as to diminish time [$\sqrt{W} \propto t^2 \propto V \propto S$], – therefore the time of motion through AD has to the time of motion through AC that ratio which is the square root of the ratio that distance AD has to distance AC [$t \propto \sqrt{S} \propto \sqrt{V} \propto v$]. The distances, then, from the beginning of motion are as the squares of the times [$S \propto t^2$]; and dividing [into equal times] the spaces passed in equal times are as the odd numbers from unity – which corresponds to what I have said all along, and have observed by experiments. And thus all truths agree with one another.[9]

As also many fictions do, alas.

V

The "contrary proportionality" of overall speeds to times, necessary to intro-duce times into this purported demonstration in order to reach the desired con-clusion, has been as great a source of perplexity to historians as the seemingly false assumption at the beginning. That the assumption was not false, but only misleading to modern readers, was seen when attention was paid to Galileo's remark about machines that act by striking. In effect he had decided to define *velocità* by impact effects rather than as we do, as a sort of ratio of distances to times. Such "ratios" of unlike magnitudes were not permitted in Euclidean mathematics, nor were any means available to Galileo for comparing speeds at points in fall physically, other than impact effects. He was free to define the term in any way he liked, provided only that it did not violate common usage at the time, or the Aristotelian rules for comparing speeds. That it now violates common usage, and even that it violates Galileo's usage after 1609, is beside the point. The fact is that no deduction occurs in Galileo's notes on motion, including those before the 1604 demonstration, in which speeds in fall were treated by him as literally proportional to distances. Even in the false hypothesis that opened f. 152r, the ratio of assumed speeds was 10:15 and that of distances was 4:9. Only in his early attempts to find a logical basis for his theorems is the specious assumption to be found; the deductions and conclusions always accord with the times-squared law after 1604, and they do not need the word "speed", since Galileo was concerned only with relations between the measurable enti-ties distance and time.

The "contrary proportionality" of speeds and times, however, was more diffi-cult to understand. Most historians supposed that Galileo's phrase referred to inverse proportionality, a concept not found in Euclid and not widely in use before the seventeenth century. The principal exception was in the case of the law of the lever, which was usually expressed by setting forth a direct proportionality and then adding the phrase "taken inversely" which applied to the second ratio, since the definition of "inverse ratio" is included in Euclid. The added phrase was sometimes "taken contrariwise," but I have not come across the words "contrary proportion" before 1604. What Galileo meant by it was clear from the context, but as Rupert Hall long ago point out, that was not a recognized mathematical operation; namely, taking on the one hand the squares, and on the other hand the square roots. Hall further pointed out the implication that overall speeds would be as the fourth powers of the times in Galileo's 1604 demonstration.[10]

I offered a rationalization of Galileo's concept in an earlier paper, and remarked that he would not have felt compelled to accept the fourth-power implication, since he was concerned only with ratios and would look upon x^4 : x^2 as indistinguishable from x^2 : x. Still, his invoking this "contrary proportionality" appeared to me strictly *ad hoc*, as a device that would get him to a result he knew on other grounds (whatever they were) to be true. It did not occur to me that it might have been precisely in order to *avoid* such an implication as that which Hall noted, that Galileo adopted the form of demonstration given above. That is what I think now, for though it does not save him in these days of algebra, it did in his era of Eudoxian proportion theory.

In 1604, Galileo needed the triangular diagram in order to relate "overall speeds" to changing speeds, which were represented as lines. His only alternative would have been to adopt some form of mean-speed concept, with a privileged member of an assemblage to represent the entire assemblage. There is no evidence among his notes that he ever tried such a device, or ever thought of it at all. Overall speed in accelerated motion was long treated by him as some kind of summation of separate speeds. To begin with, on f. 121v and f. 152r, he used numerical sums of integers; on f. 107v and thereafter he used areas to sum the lines. Galileo had his own doubts about that concept, calling this approach only probable reasoning and not mathematical proof in his *Dialogue* of 1632.[11] He was able to discard it only when he finally hit upon the one theorem that can be rigorously established without the assumption of either a mean speed or some kind of addition of infinitesimals or indivisibles.[12] This came to him only shortly before 1638, and even then, in the *Two New Sciences*, a trace of the earlier thought survives.[13] Galileo had apparently retained that in his Theorem I on accelerated motion almost up to the very time of publication.

NOTES

1 The 1604 demonstration is translated in section V, below. It bears a watermark unique among Galileo's notes in volume 72, but found also on the cover sheet of his letter to Sarpi, now preserved at the Domus Galilaeana in Pisa. The wording, con-

tent, and handwriting also show that the letter and demonstration are of like date.

2 The most widespread view now appears to be that of Alexandre Koyré, that (1) medieval tradition supported distances as the measure of speeds in free fall, and (2) geometric representation is more natural for distances than for times. Cf. A. Koyré, *Etudes Galiléennes* (Paris 1939), II-21 ff. It is at least questionable whether in fact any writer before Michel Varro, in 1584, specifically asserted that speeds in fall are proportional to distances *from rest*; see S. Drake, "Impetus Theory Reappraised," *Journal of the History of Ideas*. 36, 1975, n. 1, pp. 38–41. Cf. *Physis*. XVI, 1. pp. 47ff.

3 W. Humphreys, "Galileo, Falling Bodies, and Inclined Planes," *British Journal of History of Science*, *3* (1966–7), pp. 230 ff. A second possibility was advanced in S. Drake, "Galileo's 1604 Fragment on Falling Bodies," *BJHS*, *4* (1969), pp. 349–50.

4 It is this last that now unexpectedly appears to have been the case; see section II below.

5 S. Drake, "Galileo's Discovery of the Law of Free Fall," *Scientific American*, May, 1973, p. 84; Italian trans. in *Le Scienze*, July, 1973, p. 36. Cf. Galileo, *Opere* viii (Florence 1933), pp. 426–7.

6 Cf. S. Drake, "Uniform Acceleration, Space, and Time." *BJHS* 5 (1970), pp. 23–4. Not having seen the manuscript at that time, I dated it as probably 1609. It is earlier, possibly 1606.

7 For a reconstruction of Galileo's proof in 1602 see S. Drake, "Mathematics and Discovery in Galileo's Physics," *Historia Mathematica, 1* (1974), pp. 136–7: on p. 140 is a facsimile of f. 189*r*.

8 *Opere* viii, pp. 378–9, 387–8.

9 *Opere* viii, pp. 373–4. Galileo's correspondence with Sarpi is translated in the second paper cited in note 3 above.

10 A.R. Hall, "Another Galilean Error," *Isis, 50* (1959), pp. 261–2; see also the paper cited in the previous note, pp. 344–7.

11 *Opere* vii, p. 256; Galileo, *Dialogue*, tr. S. Drake (Berkeley 1953), p. 229.

12 Cf. S. Drake, "The Uniform Motion Equivalent to a Uniformly Accelerated Motion from Rest," *Isis, 63* (1972), pp. 28–38.

13 In the scholium to Proposition 23.

12

Mathematics and Discovery in
Galileo's Physics

I

Truly I begin to understand that although logic is a most excellent instrument to govern our reasoning, it does not compare with the sharpness of geometry in awakening the mind to discovery. [Galileo 1973, 133]

So said Simplicio, normally a spokesman for Aristotle, in the *Two New Sciences*, the book in which Galileo presented the mathematical theory of freely falling bodies which he had worked out some thirty years earlier. That theory was published in deductive form, starting from a single definition (of uniform acceleration) and a single postulate (that the same speed is attained in fall from rest through the same vertical height along any inclined plane). This orderly unfolding of results affords no clue to the procedures by which Galileo had in fact been led to them in the first place. When we reconstruct his steps from his own rough notes, we find that mathematics was indeed his most fertile source of discovery; hence it was natural for him to have one of his interlocutors express the above view. And since Galileo had begun his investigations of motion along conventional logical lines which had led him into many fallacies and errors, it was suitable to place the remark in the mouth of an Aristotelian philosopher.

Past attempts to reconstruct Galileo's procedures in discovering his new science of motion have made little use of his manuscript notes. These are bound in haphazard order in volume 72 of the Galilean manuscripts preserved at the National Library in Florence, and many of them consist only of diagrams and calculations with little to identify their nature and purpose. Indeed, as might be

Reprinted from *Historia Mathematica* 1 (1974), 129–50, by permission.

expected, the individual sheets that most likely record significant discoveries are characteristically chaotic in appearance. The orderly development of implications of each basic discovery is less difficult to trace among these notes, but it is also much less interesting than the identification of probable discovery documents.

Once a discovery has been made, the use of mathematics to develop its implications is almost routine, at least to us, and it was hardly less so to Galileo, though his methods were very elementary compared to ours in this regard. There are in fact two sorts of mathematical discovery in physics. One sort consists in the following out, systematically, of implications, and some of these may be so surprising as to be entitled to be called "discoveries," in the sense that they were unforeseen by the investigator. They were, however, implicit in what had gone before, and any mathematician would have been perfectly capable of finding them. The other sort of discovery is not a rigorous consequence of what has gone before, though it may have been suggested by that; it consists in the perception that a certain mathematical relationship holds for physical phenomena considered in a certain way. These two types of discovery in mathematical physics probably do much to account for the historical fact that progress seems to be jerky; the consequences of a discovery of the second type are usually exhausted in a generation or two, with innumerable discoveries of the first type, whereas centuries may elapse between bona fide discoveries of the second type. It is mainly with the latter that I shall be concerned in this paper.

The popularly offered reconstructions of Galileo's procedures in establishing his new science of motion are certainly mistaken with respect to the role of mathematics in them. It is quite true that if Galileo had started out with a correct definition of uniform acceleration, as he did in his final published book, he would have been led ineluctably to his conclusions; and it is also true that such a definition had been given in the Middle Ages. All he would have needed to do would be to have applied this definition to the case of free fall, and of course to have added the postulate concerning speeds at the ends of inclined planes, which seems really to have been rather trivial and easy. And it is thus that Galileo's work is presented in textbooks, as a rather humdrum extension of medieval analyses of motion.

One trouble with that account is that Galileo's first treatise on motion, far from including a correct definition of uniform acceleration, does not mention that concept at all; and indeed, it treats acceleration as essentially irrelevant to the mathematics of free fall and as entirely irrelevant to motion along inclined planes. Another trouble is that in 1604, nearly a quarter-century later, Galileo seems to have adopted a quite erroneous rule for speeds in free fall as a basis for deriving the times-squared law, which in fact follows directly from the correct definition. A third difficulty is that the fundamental concept employed by all medieval analysts, that of the mean speed, appears nowhere in Galileo's published works nor even in any of his private notes on motion, in which his final theorems were worked out. Hence the simplistic historical theory that Galileo's science of motion was worked out as an extension of medieval results is quite false, though it remains true that any competent mathematician could have so extended those results, given the idea that free fall is in fact a case of uniform acceleration, and given the postulate about inclined planes.

There is a very good reason that no mathematical physicist had done this. Or rather, there are two such reasons, one of them mathematical and the other physical. The mathematical reason is that it was only shortly before the time of Galileo that the Eudoxian theory of proportion, embodied in the fifth book of Euclid's *Elements*, became available again to European mathematicians. It was that theory which Galileo applied to free fall, treating the growth of speed as continuous in the modern mathematical sense, and all his results depended on that treatment. The physical reason is that the causal approach to acceleration in free fall, demanded by Aristotelian principles, could not allow this to be rigorously continuous. No one before Galileo had been willing to abandon the idea of cause in physics, and in fact many of his younger contemporaries, including Descartes, rejected his assumption that in order to reach any speed from rest, a body must first have passed through every possible lesser speed. The opponents of Galileo preferred hypothetical causes to the principle of sufficient reason. But these topics must not detain us; our purpose is to reconstruct Galileo's procedures, in which mathematics replaced philosophy and led on to his discoveries concerning free fall.

II

In his first treatise on motion, composed at Pisa about 1590, Galileo reduced the conditions of equilibrium on inclined planes to the law of the lever [Drabkin and Drake 1960, 63–69]. Since the topic of his treatise was motion, he tried to find from this a rule for the ratio of speeds along inclined planes. He reasoned that since weight was the cause of downward motion, it was also the cause of speed; and since speed varies with the slope of the plane, he assumed that speeds along planes of equal height should be inversely proportional to the length of those planes. He noted that such ratios were not borne out by actual trial, but among the reasons he listed for this he did not mention acceleration, which at the time he considered to be only a negligible effect at the very beginning of fall.

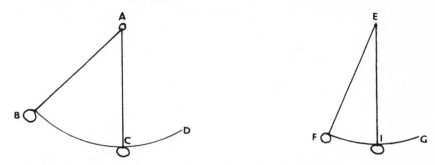

Shortly after moving to Padua in 1592, Galileo composed a treatise on mechanics. In an expanded version of this, probably about 1600, he refined his derivation of equilibrium conditions on inclined planes. This time he related the tendency to motion along an inclined plane to that of fall along a vertical circle tangent to the plane, whether the body was supported by an arc or was suspended from the center of the circle. [Drabkin and Drake 1960, 173] Though he was careful to note that this applied to the tendency at the initial point of fall, he still did not allude to acceleration.

By the year 1602, Galileo had concluded that descent along any chord to the lowest point of a vertical circle was made in equal time, and also that descent was swifter along two conjugate chords than along the single chord determined by them. He also conjectured that descents along all arcs of the lower quadrant were isochronous. Guidobaldo del Monte, to whom Galileo communicated these ideas, replied that they were implausible and were not borne out by experiments in which a ball was dropped along the inner surface of a large hoop. The earlier letters are lost, but Galileo's reply (dated 29 November 1602) reads as follows [Favaro 1934, Vol. 10, 97–100]:

You must excuse my importunity if I persist in trying to persuade you of the truth of the proposition that motions within the same quarter-circle are made in equal times. For this having always appeared to me remarkable, it now seems even more so that you have come to regard it as impossible. Hence I should deem it a great error and fault in myself if I should permit this to be repudiated by your theory as something false; for it does not deserve that censure, nor yet to be banished from your mind – which better that any other will be able to keep it the more readily from exile by the minds of others. And since the experience by which the truth has been made clear to me is so certain – however confusedly it may have been explained in my other [letter] – I shall repeat this more clearly so that you, too, by making this [experiment], may be assured of this truth.

Therefore take two slender threads of equal length, each being two or three braccia long; let these be AB and EF. Hang A and E from two nails, and at the other ends tie two

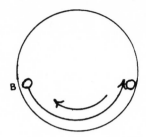

equal lead balls (though it makes no difference if they are unequal). Then, removing both threads from the vertical, one of them very much, as by the arc CB, and the other but little, as by the arc IF, let them go free at the same moment of time. One will begin to describe large arcs like BCD, while the other describes small ones like FIG. Yet in this way the moveable B will not consume more time in passing the whole arc BCD than that which is used by the other moveable F in passing the arc FIG. Of this I am rendered quite certain, as follows.

The moveable B passes through the large arc BCD and returns by the same DCB and then goes back toward D, and it goes 500 or 1000 times repeating its oscillations. The other goes likewise from F to G and then returns to F, and similarly will make many oscillations; and in the time that I count, say, the first 100 large oscillations BCD, DCB, and so on, another observer counts 100 of the other oscillations through FIG, very small, and he does not count even one more – a most evident sign that one of these large arcs BCD consumes as much time as each one for the small ones FIG. Now if all BCD is passed in as much time [as that] in which FIG [is passed], though [the latter is] but one-half thereof, these being descents through unequal arcs of the same quadrant, they will be made in equal times. But even without troubling to count many, you will see that moveable F will not make its small oscillations more frequently than moveable B makes its large ones; they will always go together.

The experiment you tell me you made in the [rim of a vertical] sieve may be very inconclusive, perhaps by reason of the surface being not perfectly circular, and again because in a single passage one cannot well observe the precise beginning of motion. But if you will take the same concave surface, and let ball B go freely from a great distance, as at the point B, it will go through a large distance at the beginning of its oscillations, and a small one at the end of these; yet it will not on that account make the latter more frequently than the former. Then as to its appearing unreasonable that given a quadrant 100 miles long, one of two equal moveables might traverse the whole, and the other but a single span [in the same time], I say that it is true that this contains something of the wonderful. But our wonder will cease if we consider that there could exist a plane as little tilted as that of the surface of a very slowly running river, so that in this [plane] a moveable will not have moved naturally more than one span in the time that on another plane, steeply tilted (or coupled with a great [initial] impetus even on a small incline), it will have passed 100 miles. Perhaps the proposition has inherently no greater improbability than that triangles between the same parallels and on equal bases are always equal, though one may be quite short, and the other a thousand miles long. But sticking to our subject, I believe I have demonstrated that the one conclusion is no less thinkable than the other.

Let BA be the diameter of circle BDA erect to the horizontal, and from point A out to the circumference draw any lines AF, AE, AD, and AC. I show that equal moveables fall in equal times, whether through the vertical BA or through the inclined planes along lines CA, DA, EA, and FA. Thus, leaving at the same moment from points B, C, D, E, and F, they arrive at the same moment at terminus A; and line FA may be as small as you wish.

And perhaps even more surprising will this appear, [which is] also demonstrated by me: that line SA being not greater than the chord of a quadrant, and lines SI and IA being any [two chords conjugate to SA] whatever, the same moveable leaving from S will make its journey SIA more swiftly than just the trip IA starting at I.

I have demonstrated this much without transgressing the bounds of mechanics. But I cannot manage to demonstrate that the arcs SIA and IA are passed in equal times, which is what I seek.

Do me the favor of conveying my greetings to Sig. Francesco [del Monte?], and tell him that when I have a little leisure I shall write to him of an experiment that has come to my mind for measuring the force of percussion. And as to his question, I think that what you say about it is well put, and that when we commence to deal with matter, [then] by reason of its accidental properties the propositions abstractly considered in geometry commence to be altered – from which, thus perturbed, no certain science can be assigned, though the mathematician is so absolute about them in theory. I have been too long and tedious with you; please pardon me, and love me as your most devoted servitor.

III

Of all the interesting points raised by this letter of 1602, we are concerned only with the nature of Galileo's proofs of the two propositions about descent along chords of a vertical circle. He considered those proofs to belong to mechanics, and he did not mention acceleration. Those clues suffice for reconstruction of his probable proofs in 1602, which do not survive in their original form among the surviving notes. Previous speculations about them have assumed that Galileo's proofs followed from correct premises. I shall take nothing for granted except Galileo's knowledge of geometry and what he is known to have assumed earlier.

One of the problems posed by Galileo in his *De motu* of 1590 was that of finding two planes of equal height along which the speeds would be those of two bodies having different "natural" speeds in free fall. (The weights of bodies in air were supposed to affect their speeds.) His idea was thus to equalize the times by dropping the faster body along the longer plane. A logical variant of this problem would be to seek two distances along different planes which would be traversed in the same time. Under the mistaken assumption of uniform speeds determined by slopes in accordance with equilibrium conditions, Galileo's theorem of 1602 is evident from the following diagram:

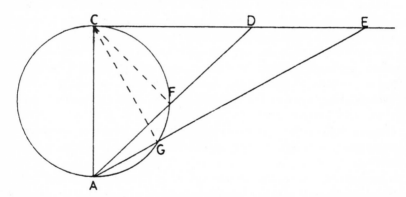

By Galileo's mistaken premise, the speed along DA is to that along EA as EA is to DA. Since CA is the mean proportional between AF and AD, and also between AG and AE, it is evident that EA is to DA as FA is to GA. (EA is to DA as sin CDA is to sin CEA; angles CDA and ACF are equal, as are angles CEA and ACG; hence AF and AG are inversely proportional to AD and EA.)

Hence that the overall speeds along the chords are as the chords follows as a consequence of the mistaken assumptions. But when overall speeds are as the distances traversed, the times consumed are equal; this rule is valid for all motions, whether uniform or not. It can be proved from Aristotle's definitions of equal and greater speed by using the Eudoxian definition of "same ratio," as

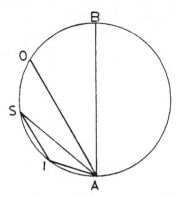

shown later by Galileo in the *Two New Sciences*. [cf. Drake 1973b] Galileo's reasoning at the time of writing to Guidobaldo was doubtless of this form.

Thus this striking result, that the times of descent are equal through all chords to the bottom of a vertical circle, was discovered mathematically from a false premise. The result would be easy to test by releasing two balls simultaneously from points properly marked on facing planes of different slope, or by propping two boards against the sides and bottom of a large circular frame. If Galileo made such a test, however, it would have further confirmed him in his mistaken belief that "speeds" along planes of the same height are inverse to the lengths of planes. At the same time, it would make the experimental failure of his earlier (and illusory) ratios even more puzzling to account for.

The second proposition likewise followed from the same false premise. Chords SA and IA are traversed in equal times, by the foregoing theorem, and the motion along SA is the swifter, by Aristotle's definition (more distance covered in the same time). Then motion along SI is still swifter, since at this speed it is OA, parallel to SI, that would be traversed in the same time as SA, again by the foregoing theorem. But the motion in IA ensuing after motion through SI cannot be slower than the motion in SI alone, since in this case the body starts with that speed from I; and even if IA were level, motion through it would not be less swift than that with which it started. Hence the motion in every part of descent SI-IA is swifter than that in SA, and is at least swift enough to traverse OA in the time of SA or IA. It follows that descent along SI-IA is completed in less time than descent through SA, and hence than descent through IA alone. [Favaro 1934, 213]

All these things are in fact true, and Galileo was later to prove them correctly, after he had discovered the law of acceleration in free fall and extended it to inclined planes. But here it will be good to note that these correct results are much easier to reach from Galileo's mistaken assumption of constant speed depending only on slope of plane than they were for him to deduce from the law of acceleration. Thus, if it is assumed that the appropriate speed is gained immediately, without acceleration, on the steeper plane, no problem arises when SI is very short; but it is not equally obvious, under the assumption of acceleration, that an overall gain in speed will more than make up for the increased length of descent along conjugate chords. The analysis gave Galileo a good deal of trouble later, when he tried to prove his earlier theorems by means of the law of free fall; indeed, his various attacks on propositions 6 (equal times along equal chords) and 36 (time is shorter along conjugate chords) are among the hardest of the fragments to arrange in chronological order.

Both these theorems, discovered mathematically, involved the comparison of "speeds," though there had been no prior attempt to define that concept. The existence of a *ratio* of speeds had been assumed from a well-defined concept in mechanics, that of a ratio of weights. Galileo later compared speeds by confining himself to time-ratios and distance-ratios, virtually eliminating the need for our speed concept. Under Euclidean rules no ratio could be formed of magnitudes unlike in kind, such as distance and time; hence speed could be rigorously treated only by proportionalities.

IV

The importance of acceleration in free fall probably did not become apparent to Galileo until Guidobaldo's objection to his theorems in 1602 stimulated him to suggest the use of pendulums in place of circular surfaces. Observations of long pendulums call attention to acceleration. The bob of a Foucault pendulum is visibly accelerated on each downswing, as one of three braccia (about six feet) long would be. Having in mind the association in his *Mechanics* between initial speeds on an inclined plane and along the tangent circle, Galileo may well have speculated that a similar, but even faster, increase of speed ought to occur along a plane of fixed slope, since the successive tangent planes become less and less tilted as the pendulum approaches the lowest point. At any rate, Galileo did turn his attention to the question of acceleration as such during 1603–04, and successfully searched for a rule linking distances, speeds, and times in free fall.

The discovery of this rule in turn was mathematical in character, and it also started from an initial false premise, though in this instance the error was immediately corrected in the process of discovery. The document recording this event (f. 152r) has recently been reproduced and analyzed. [Drake 1973a] Here I shall include only an English transcription of it (Plate 6 with transcription in Plate 7) and a brief comment. It is a typical discovery document, written partly in Italian and partly in Latin; it starts with a common but false hypothesis of medieval origin, and ends with the correct result. A mistaken attempt to apply compound ratios of time and distance, at the upper right, is related to still another fragment, probably of 1601. To obviate conflicts of ratios, Galileo adopted the mean proportional, and this put the times-squared law into his hands.

Galileo's first move after obtaining the law of free fall was to return to his investigations of motion along inclined planes and to test its applicability to them. The principal document here is f. 189r (Plate 8), linked to f. 152r by the parabola roughly sketched at the left. This sheet is also linked to Galileo's letter to Paolo Sarpi in October, 1604, and to the important demonstration written out for him at the same time, by another figure, probably drawn with the sheet turned clockwise through 90°, which shows a parabola in a right triangle. This diagram marks Galileo's reduction of the inconvenient parabola of speeds to a simple triangle by the simple device of squaring the abscissae.[2] In the proof written out for Sarpi, he dealt with "speed" as the square of the value we now use, which value Galileo himself adopted in 1609. That f. 189r comes very early among Galileo's developments of the law of free fall is also shown by the erroneous value adopted for $\sqrt{2}$, corrected in calculations belonging to 1605 that will be shown later. Here Galileo wrote 7071 as a first approximation for $\sqrt{2}/2$ and then mistakenly wrote 70771 (for 70711) as the next approximation.

The calculations made on f. 189r confirmed that the new law of free fall could be applied to inclined planes consistently, and that the theorems previously sent to Guidobaldo del Monte were consistent with this treatment. Since the reproduction of the manuscript is hard to read, a partial transcription (Plate 12) is provided for use in following the reconstruction of Galileo's procedure.

Galileo's procedure was the following:

1   He drew the circle with quadrant chord CF and its two equal conjugate chords CE and EF, extending the latter out to intersect the horizontal ABC, produced. The radius BC was taken as 100000, later reduced to 100 in order to simplify the calculations.

2   The calculated total length of FD (using the erroneous value for $\sqrt{2}/2$ previously mentioned) was 261761 (or 262); the length of ED was obtained by subtracting from this the length of FE, 76536, giving 185225. Calculation of the time through ED by the mean-proportional rule relating times to distances from rest was then begun just below and to the right of the diagram, but was broken off in favor of using only three-digit figures.

3   Taking the length FD as a measure also of the time from rest through that distance, T_{DE} was computed from the proportionality $T_{DE} : T_{DF} :: \sqrt{185} \times 262$; this gave $T_{DE} = 220$.

4   The next step was of capital importance, since it marks Galileo's abandon-ment of his old mistaken assumption that speeds should be inverse to lengths of plane and his adoption of the direct proportionality of times to lengths of planes of the same height. Here Galileo assumes that $T_{DE}:T_{CE} :: DE:CE$; this enables him to calculate $T_{CE} = 90$ (immediately below the previous work).

5   Since the correct new assumption makes the times to F from rest at any points along the horizontal BCD proportional to the lengths of the correspond-ing planes, the time of descent along EF after fall from D along DE can be obtained by simple subtraction, and thus $T_{DF} - T_{DE} = {}_DT_{EF} = 42$. (The time through a distance in motion from rest at any point other than the initial point of that distance will be designated by showing the point of rest as a preceding sub-script; thus ${}_DT_{EF}$ means the time through EF in motion starting from rest at D.)

6   Galileo's next assumption is in effect that of the single postulate adopted in the *Two New Sciences*. He supposes that the time along EF after fall from rest at a given height remains the same, whether fall from that height to E is vertical, or is supported on the incline. This amounts to assuming that the same speed is acquired in a given vertical fall from rest, regardless of the path of fall. This assumption, at the time of writing f. 189r, probably had no other basis than that it made possible a very simple calculation. Here it gave the total time along the conjugate chords CE-EF as 90 + 42, or 132. This is a shorter time than $T_{CF} = 141$, confirming Galileo's earlier theorem that the time along conjugate chords is shorter than that along their corresponding single chord. And for the first time it indicated a quantitative measure of these times and their difference.

. In all the above, the new assumptions required had their basis in mathemati-cal simplicity alone, though they were in a way not devoid of experimental con-firmation, since the results conformed to two theorems that Galileo already knew to be borne out by test. A rough direct test would also have been possible for the assumption that speeds are the same at the ends of two planes of equal height; Galileo would only have had to watch balls rolling along a level surface after leaving the planes. But whether or not Galileo made such a test, it is evi-dent from the work on f. 189r that he did not reach his postulate on that basis.

Two other investigations are found on f. 189v, one of which relates to the general relation of motions on planes differing in both height and slope, but these probably belong to a somewhat later date.[3]

v

The next important fragments in point of time are ff. 166 and 183, which show a very elaborate diagram and a related tabulation of distances and times. Across the face of the diagram, after making the tabulation, Galileo wrote a number of notes. The diagram is accordingly reproduced [Plates 9, 13, and 14] from Favaro's printed edition, where the notes were placed separately and do not obscure the diagram. Many of the calculations relating to these sheets are preserved on ff. 184 and 192. All these sheets bear watermarks related to one another and to dated letters of Galileo's, placing them in the year 1605.

The purpose of the diagram and tabulation is evident. Having found the law of free fall and related it to motion on inclined planes and broken lines, Galileo was in a position to resume his earlier inquiry into the question of fall along circular arcs. He now approached this through an analysis of the time of fall along the chord of a quadrant as compared with the time along two equal conjugate chords; that is, two sides of the inscribed octagon, then along four sides of the 16-gon, eight sides of the 32-gon, and so on. A relation discovered among the ratios of those times would lead to the arc as a limit. Some notations written over the diagram embody such conjectures; but it appears that Galileo then abandoned this project in favor of further explorations of fall along straight lines in various combinations. Of special interest is his concept of *casus*, discussed below.

Galileo's diagram (Plate 9) does not show the circle of which ac is the chord of a lower quadrant; this circle would pass through points a–f–e–g–8–c.[4] To compute the figures in the tabulation requires only the assumptions previously made on f. 189r – that the times along inclined planes of equal height are proportional to the lengths of plane, and that the times from rest to two points along a given plane are as the shorter distance is to the mean proportional between the two distances.

The letter x, seen along the vertical ad in Plate 9, indicates the point of intersection of the designated line with the horizontal ba, extended. Zodiacal symbols near ad likewise refer to points of intersection with the upper horizontal line. Galileo's procedure in the calculations was to apply the mean-proportional rule to vertical distances, and then to adjust them to the incline by recourse to similarity of triangles. Thus the mean proportional between ad and te is 84090, which is multiplied by 76536/70711 to get 91707, the time through ae.

A notation across the diagram to which I alluded earlier reads as follows: "The time through the two [planes] eg–gc, from rest at e, is 66236; it should be 71757 if [that] fall (*casus*) had the same ratio to fall through ec as fall through ae–ec has to fall through ac." Here the word *casus* means "time of descent in units in which the time through ad is 100000." This meaning can be ascertained by deriving Galileo's figure of 66236 (or rather, 66326, since it appears that he made a transposition in writing this). The derivation, outlined below, shows that he was able not only to calculate the time along a broken line of any number of sections, but that he could do this in terms of a single standard unit of measure. Since T_{ad} and ad had both been taken as 100000, Galileo had in effect a unit speed, giving him a means of comparing speeds without violating the Euclidean rule against forming ratios between magnitudes unlike in kind. The modern reader may object that the concept of "unit speed" already implies such a ratio. I shall not argue this point here, but am content to point out that the calculations outlined below foreshadow Galileo's later device of taking a selected line to represent both a distance and the time of fall from rest through that distance, which device was used frequently in the theorems of the *Two New Sciences*, and in his notes.[5]

The table on f. 183 (Plate 14) gave the time of fall through eg–gc, but only after initial fall from rest through ae; that is, it gave not T_{eg-gc}, but rather $_aT_{eg-gc} = 21657 + 19896 = 41553$. The new problem was to find T_{eg-gc}, and to find this in units such that $T_{ad} = 100000$. Galileo's procedure was a direct extension of what had been done on 189r.

Since $ad = 100000$, $ec = 76536$. Let $T_{ec} = ec$; then $T_{e\theta} = e\theta = 29289$. Now, $e\omega = vg - te = 21677$, the difference between verticals dropped from ab to g and to e. Hence $T_{e\omega}$ is 25197, mean proportional of $e\omega$, $e\theta$. It follows that $T_{eg} = 25197(eg/e\omega) = 25197(39017/21677) = 45353$. Likewise $_eT_{gc} = _eT_{\omega\theta}(gc/\omega\theta) = 29289 - 25197)(39017/7612) = 20974$. And $T_{eg-gc} = T_{eg} + _eT_{gc} = 45353 + 20974 = 66327$.

Thus Galileo had developed a systematic procedure for the calculation of times of fall along any broken lines in terms of the time of fall through a standard vertical distance. (It should be noted, however, that these comparisons would not be borne out by experiment because of the factor of 5/7 for inertial moment in rolling as against the rate of acceleration in free fall.)

One other notation on f. 166 (Plate 13) also deserves comment: "Let ad be 100 *punti* long; let the time of fall through this be 180 minutes, and through both ad–dc, 270 minutes." This implies the double-distance rule, to which Galileo may have been led by noting the numerical relations as the inclined planes of f. 166 were approaching the horizontal. The rule was proved by Galileo later, probably in 1607, in a memorandum establishing one-to-one correspondence between speeds in uniform and in uniformly accelerated motions.

VI

The foregoing documents show the manner in which purely mathematical considerations entered into Galileo's basic discoveries concerning free fall. The procedures seem to me rather different from Plato's demonstration that the triangle and the number 3 are the cause of fire, and from Kepler's proof that the number of planets must be six in order to accommodate the five Platonic solids each once and only once. Nor do they particularly resemble the calculations of Swineshead, Heytesbury's proof that uniform acceleration from rest is possible, or the famed triangular proof of Oresme that such acceleration is represented by its middle speed.

Galileo's first correct theorem concerning fall was that of the equality of times of descent along chords of a vertical circle to its lowest point. He reached that theorem by valid mathematical reasoning from a false assumption about ratios of speeds and without considering the role of acceleration at all. His attempt to persuade a friend that the conclusion was correct appears to have called his own attention to the importance of acceleration. An attempt to discover consistent ratios for accelerated motion, starting from an erroneous legacy of the Middle Ages, was successful through the adoption of an arbitrary mathematical device. That success gave him the law of free fall, but in a form that tempted him to define "velocity" in a bizarre way and then to abandon that concept for several years in favor of time and distance ratios. By means of these, he was able to discover many theorems concerning accelerated motion without formalizing their conceptual basis separately from the mathematics employed in them. It was only toward the end of his life that he turned to that task.

The profound difference between medieval and Galilean physics involves many things. Not least among these was the restoration of Eudoxian proportion theory in the sixteenth century, an event of great importance to the history of mathematics quite apart from the work of Galileo. An excessive concern with the history of philosophy on the part of historians of science has tended to conceal this fact and to create an illusion of greater continuity between the fourteenth century and the seventeenth than is justified by the events. It is probably true that given the correct text of Euclid and the authentic works of Archimedes, Galileo needed nothing from the Middle Ages for his work on motion. It is probably also true that given only the medieval Euclid, all the works of Jordanus Nemorarius, Thomas Bradwardine, William Heytesbury, Jean Buridan, and Albert of Saxony, this would not have enabled Galileo to go even as far as Nicole Oresme in the approach to a valid mathematical physics in the modern sense of that phrase.

BIBLIOGRAPHY

Carugo, A. and L. Geymonat, ed. 1958 *Galileo Galilei: Discorsi e Dimostrazioni Matematiche Intorno a Due Nuove Scienze*. Paolo Boringhieri, Torino.
Drabkin, I.E. and S. Drake 1960 *Galileo on Motion and on Mechanics*. University of Wisconsin Press, Madison.

Drake, S. 1973a "Galileo's discovery of the law of free fall," *Scientific American* 228(5), 84–92.

Drake, S. 1973b "Velocity and Eudoxian proportion theory," *Physis* (in press).

Favaro, A. 1934 *Le Opere di Galileo Galilei.* Vol. 8 unless otherwise indicated. G.R. Barbèra, Florence.

Galileo 1973 *Two New Sciences.* Translated by S. Drake. University of Wisconsin Press, Madison.

NOTES

1 A proof on mechanical assumptions is found on f. 151, but this is probably of later date. A similar proof was given as the second of three under Theorem 6 on accelerated motion in the *Two New Sciences*; this resembles the proof on f. 160.

2 Explanation of the proof for Sarpi was given in Drake 1973a, 91–2, on the basis that Galileo meant by *velocità* what we call v^2, before this diagram was noted.

3 The page was folded and the other side used with the paper turned at right angles to the position of f. 189r at one time, and then reversed at another. The two portions are unrelated, and the writing is smaller.

4 Galileo ran out of letters for this diagram and continued with numbers and conventional symbols; see below.

5 Galileo did not explicitly justify the procedure when he introduced it in the Third Day of the *Two New Sciences*. It was also used in the Fourth Day for "impetus," and explained in the discussion there. He would not have considered time-distance fractions legitimate, as we do, though his contemporary Marin Mersenne moved in that direction.

13

The Role of Music in Galileo's Experiments

Einstein saw Galileo as a kindred spirit who had faced problems not very different from his own. The testing of relativity theory pressed the limits of observational accuracy, and so had Galileo's experiments on free fall three centuries earlier. Perhaps Einstein had that in mind when he wrote, in the foreword to my English translation of Galileo's *Dialogue Concerning the Two Chief World Systems*: "The experimental methods at Galileo's disposal were so imperfect that only the boldest speculation could possibly bridge the gaps between empirical data. For example, there existed no means to measure times shorter than a second."

It is true that in Galileo's day there were no reliable clocks or watches, let alone the instruments and techniques that are now available for measuring extremely short times. It does not follow, however, that Galileo lacked any means by which he could divide time into equal intervals shorter than a second. We have grown so accustomed to precise instruments calibrated in standard units that we may think they are necessary for exact experimentation. That is not the case, however; if it were, modern physics could never have got started. Bold speculation has often advanced physical science, but it alone did not suffice to inaugurate modern physics. Neither classical antiquity nor the Middle Ages lacked for brilliant speculations, but for 2,000 years they did nothing useful to bridge the gaps in the empirical data. The modern era began with Galileo's law of falling bodies, and if Galileo had had no way of dividing time into intervals of less than a second, he could never have established that law firmly enough for it to deserve acceptance.

Reprinted from *Scientific American* 232, no. 6 (June 1975), 98–104. Reprinted with permission.

The phrase "measure time" makes us think at once of some standard unit such as the astronomical second. Galileo could not measure time with that kind of accuracy. His mathematical physics was based entirely on ratios, not on standard units as such. In order to compare ratios of times it is necessary only to be able to divide time equally; it is not necessary to name the units, let alone to measure them in seconds. The conductor of an orchestra, moving his baton, divides time evenly with great precision over long periods without thinking of seconds or any other standard unit. He maintains a certain even beat according to an internal rhythm, and he can divide that beat in half again and again with an accuracy rivaling that of any mechanical instrument. If the cymbalist in the orchestra were to miss his entry by a tiny fraction of a second, say by a 64th note in the music, everyone would notice it, not just the conductor. Professional cymbalists have virtually perfect timing, although many of us may feel that theirs is the one place in the orchestra we could learn to fill. In that we are mistaken, but the point is that we would not be tempted to think we could unless we believed it is quite easy for anyone to divide time accurately into very small intervals.

In fact, that is what we do whenever we dance or sing. A dance step may take half a second, but an error of much less than half a second is quite noticeable and even unbearable. Few people can stand the irregularity of the ticks when a metronome or a pendulum clock is slightly tilted. Most of us can sing to an established beat and would wince if a pianist played a grace note where the principal note ought to be (that is, a hundredth of a second too late). And it is not difficult for us to tell, when we hear two notes close together, which one we heard first.

I have been describing equipment that was second nature to Galileo, whose father and brother were professional musicians and who himself composed creditable music for the lute, an instrument he is said to have played quite well. We may think such equipment is not very suitable for scientific experiments, but that is mainly because we no longer need it for them. It was certainly not suitable for a published experiment designed to convince Galileo's readers; even in his day it would have been foolish to write, "I tested this law by singing a song while a ball was rolling down a plane, and it proved quite exact." As a result the experiments later described in print by Galileo were different. Instead of dividing the time equally he fixed the distances in established ratios and then measured the times over the distances by weighing on a delicate balance the water that had flowed from a large container through a narrow orifice. The ratios of distances agreed with the square of the ratio of the times thus measured – within a tenth of a pulse beat, Galileo declared. His published experiments were duplicated in 1960 by the historian of science Thomas B. Settle, who found that Galileo's stated accuracy was more than confirmed.

Galileo did not say, however, that he first obtained his law by performing experiments. All he said about the discovery was that nature had led him almost by the hand to his rule that in natural motion equal speeds are added in equal times. Historians have poked fun at Galileo for that picturesque statement. In October, 1604, Galileo wrote to his friend Paolo Sarpi that the times-squared law followed from a different assumption: that "speeds" in fall are proportional to the distances fallen from rest. No evidence was forthcoming that nature's guiding hand had led Galileo out of his apparent error during at least the next four years, and possibly longer.

Recently, however, things have begun to look different. In an earlier article in this magazine I presented evidence that Galileo's word for "speed" in his letter to Sarpi did not mean our v but rather what we write as v^2 [see "Galileo's Discovery of the Law of Free Fall," p. 248 in this volume]. For the first time a certain page of Galileo's notes on motion (designated $f.$ $152r$ (Plate 6 with transcription in Plate 7) in Volume 72 of the Galilean manuscripts preserved at the National Central Library in Florence) was published in full. The proportionality of speeds in acceleration to times elapsed from rest was already implied in that document, which was almost certainly written before Galileo's letter to Sarpi. In his attempt to derive the law of free fall mathematically from a physical notion, however, he had decided to define "speed" as something that is in fact proportional to distance. His reasons can now be filled in from marginal notations on $f.$ $107v$ (Plate 10 with transcription in Plate 11), another previously unpublished page of Galileo's notes in the same volume.

First, from $f.$ $107v$ we can reconstruct an experiment that probably took place in 1604, after Galileo had abandoned his first hypothesis on $f.$ $152r$ and before he went on to establish a correct rule for speeds and distances from rest. This new document looks singularly unpromising at first glance, but it records an actual experiment. The figures listed in the third column at the top left represent very nearly the distances from rest of a ball rolling down an inclined plane at the ends of eight equal times. The distances were actual, not theoretical, since they differ slightly from the products of the first number, 33, by the square numbers 1, 4, 9, ..., 64. Those numbers in the first column, barely legible in reproduction, were added later with a different pen and ink. Galileo's figures in the third column are not exact multiples of squares, and neither could they have been obtained by using water-timing as in Galileo's published experiment. To get such accurate figures a single time had to be divided into eight equal intervals. I believe this was done as outlined below, which will explain why in spite of this experiment's interest and historical importance it was not published by Galileo.

Place a grooved inclined plane about 6½ feet long at an angle of about 1.7 degrees. Fit a stop at the higher end, against which a steel ball can be held by resting a finger on it lightly. Now sing a simple march such as "Onward, Christian Soldiers" at a tempo of about two notes per second, very crisply. When the tempo is established, release the ball at some note and mark with chalk the position of the ball at other notes (half-second intervals). In three or four runs eight positions can be marked; put a rubber band around the plane at each mark. (Galileo would have tied gut frets at these places, in the way that adjustable frets were tied around the neck of a lute, but of course today rubber bands are easier to place and adjust.) Then, making many repeated runs, adjust the rubber bands so that the audible bump made by the ball in passing each one always coincides exactly with a note of the march. When the inclined plane has thus become a kind of metronome, measure the distance in millimeters from the resting position of the ball to the bottom of each rubber band. The figures should agree tolerably well with those in the third column on *f. 107v.*

Galileo's calculations to the right show how he actually got the numbers entered in the third column. In each calculation he multiplied 60 by some integer and then added a number less than 60. That is just what someone would do if he were measuring distances with a short ruler divided into 60 equal parts. Galileo certainly had such a ruler, because several diagrams in his notes on motion were drawn so that a principal line contained 60 times the number of units in which all the other lines were measured. In one note he gave "180 points" as the length of a line I measured as 174 millimeters, so that we may take Galileo's unit as being 29/30ths of a millimeter. That is almost exactly the shortest graduation (which he called a "point") measured on his own proportional compass at the Museum of the History of Science in Florence.

Since Galileo actually measured the distances listed on *f. 107v,* it is safe to say that when he started, he did not yet know the times-squared law. Otherwise his natural procedure would have been to tie frets at appropriately calculated distances, saving many runs of the ball and a lot of adjusting. In that way a single run would have confirmed the law and there would have been nothing to write down. Hence the actual calculations, combined with evidence that the first column was written after the third, indicate that Galileo did not have the law of free fall when he began *f. 107v.*

Still further evidence is provided by the series of alternate odd numbers (1, 5, 9, 13, 17, 21) he wrote down on *f. 107v* and then canceled. That represents his first guess at a rule for the growth of the third column. It may seem a strange guess, but (as I shall explain below) Galileo had good reason to expect some kind of regular additions to the speeds and distances in successive equal times.

Time	Time²	Distance (centimeters)	Distance (points)	Figures in f. 107v	Difference (points)	Distance in 1/64 second (points)
.55	.30	3.176	32.9	33	−.1	1.8
1.10	1.21	12.705	131.4	130	−1.4	3.7
1.65	2.72	28.59	295.7	298	−2.3	5.6
2.20	4.84	50.82	525.7	526	−.3	7.4
2.75	7.56	79.41	821.5	824	−2.5	9.3
3.30	10.89	114.3	1,182.4	1,192	−9.6	11.2
3.85	14.82	155.6	1,609.8	1,620	−10.2	13.1
4.40	19.36	203.3	2,103.1	2,123 [2,104]	−20.1 [− .9]	14.9

CLOSE AGREEMENT is evident between Galileo's figures and the theoretical figures for the experiment assumed by the author and his colleagues to underlie the f. 107v data. The distances passed by a ball rolling on the inclined plane in eight equal times were calculated both in centimeters and in Galilean "points." (A "point" in Galileo's system of measurement equaled 29/30ths of a millimeter.) The plane was assumed to have an elevation of 60 points in 2,000, giving a theoretical acceleration of 21 centimeters per second squared. Two figures are given in the table for Galileo's final distance, the first being the one he originally measured and the second being the altered figure he entered in the third column of f. 107v. Before correction this figure had been the only one showing a difference greater than can be accounted for by assuming a tolerance equal to the distance run by the ball in a 64th of a second at each speed; calculated tolerances are indicated in last column at right.

Since the actual rule of growth is 1, 4, 9, 16, ..., Galileo's canceled figures were not far wrong up to the fifth, and he stopped at the sixth.

If my reconstruction is correct, this experiment will surprise other historians of science as much as it did me. We all believed that until considerably later than Galileo's lifetime, experiments were made only to test some preconceived rule, not to seek a rule from measurements. The only exception previously known to me was a series of experiments by Galileo's father, Vincenzio Galilei. He discovered by measurement that the weight loading similar strings of equal length that sound a given musical interval are as the inverse squares of the string lengths sounding that same interval under equal load. Music thus appears to have had a dual role in the beginnings of experimental physics, since both pitch and time played their part.

The slope of plane and unit of time Galileo used to get the data recorded on f. 107v cannot be exactly specified, but there are some practical limitations on both. A plane much steeper than two degrees will produce an inconveniently high speed too quickly, whereas a much gentler slope causes the ball to hesitate at the first few frets. Intervals of less than half a second or more than a second cannot be judged to be equal as accurately as those within that range, perhaps because the heart beat (and its half) supply unconscious rhythms. A probable incline for the experiment is reasonably suggested by the length of the plane used (2,100 points) and the length of Galileo's ruler (60 points). The same plane, but one more steeply tilted, was probably again used for Galileo's experiments leading to the discovery of the parabolic trajectory, in which the greatest length needed was 2,000 points [see "Galileo's Discovery of the Parabolic Trajectory," by Stillman Drake and James MacLachlan, p. 160 in this volume]. A ball rolling on a plane elevated 60 points in 2,000 has an acceleration of 21 centimeters per second squared, for which the distances passed in eight equal times out to 2,100 points agree closely with Galileo's data. The distance run in a 64th of a second at each speed yields a probable variance from the ideal placement of frets.

Two figures are given in the table for the final distance, the first being the one Galileo originally measured and the second being his altered figure in the third column. Before correction this distance had been the only one showing a difference greater than can be accounted for by a tolerance of a 64th of a second. It is interesting that after Galileo's correction the last fret was almost exactly in place, whereas the two preceding frets were left about a centimeter off, although they were still within the assumed range of tolerance. Probably after the frets had been placed to Galileo's satisfaction and the distances had been measured and recorded he made a final run or two and decided that the last bump was definitely late. After two of the other figures he wrote a plus mark and after two a minus mark, probably indicating that those sounds seemed a little early or late but not enough to require further adjustments.

Consistently positive deviations (except for the second fret) are not surprising. A trained observer will judge the instant a star crosses the meridian with great consistency among his own data, but his data will not be identical with those of another equally skilled observer. Astronomers have long known of this phenomenon, which is technically called the personal equation. If my reconstruction happens to be right for the slope and time used by Galileo, it means that he judged each beat to be simultaneous with a bump when the bump was in fact a little after the beat. (A different slope or time would at most reverse the sign, without changing the explanation.) The negative difference in placing the second fret, on the other hand, has a physical explanation rather than a psychological one. In practice unless the ball is very massive it is delayed in climbing the first fret, so that the second fret must be moved a little closer than it is in theory in order to get the second time equal to the first. No such delay is noticeable after the ball is really under way.

Thus everything about the data on *f. 107v* conforms with the foregoing experimental procedure, except perhaps that the high degree of accuracy seems incredible. Yet once one abandons the notion that precisely calibrated instruments are absolutely necessary it is apparent that Galileo had more than one means available to him for the equal division of small times. He knew about the isochronism of the pendulum by 1602, so that he could have used returns to the vertical of a pendulum hung from a point on a vertical line. Or he could have used the sound of drops of water falling into a resonant basin from a small hole in a large tank. In practice, however, it is harder to judge the agreement of a sound with a position of a moving pendulum, or even the agreement of two different sounds coming from outside sources, than it is to decide when a single set of external sounds agrees with a strong internal rhythm. Those who have studied music with a professional performer know how remarkably the last ability can be developed. That is why the very accuracy of Galileo's data persuades me that his experiment was of the kind I first described.

So much for the musical and experimental implications of *f. 107v*. Its other parts – the notations in the right-hand margin – are equally interesting for the light they shed on Galileo's return to *f. 152r* (Plate 6) for its successful completion and on his first attempted logical derivation of the law of free fall. Those matters were discussed in my 1973 article, but on the assumption that mathematical persistence and ingenuity alone accounted for Galileo's discovery of the times-squared law for distances in free fall. With a little repetition of what was said there, more steps can now be filled in from *f. 107v*, as follows.

From the Middle Ages to the time of Galileo the usual way of thinking about actual acceleration in free fall was in terms of a series of small and rapid spurts of uniform speed, accumulating as though by quantum jumps. The reason for this tradition was that early physicists were primarily interested in causal explanations. Increments of added impetus provided a satisfactory cause for successive acceleration, but any truly continuous change of speed seemed to require different effects within a single cause, contrary to certain of Aristotle's principles. (It made sense for the early physicists to study change, but not change of change.) When Galileo began his search for the mathematics of acceleration on *f. 152r*, he likewise assumed successive increments, and he sought the "overall speed" from rest by adding up separate "degrees of speed," putting $1 + 2 + 3 + 4 = 10$ "degrees" as the overall speed for four units of time and $1 + 2 + 3 + 4 + 5 = 15$ "degrees" for five units of time. This hypothesis led him to contradictory meanings for "15 degrees of speed," however, and so he put *f. 152r* aside for a time.

The experiment of *f. 107v* left Galileo no room for doubt that distances from rest are related as the squares of the times. He went on to reason that since in this experiment the individual times were equal, the successive speeds (each of these being thought of as a brief overall speed) must be proportional to the successive distances passed. And since the overall distances from rest went up as the square numbers, the successive partial distances must go up as the odd numbers. (Mathematicians used to call the odd numbers the "gnomon" of the squares, because when one adds in turn 1, +3, +5, +7, ..., one gets 1, 4, 9, 16, ...) Therefore in order to represent the increasing speeds Galileo wrote the series of odd numbers seen at the right-hand margin. To the left he drew a series of uniformly increasing parallel vertical lines, in the way that Archimedes always represented an arithmetical progression.

The series of odd numbers has the unique property among arithmetical progressions that the ratio of the first number to the second is also the ratio of the first two numbers to the second two, and so on. Thus 1 is to 3 as 1 + 3 is to 5 + 7 and as 1 + 3 + 5 is to 7 + 9 + 11. This means that if the initial time unit had been any multiple of the one used, the relations would still have been the same. That thought had probably struck Galileo when he wrote down a 3 over a 1 near the odd numbers on *f. 107v*. Then he drew the little triangle at the right to represent overall speeds as being built up of successive speeds in small equal times, which speeds he represented by lines parallel to the base. In effect this was merely enclosing in a triangle the Archimedean diagram for an arithmetical progression. There is a logical oversight in this step, however, because the triangle actually comes to a point, whereas the other diagram always begins with a line, no matter how short a line. This particular triangular diagram was accordingly destined to lead Galileo astray several times in the ensuing years, whenever he sought a purely logical foundation for the law of free fall. Having represented speeds by lines parallel to the base and distances of fall by the altitude, he thought he had somehow to make "speeds" proportional to distances from rest. Actually what we call speeds are proportional to the square roots of those distances. Galileo eventually treated them as such, but not in 1604.

If I am correct about the order in which Galileo's notes were written, he quickly derived the square-root rule for speeds and almost as promptly rejected it. This rule was derived on *f. 152r* from the simple (and correct) assumption that the speeds at two given points in fall are proportional to the overall speeds of a body in reaching those points from rest. The two distances Galileo happened to select for his original hypothesis on *f. 152r* were 4 and 9, and by a coincidence the two overall speeds he had then assigned were 10 and 15, as I have explained above. The square roots of those distances are in the same ratio as the next two "triangular" numbers used for speeds, but that is not usually the case. (For example, the roots of 9 and 16 are not in the same ratio as 15 and 21, the next two triangular numbers.) Galileo had before him two "overall speeds" based on the traditional discrete-addition concept, which happened to fit the square-root relation that was really valid only for continuous acceleration from rest. He duly noted on *f. 152r* that the square-root relation would make the speeds plotted against distances fall along a parabola. Since there was no way to reconcile that with the triangle of "speeds" already drawn on *f. 107v*, a choice had to be made.

It seems that Galileo put his faith in the diagram he had arrived at (by a logical oversight) from his experiment on *f. 107v* rather than in the correct assumption he had made on *f. 152r*. Thus in the demonstration of the times-squared law that he went on to work out for Sarpi he decided to make his "overall speeds" proportional not to the speeds acquired but to their squares. He made the acquired "degrees of speed" in turn proportional to the distances from rest. That came about because when Galileo sought a physical meaning for "speed," he hit on "the effects of machines that work by striking," as he said in the 1604 derivation. A pile driver doubles its effect by dropping the weight from a doubled height. Galileo reasoned that the weight stays the same and all that changes is the speed, so that it was reasonable to make "speed" proportional to distance fallen. Galileo changed this to time of fall when he discovered the parabolic path of projectiles five years later.

Einstein might have been surprised at the role played by precise experiment in Galileo's discovery of the law of free fall, although as a musician Einstein would have particularly liked its origin. Galileo's boldest speculation bridged the gap not between empirical data for lack of means to deal with times shorter than a second but between the discrete and the continuous. It occurred to Galileo only after he had made several more attempts to reconcile his first notion of additive speeds with mathematically continuous acceleration from rest. We still cannot prove experimentally, any more than Galileo could, that continuous change really underlies the changes we can measure only in discrete units of some kind. That assumption took an act of faith and resulted in a bold speculation that put Galileo at odds with some of his ablest contemporaries. In accepting a mathematically continuous change of speed from rest, Galileo consciously abandoned the traditional search for a physical cause of acceleration in free fall and contented himself with its mathematical description. The elimination of cause as a necessary concept in physics was as distasteful to most of Galileo's contemporaries as the elimination of force was later to many of Einstein's.

14

New Light on a Galilean Claim about Pendulums

Perhaps the best single reason for doubts that Galileo attached importance to precise experiment has been a passage in the Fourth Day of the *Two New Sciences* concerning two equal pendulums swinging through widely differing arcs. Taken in isolation from the rest of the discussion, the passage as I translated it two years ago is this:

> But experiment contradicts this, for if two friends shall set themselves to count the oscillations, one counting the wide ones and the other the narrow ones, they will see that they may count not just tens, but hundreds, without disagreeing by even one, or part of one.[1]

This appears to be such a palpable exaggeration that I appended two notes, saying first that "The statement is false, and the experiment adduced in its support is fictitious. Galileo probably deduced the result from his mistaken assumption that resistance of the medium is proportional to velocity ... ," and second that "A disagreement of about one beat in thirty should occur with pendulums of the length and amplitudes here described as isochronous. Writing this passage in his old age, Galileo may have recalled his valid experiments with pendulums weighted with cork and lead as described in the First Day, and confused them with the quite different result incorrectly asserted here."

My translation and notes have now appeared in print, but I no longer believe them to be right. The final four words – "or part of one" – are not what Galileo said; yet, like every other translator from 1730 on, I saw no viable alternative. Weston, in 1730, gave "nay, not of the least part of one"; Crew and De Salvio, in 1914, wrote "not even by a fraction of one;" and modern German and French

Reprinted from *Isis* 66 (1975), 92–5, by permission of the University of Chicago Press. Copyright 1975 by the History of Science Society, Inc.

translations have the equivalent. Even the usually accurate Spanish version reads *ni siquiera en un solo punto*. Thomas Salusbury, in 1665, stayed closest to Galileo's text with "yea, or one sole point"; yet the sense of every rendering, even those which kept "point" and did not introduce "part," was to convey the idea that the two friends would call out the same number at the same time throughout, the two pendulums being always at the same relative point of swing. Yet any experimental test shows from the very start that that is not true, and thus for Galileo to have added it would seem to have been a pointless exaggeration added to an initial misstatement.

Galileo's actual words are "... *senza discordar d'una sola [vibrazione], anzi d'un sol punto.*"[2] Now, *punto* does not mean "part," and still less, "fraction"; on the other hand, if we keep "point," *anzi* does not mean "yea, or," and still less, "not even," which is the sense of *ni siquiera*. What *anzi* does mean is "rather," in the sense of "or rather," when the speaker adds a more precise or even a pedantic correction either because he wants to avoid a carping criticism or because something of interest would otherwise be missed by the hearer. Thus I might say, "Euclid's fifth definition in Book V defines equal ratios; or rather, 'the same ratio.'" In this way I speak first for the casual hearer and then fend off a sophisticated critic, at the same time suggesting to others that an interesting distinction was once made that we now ignore.

Galileo's text is accordingly to be read: "... they may count not just tens, but even hundreds, without disagreeing by even one: or rather, by one count alone."

Among the meanings for *punto* listed in an enlarged Crusca dictionary in 1833 is that of a numerical count, as "point" in a game score. This was a poor choice by Galileo, as things turned out, but it need no longer interfere with our understanding of what he meant. The terminal correction above was meant by him not only to fend off pedantic critics, but also to disclose a further very interesting fact revealed by experiment; namely, that the two friends will for a time remain in agreement, and thereafter will differ by only a single count no matter how long they continue the observations.

My colleague James MacLachlan has conducted an actual test using pendulums 110 inches long, which is within the range specified by Galileo (4 to 5 *braccia*). One pendulum was started through a vertical angle of 80° and the other through 10°. Agreement in count continued through 30 swings, which took 1¾ minutes. During the 31st swing of the smaller, the person counting that advanced one count, which remained the difference after 150 swings, or 8 minutes. By that time the oscillations of both were through arcs so small that no greater difference in tally was ever to be expected.[3]

MacLachlan's observations convince me that Galileo's *anzi* was inserted to distinguish the first three tens from further tens and hundreds; that is, no difference in count beyond a difference of one. The Italian *anzi* is identified with Latin *immo* in the same Crusca dictionary previously mentioned. *Immo* means "yea rather" or "nay, but rather;" it can mean "indeed, at best," but it never has the sense of "not even." Now, it is of interest that the one translation which may be of considerable authority is the very accurate Latin version published in 1699–1700 at Leiden, which reads: *se non tantum decades, sed etiam centenarios, absque unius vibrationis, imo unius puncti differentia esse numeraturos.*[4] Here the doubtful phrase is definitely grouped with the counting, eliminating any possibility of an elliptical reference of *puncti* back to the relative positions of the pendulums. The Latin version was published anonymously at Leiden, at a time when a new translation of Galileo's book would have met with small demand. Hence I think it likely to have been the translation made under Galileo's own supervision and intended for the collected works in Latin that he negotiated with the Elzevirs at Leiden, but which did not appear because of the slow sale of the works they had previously published for him.[5]

The failure of modern translators, including myself, to catch Galileo's meaning in this passage is only partly due to the odd word *punto* for "a count," with the related modification of his word *anzi* in order to make some sense of the concluding phrase. Part of the trouble has been a tendency also to read his use of *etc.* in the description of the experiment as disjunctive, as though we were to compare repeated swings through some large arc of fixed size with repeated swings through a small arc. No such experiment is even possible: but a habit of discrediting Galileo's descriptions as thought-experiments tempts us to dismiss that fact. The experiment Galileo described was clearly one in which two equal pendulums were started through very different arcs and were then allowed to swing freely without interference of any kind. Galileo's *etc.* simply indicates (as he says) the gradual diminution of arc. Here is the full context:

As to the other point, we must show that the impediment received from the air by the same moveable when moved with great speed is not very much more than that which the air opposes to it in slow motion. The following experiment gives firm assurance of this.

Suspend two equal leads balls from two equal threads four or five braccia long. The threads being attached above, remove both balls from the vertical, one of them by eighty degrees or more, and the other by no more than four or five degrees, and set them free. The former descends, and passing the vertical describes very large [total] arcs of 160°, 150°, 140°, etc., which gradually diminish. The other, swinging freely, passes through small arcs of 10°, 8°, 6°, etc., these also diminishing bit by bit.

I say, first, that in the time the one passes its 180°, 160°, etc., the other will pass its 10°, 8°, etc. From this it is evident that the [overall] speed of the first ball will be 16 or 18 times as great as the [overall] speed of the second; and if the greater speed were to be impeded by the air more than the lesser, the oscillations in arcs of 180°, 160°, etc. should be less frequent than those in the small arcs of 10°, 8°, 4°, and even 2° or one degree. But experiment contradicts this, for if two friends shall set themselves to count the oscillations, one counting the wide ones and the other the narrow ones, they will see that they may count not just tens, but even hundreds, without disagreeing by even one; or rather, by one count alone.

I believe that Galileo never deserved the charge that has been made against him on the basis of this passage: my former opinion that the whole experiment was fictitious and that the claimed result was false rested on a simple misapprehension of his words. Nevertheless, Galileo's argument was invalid at this point for a reason that has remained buried under the misdirected criticism.

The pendulum experiment was intended to show that for reasonable speeds through air (bullets and cannonballs having been excluded), resistance of the medium is not appreciably different for swift and slow motions. Galileo remarked that the speed of a long pendulum swinging through a large arc is indeed very considerable at times, and much greater than that at any time in a small total arc. He then argued that comparison of the times elapsed during a great many such swings supported a proportionality of air resistance to speed.

But it is clear that for the purposes that led Galileo to describe this experiment only those swings in which one pendulum moves quite swiftly with respect to the other could be relevant. The mere accumulation of large numbers of swings added nothing, and indeed this distracted attention from the important early swings. Only in the first ten swings are the top speeds vastly different, and even then both pendulums spend an appreciable part of the time in motions that are slow and virtually equal for both. Now, in the early swings, the total time consumed from the beginning is conspicuously different between the two, and it is only by running the count into the hundreds that the difference in total time is made relatively small. Galileo's argument as a whole is therefore deceptive, even though his description of the experimental facts is vindicated.

NOTES

1 Galileo, *Two New Sciences*, trans. Stillman Drake (Madison: University of Wisconsin Press, 1974), pp. 226–227. The entire passage is cited below.
2 *Le opere di Galileo Galilei*, ed. Antonio Favaro, Vol. VIII (Florence: G. Barbèra, 1933), p. 277.
3 Lead bobs of great size would not be reduced to virtual isochronism within 60 swings, as are small ones. On the other hand, large bobs of small density also exhibit the effect

described. Thus it holds for a considerable range of materials, though not for all pendulums however weighted.

4 Galileo, *Discursus et demonstrationes mathematicae* ... (Leiden, 1699), p. 224. This was published together with the fourth printing of Bernegger's Latin translation of the *Dialogo*, for which the engraved title page is dated 1700.

5 Cf. *Opere*, Vol. VIII, p. 204, Elia Diodati to Galileo, June 15, 1640: "The Elzevirs write to me that they are deferring for some time the printing of the Latin translation of your works, until they have sold a greater number of those already printed by them, more than 500 copies of each being still on hand"; that is, of the *Dialogue*, the *Letter to Christina*, and the *Two New Sciences*, of 1635, 1636, and 1638, respectively.

15

Galileo's Accuracy in Measuring Horizontal Projections

In 1973 I published f. 116v (Plate 3 with transcription in Plate 4) of Galileo's notes on motion, a record of measurements of projections of a ball rolled down an inclined plane, deflected horizontally, and then falling from a table to the floor.[1] Galileo's unit was identified by a line drawn on f. 166r (Plate 9), and my analysis at the time implied his measurements to have been accurate within ±3½%. That surprised me, as did his use of a unit as small as 0.94 mm. Only three years before, I shared with other specialists the opinion that Galileo had not discovered by carefully controlled measurements the results he later confirmed in that way as described in his last book.[2] The previously unpublished f. 116v convinced me that so far as discovery of the parabolic trajectory was concerned, the prevailing opinion was mistaken, though Galileo's more fundamental discovery of the times-squared law of distances in fall from rest remained to be explained.

In 1975 I published f. 107v (Plate 10 with transcription in Plate 11), which recorded a series of measurements in the same unit, accurate within 1/64th of a second for descents at the end of each of eight successive equal time intervals.[3] That convinced me that the law of fall had also been discovered from carefully controlled measurements, though the only explicit entry on f. 107v recording square numbers had been added in a margin, in ink different from the rest and presumably a bit later. Not until the end of 1982 did I succeed in identifying f. 189v1 (Plate 15), rather than f. 107v, as the page on which Galileo had been writing when he first recognized the times-squared law.[4] He then entered the square numbers in the margin of f. 107v to confirm that the law held also for descent along an inclined plane, Galileo having discovered it in the course of calculations involving his timing of free fall through 3.76 meters (4,000 *punti* in

Reprinted from *Annali dell'Istituto e Museo di storia della scienze di Firenze* 10 (1985), 3–14, by permission.

his unit). That timing was accurate to 1/20th of a second, while two others (of a 2,000-*punto* pendulum and of fall through 2,000 *punti*) were virtually exact. No room was left for doubt that the opinion I had shared with other specialists, before examining Galileo's working notes on motion, was quite incorrect.

Analysis of Galileo's measurements of downward oblique projections, on ff. 114v (Plate 5) and 81r (Volume 1, Plate 1), made it possible to describe in fair detail the apparatus and procedures he had used.[5] His inclined plane was grooved to guide the ball, as described in *Two New Sciences*, width of groove 8½ *punti* being such as to reduce linear acceleration down the plane by 5.75%, the ball used being of radius 10 *punti*. For f. 114v the angle of the plane was set so that horizontal advance was double the vertical descent; that is, at $\tan^{-1}$ 1/2, or 26°.565. For f. 81r, the angles were $\sin^{-1}$ 1/3, 1/6, and 1/12, Galileo's measurements of projections on both documents were accurate within 4 *punti*.

This information, gleaned over a period of years, made it no longer tenable to suppose that on f. 116v Galileo had measured incorrectly by as much as 40 *punti*, as seemed to be implied by calculations of his own on that page. A skilled experimentalist capable of holding his results within a variance of four units would not err by ten times that amount in two out of five recorded measurements. Accordingly I have re-examined f. 116v to find, if possible, just what Galileo had measured before making his *doveria* calculations.[6] Using the information gained from other documents about his apparatus and procedures, I concluded that Galileo made those calculations not in the expectation of their agreeing with the measurements he compared with them, but in hope of obtaining further information about inter-relations among vertical descents (and rolls) along the plane, distances of fall during projection, and lengths of projections as such.

That pattern of search was the same, in the case of horizontal projections, that Galileo later adopted in his study of oblique projections downward, without deflection at the end of rolls on the plane. The great difference was that the former led to notable success, whereas the latter ended fruitlessly. An apparent difference was that vertical descents on the plane were emphasized in the first, and distances of roll along the plane in the second, but in fact both entered into each investigation. That is not immediately evident from the documents, but emerges from their analyses. It is important to remember that the notes we have were intended only for Galileo's own use, so that he did not spell out on them everything that he noticed and put to use.

The calculations I have called *doveria*[7] above had a meaning that Galileo did spell out in the first instance, thereafter abbreviating to the single word *doveria*, "it should be," or better, "it would have to be." In its first use, Galileo wrote out "it would have to be, to correspond with the first," alluding to projection of 800 *punti* after vertical descent of 300 along the plane, horizontal deflection, and fall through a distance that I, and others, had assumed to have been 828 *punti*.[8] The assumption was seemingly safe, because the verbal statement on f. 116v is "828 *punti* height of the table," and terminal fall was from the table. No one noticed that Galileo used a dotted line to indicate that height; at least, none regarded that as significant. Pierre Costabel noticed another dotted line on f. 116v, representing the path of the ball from end of the table to projection marked 800 along the floor,[9] but he did not notice or comment on the fact that there is also a solid line (as in all the other cases) that met the floor just a trifle beyond 800. That probably represented 803½ *punti*, as will be explained. For his *doveria* calculations Galileo used 800, but because that was not the exact distance 803½ which he had measured, he connected 800 to the table with a dotted line. I shall show why I believe the other dotted line, for table height of 828 *punti*, represented a distance used only in one *doveria* calculation and not in any of the recorded measurements. How Galileo arrived at 828 will be explained in due course, as well as what distracted him into a different (and more fruitful) line of inquiry than further alteration of the height of his table.

I regret the necessity of postponing matters of interest, but I must first return to the meaning of *doveria*. To "correspond with the first," in Galileo's sense, meant to correspond with vertical descent of 300, deflection in a certain way, and projection of 800 during fall of 800 – not fall of 828, as we all assumed. That exact projection with Galileo's apparatus would have required a change of procedure, undesirable for other reasons, but was quite close to his initial experimental situation for f. 116v. With a plane grooved, in relation to the ball used, so as to reduce acceleration by 5.75% from that of roll along a smooth plane, projection at Padua after vertical descent of 300 and fall through 800 would be 803.6 *punti* by calculation, taking $g = 980.7$ cm/sec^2, and projection being made at the speed reached during roll. That is, so to speak, the minimum projection with Galileo's apparatus for the stated vertical descent and terminal fall, assuming no loss of energy in the process of deflection.

To achieve horizontal projection at the speed reached by the ball at the end of the plane, a simple procedure is to place two parallel boards to receive the ball smoothly from the end of the groove. A grooved incline resting on a table necessarily has its lowest point of top surface somewhat above the table, and that height determines the thickness of the boards which, separated by the width of the groove, will receive the ball smoothly. Rolling radius of the ball being unchanged, speed of projection will be the speed reached at the end of descent.

At least five projections noted on f. 116v, three of which have not been previously identified or explained, were made in the above way. The first was noted only by intersection of the solid line with the floor, just beyond the dotted line, at 804 to the nearest *punto*. The second, recorded as 1,328, had 820 *punti* as either vertical descent or terminal fall, the other being 800. (Vertical descent and terminal fall are interchangeable for a given mode of projection; that is, descent of x and fall through y gives the same projection as descent of y and fall through x.) Three other projections made in this manner are noted at lower center of f. 116v, the page having been rotated; they are 1,394, 1,603, and 1,866, for terminal fall of 800 *punti* after vertical descents of 900, 1,200, and 1,600. They will be mentioned again later.

Now, the ball can be received smoothly at the end of its roll in another way, and it will be projected farther. Instead of placing parallel boards to take the ball from the groove and continue its motion along an equal horizontal groove, a single board not quite so thick will receive smoothly the lowest point of the ball. It will then be projected at higher speed than before, almost exactly 2% higher with Galileo's apparatus. The reason for this is easily seen, though the calculation is rather complex. Along the grooved plane, the ball acquires a certain speed of rotation while rolling on a reduced radius, part of that being always below the surface of the plane. If that rate of angular rotation were conserved unaltered, any roll on the full radius before projection would increase the lineal speed of the ball.[10] The ball which previously was projected 803.5 *punti* would travel 819.7 *punti*, or 820 to the nearest *punto*. That is the new figure I introduced above without explanation. It was taken from f. 116v, at lower right, where Galileo made a calculation with 820 precisely the same in form as his first *doveria* calculation, with 800, at top center. Either calculation gives projection after double the vertical descent for which projection is already known, in this case for vertical descent of 600. We would simply multiply by $\sqrt{2}$; Galileo multiplied the number by its double and then extracted the square root of the product.[11]

Galileo's calculation shows that he had measured the projection of 820 *punti*, implying his use of the second mode of deflection in which the ball rolls on its full radius before it enters into fall. That mode of deflection was also used for the longest projection, 1,500 as recorded by Galileo. Calculation puts this at 1,496, for vertical descent of 1,000 and terminal fall of 800 *punti*, the ball having been received on its lowest point. For vertical descent 800 and terminal fall 800, calculation gives projection of 1,338½ in this mode; Galileo wrote 1,340. That number is of interest in more than one way, Galileo's *doveria* calculation for comparison with it having been the only one to use 828, so before proceeding to the remaining measured projection (1,172) I shall say more of it.

The number 1,340 appears to have been written in after 1,328 and 1,500 were already in place below the horizontal line representing the floor. At any rate Galileo placed it above that line. The only clue to is origin, in 1973, seemed to be Galileo's use of 828 in the place occupied, in all other *doveria* calculations, by a vertical descent. Accordingly I assumed that Galileo had belatedly tried vertical descent of 828 *punti*, without entering that number on the vertical line above the table in his diagram, where I supplied it in square brackets. I went on to offer an explanation for the added trial, derided by critics (who offered no alternative explanation whatever).[12] The new explanation which follows now seems to me better.

Although the angle of the plane does not matter in analyzing horizontal projections, it is convenient in calculation to assume some particular angle; and, knowing that for f. 114v Galileo used $\tan^{-1}$ 1/2, I chose that for my calculations. Now, at that angle, roll along the plane for vertical descent of 600 is 1,341.6. Another previously unexplained notation at the bottom of f. 116v, which will be discussed later, inclines me to believe that Galileo did put his plane at the same angle for both ff. 114v and 116v. If so, the origin of the projection 1,340 may have been Galileo's interest in finding a projection that was equal to a length of roll used in obtaining another projection. That would account for the lack of any new vertical descent associated with this projection, along the line above the table in the diagram, since Galileo found that descent of 800 and fall of 800 gave the projection he measured as 1,340, when the ball was received on its lowest point.

The remaining projection, 1,172, corresponds very nearly to that given by the second mode of deflection when vertical descent is 600 and terminal fall is through 820 *punti*. Calculation gives this as 1,173½. The previous appearance of 820 was ambiguous, as I remarked, though again the absence of any such designated vertical descent weighs in favour of its having been a terminal fall. The present instance seems to me decisive, it being more probable that Galileo raised his table height from 800 to 820 than that he lowered it to 600 *punti*. Whether or not he raised it again, to 828 *punti*, can be left to consider after Galileo's source for the number 828 has been explained, something not necessary in accounting for the recorded measurements on f. 116v as above.

Deflection in the first manner gives, so to speak, the minimum projection for the grooved plane in use. Deflection in the second manner yields a longer projection, probably to Galileo's surprise. To wander about the maximum projection would then be natural. For vertical descent of 300 and fall of 800, the first mode of deflection gave projection of 803½; the second mode, 820. Maximum projection is 828, as the reader may have guessed. It is achieved by having the ball roll on its full radius at all times. At first sight, that seems to mean that the ball must remain unguided during roll along the plane. Certainly rolling it along the ungrooved edge of Galileo's plane would be to invite mishaps. What Galileo did is hinted in *Two New Sciences*. Over the groove he glued limp vellum,

then in common use by bookbinders.[13] Weight of the ball indented this sufficiently for straight guidance, while contact by the lowest point of the ball
assured rolling without slippage on the full radius.

Returning now to the projection of 1,172, it was noted by Pierre Costabel that
the number is very nearly $828 \sqrt{2}$. But that is 1,170.97, and if Galileo had
merely calculated from 828, he would have written 1,171, not 1,172. That was
definitely a measured projection. Its variance from my calculated 1,173½ may
be ascribed to a small error of measurement, or to some loss of energy when the
ball underwent friction during its acceleration to higher speed by rolling on its
full radius. When Galileo found that 828 was the projection for 300 vertical
descent on the vellum-covered groove, it is likely that he tried 600 descent and
measured close agreement with the projection of 1,172 already noted; but it is
not likely that he made the calculation elsewhere, nor is it seen on f. 116v.

Galileo's next move, as with 820 earlier, would be to test projection of 828 as
a terminal fall. He recorded this as "height of the table," drew the dotted line,
and might have gone on had not his *doveria* calculations at this time led him to
a conjecture and a new line of inquiry that turned out to be very fruitful.[14]

As I wrote in 1973, the work behind f. 116v began as a test of a conception
Galileo had reached earlier by reasoning: horizontal motion, in the absence of
impediments, should continue at uniform speed. The way to an accurate quantitative test of this was opened by a proposition that Galileo wrote out on f. 164r
about the end of 1607 (as I did not know in 1973): speeds in descent are as the
square roots of the vertical distances passed from rest. That became the basis of
Galileo's *doveria* calculations; but before he began writing f. 116v, he had
probably already verified his proposition by measurements differing from calculation no more that one or two *punti*. For such verification, the first mode of
deflection sufficed, and his preliminary work probably also led him to select
terminal fall through 800 *punti* and vertical descents of 300, 600, and 800 for
special study.[15] The path taken by the ball was not uppermost in his mind, if it
was present at all. Calculations to confirm uniform horizontal speed concerned
ratios of speeds at the end of the plane and ratios of distances of projection at
the end of the same fall; nothing more.

On f. 116v Galileo began inquiries into relations among the selected vertical
descents and the resulting projections. He found that the latter were altered by a
different mode of deflection, and then that descent outside the groove maximized
them. In the way he had already established uniformity of horizontal motion, he
made *doveria* calculation for comparing a mixed set of actual projections with an
arbitrary standard,[16] the departures might reveal some new relation.

In the course of making these calculations, Galileo was aware that two tendencies to motion, both now known to him quantitatively, were obeyed by the

ball during projection. In traditional natural philosophy, only the stronger of two tendencies simultaneously present in a body could make itself evident, and only if that one weakened in conflict with the other, until it no longer dominated, could the other produce its effect. After the birth of impetus theory in the fourteenth century, that was assumed in explaining projectiles, supposed to go straight while the "forced" motion remained stronger than the "natural" tendency to fall; the latter, being natural, remained undiminished by the conflict and at last dominated and brought the projectile straight down. Tartaglia postulated a short circular arc joining the two straight motions, but remarked that the initial motion was never truly straight. Benedetti stressed the "mixed" character of the entire motion, but lacked quantitative means to analyze it further. Galileo surmised that the two tendencies did not conflict, both being natural in his physics, and should accordingly be compounded independently of one another in the motion of projection, neither one dominating or being weakened.

Instead of proceeding with further projections utilizing a new height of terminal fall, Galileo sketched the paths of descent from table to floor, put aside f. 116v, and on f. 117 (Plate 1) proceeded to draw and analyze the parabolic paths followed by horizontally projected bodies. On f. 91v, for the first time, he correctly linked speed and time in natural descent, and on other folios he began the making of calculations and formulation of theorems concerning parabolic trajectories.

Early in 1609 Galileo wrote to Antonio de' Medici about his recent work, announcing that shots fired horizontally (point-blank, as that was called) struck the ground in the same time regardless of powder charge and distance traveled.[17] That followed from his work on f. 116v, once he had explicitly linked speed in fall with time. Originally, time as such had not entered into the calculations, but only speeds.

In the same letter Galileo announced that shots fired upward, provided that they reached the same maximum altitude during flight, also took the same time from gun to ground. That conclusion does not follow evidently from the work hitherto discussed; it came from a series of experiments for which we have only a single document as evidence. On f. 175v (Plate 2), Galileo drew a diagram of projection upward by a hook-shaped device at the end of a very steep incline. Perceiving that this had something to do with projection, I published it in reduced facsimile in 1973, and was misled into thinking that Galileo had used a steeply inclined (64°, the complement of the angle I then thought had been used for f. 114v) to obtain his projections for f. 116v. But the hook-shaped device has no utility in making horizontal projections to be accurately measured; for that, some horizontal roll must be allowed before projection.

On f. 175v a ball is shown in two positions, one at the start of a descent, with the number 3 beside it, and the other as it rises from the device. Taking the number 3 to indicate 300 *punti* above the horizontal line drawn, measurement of the trajectory sketched suggests projection of 1,000 and maximum altitude of 100. The incline with upward deflector could be turned to different angles of projection, and a series of points marked along the incline suggest that Galileo dropped a ball from them at different positions of the apparatus. I have not made the calculations thus suggested, since I believe that Galileo learned no more from this work than he wrote to Antonio de' Medici. Even to have discovered that would take a good deal of ingenious testing by Galileo, to find descents and angles that brought the ball to the same maximum altitude in projection, perhaps by placing a board above and parallel to the horizontal base and finding "shots" that caused the ball just to touch it in flight.

Pending publication of my paper on Galileo's timings recorded on ff. 154v and 189v1, readers may be interested to know that Galileo collected during a motion, and weighed in grains, the water that had flowed at a rate of 3 fluid ounces per second through a tube in the bottom of a large container – as he reported, without quantitative details, in *Two New Sciences*. To have a more convenient unit in calculating, he divided the weights by 16; this made his unit, which I call the *tempo*, equivalent to flow of 1/30th of a fluid ounce, or one gram, and to 1/90th of a second. Galileo's poorest recorded timing was accurate to 5 *tempi*, or 0.056 second. Hence he would have been able to verify equality of times for upward oblique projections reaching the same maximum altitude, in the trials that I suppose him to have made with the device shown on f. 175v.

Earlier in this paper I mentioned another set of notations on f. 116v, previously unexplained, to which I promised to return. Three numbers appear at the bottom of the page, written when it was turned as for the additional projections cited previously; these appear as 236, 247, 59. However, the page was trimmed so that only the third figure, 59, is certain, the first two having probably had an initial digit removed by the trimming. Now, 59 *tempi* is 0.656 second, and the time the ball was on Galileo's plane during vertical descent of 300 *punti* at $\tan^{-1}$ 1/2 was 0.654 second. Because no other number on f. 116v is of so small a magnitude as 59, I believe that Galileo timed that roll and noted his result. Above it is the 247 which I believe was originally 1,247; that is the square of the time, in Galileo's units, of terminal fall through 800 *punti* at Padua, or more precisely, I calculate the latter at 0.3916 second, while Galileo's finding was 0.3923 second expressed in *tempi*. (So far as I know, Galileo never related his *tempi* to the second, nor did he need to for his calculations.) The figure above 247 appears as 236; I believe it to have been originally 2,236, the length of roll at $\tan^{-1}$ 1/2 for vertical descent of 1,000 *punti*.

The additional projections noted nearby, as said earlier, imply vertical descents of 900, 1,200, and 1,600 *punti*. Those were distances of roll used in obtaining oblique projections from the plane at $\tan^{-1} 1/2$ for f. 114v. My view is that f. 114v came after f. 116v, Galileo having first confirmed uniformity of horizontal motion in the absence of impediments, and having known how to make the necessary calculations for horizontal projections. The purpose of f. 114v was to investigate a class of natural motions not reducible to horizontal projections, as I have explained elsewhere. The foregoing interpretation of f. 175v now adds to Galileo's studies of horizontal and of downward oblique projections a study of upward oblique projections, conducted before February 1609.

Probably further information about Galileo as a careful and resourceful experimental physicist, guided by accurate measurements,[18] will be found among his working papers by other students, who may appreciate knowing, if only to challenge and overthrow, the analyses I have ventured here. I am content to have shown that we are no longer obliged to suppose Galileo to have been satisfied by measurements as approximate as ±3½% when he found horizontal motion to be uniform in the absence of material impediments.

NOTES

1 S. Drake, "Galileo's Experimental Confirmation of Horizontal Inertia," *Isis* 64 (1973), 289–305. I would not now use the word "inertia," a dynamic concept, but the phrase "conservation of speed," appropriate to Galileo's kinematic viewpoint.

2 S. Drake, *Galileo Studies*, Ann Arbor, 1970, p. 218.

3 S. Drake, "The Role of Music in Galileo's Experiments," *Scientific American* (June, 1975), 102–110.

4 S. Drake, "Galileo and Mathematical Physics," in *Scienza e Filosofia*, ed. C. Mangione, Milano, 1985.

5 S. Drake, "Analysis of Galileo's Experimental Data," *Annals of Science* 39 (1982), 389–397.

6 Described physical apparatus used in described procedures can yield only specific data, which can be calculated within a specifiable range of accuracy. In this paper I present results Galileo could have obtained with apparatus and procedures described in detail. To dispute them, others should offer equally detailed specifications of other apparatus no less likely to have been used by Galileo.

7 I took the obsolete form *doveria* to correspond with modern *dovesse*. In the 1611 edition of John Florio's Italian-English dictionary, however, *doveria* was listed as an alternative form of *dovrebbe*. Florio himself said he had never seen a grammar that correctly explained the important difference between forms such as *dovesse* and *dovrebbe* (which he classified as tenses), and wrote: "There is no tence (*sic*) in the Italian tongue, wherein Englishmen commit more incongruities." Had Galileo

expected the projection he measured as 1,172 to be 1,131, he would have written *dovesse*. Knowing what he had actually measured, he wrote *doveria*, "it would have to have been." Galileo then noted the difference of 41 *punti* (and other differences in *punti*) as of possible use in assigning quantitative effects of the varied procedures described in this paper.

8 It will be shown that height of the table was 800 *punti* at the outset, and was raised to 828 *punti* only after Galileo had entered the measurements of projections shown along the floor-line.

9 P. Costabel, "Mathematics and Galileo's Inclined Plane Experiments," *Reason, Experiment, and Mysticism*, ed. M.L. Righini Bonelli and W.R. Shea, New York, 1975, p. 183.

10 What is conserved unaltered is here slightly over-simplified for ease of understanding the essential physical fact that additional lineal speed is acquired by roll on full radius.

11 Galileo's calculation differs in form only, not in basis, from his *doveria* calculations not involving the exact double.

12 W.R. Shea, N.S. Wolf, "Stillman Drake and the Archimedean Grandfather of Experimental Science," *Isis* 66 (1975), 397–400.

13 The roll for 300 *punti* vertical descent at $\tan^{-1}$ 1/2 is 671 *punti* or 63 cm. Limp vellum is pliable and easy to work with; smooth joins are often found on repaired old bindings.

14 It is probable that Galileo raised the table height from 800 to 820 and finally to 828 *punti*, and that at the final height he made a very interesting set of measurements shown in the table below. If so, the two dotted lines on f. 116v are related by Galileo's study of distances of roll, which became of prime importance as he proceeded to oblique projections on f. 114v. At full, acceleration along the plane at $\tan^{-1}$ 1/2 and terminal fall of 828 *punti*, after *rolls* of double the distances shown on the vertical line above the table, calculated projections are compared below with Galileo's *doveria* figures.

Verticals on f. 116v	Rolls assumed	Projections	Doveria figures	
300	600	797½	800	
600	1,200	1,128	1,131	
800	1,600	1,302½	1,306	
[828]	1,656	1,325	1,329	(See Galileo's calculation)
1,000	2,000	1,456½	1,460	

15 No folio bearing such preliminary measurements and calculations survives, nor would Galileo have any reason for preserving it after transferring to f. 116v those he selected for further study. His discarding of superseded calculations is illustrated by

f. 90^{b}v, a fragment of a folio of pendulum calculations pasted down in 1609 with a note on a different topic written on the side previously blank, now numbered 90^b. This discarded pendulum notation remained concealed until 1979.

16 Knowing that two modes of deflection had been used, Galileo would not expect agreement with the calculations, particularly when those were based not on an exact measurement but on an approximation to it that simplified the work of calculation.

17 *Opere di Galileo Galilei*, ed. A. Favaro, Florence, 1934, vol. x, p. 229.

18 Astronomy became the first exact science by attending closely to accurate measurements. Physics became the second, two millennia later, by doing the same thing. Galileo pioneered that move, to the exasperation of contemporary natural philosophers.

16

Galileo and Mathematical Physics

The seventeenth century witnessed a revolution in physics even more profound than the astronomical revolution, though the latter appears to have more greatly altered the course of men's thinking. Kepler's discovery of the planetary laws rested on the Copernican astronomy and on the unrivaled accuracy of Tycho Brahe's observations, both having been events of the preceding half-century. That indeed revolutionized astronomy, but it exhibits an orderly sequence of events in a science of great antiquity. The Copernican arrangement of heavenly bodies and their motions had been adumbrated by Aristarchus, even before Hipparchus and long before Ptolemy. The practice of modifying hypotheses as needed to fit more accurate observations was as old as astronomy itself. Without detracting from the merits of Kepler or of Newton, and without questioning that vehement opposition was aroused by the new astronomy, historians can reasonably view it as having evolved quite understandably, though with a great burst of energy during a relatively short period of time, often called a "revolution."

Galileo's announcement of a mathematical law of physics, exact within a specified range of experimental variance, was, in contrast, entirely unprecedented. It was simply stated – distances in fall from rest are proportional to the squares of the times – and it had been tested with no greater variance than one-tenth of a pulsebeat. In antiquity, Archimedes had reasoned mathematically about statics and hydrostatics, but not about motions, or by appealing to actual measurements. During the fourteenth century, some natural philosophers had analyzed motions mathematically, but had not applied their findings concerning uniform acceleration to the fall of bodies. Indeed, no record is known to me of

Reprinted from *Scienze e filosophia: saggi in onore di L. Geymonat*, ed. C. Mangione (Milan: Garzanti, 1985), 627–42, by permission.

any attempt before Galileo to make measurements of accelerated motions, even rough and approximate measurements. Within half a century after his *Two New Sciences* of 1638, Isaac Newton's *Principia* moved on to the law of universal gravitation, established the three laws of motion that underlay physics until our own time, and applied those principles in the first truly integrated system of the world.

Because of the importance of Galileo's law of fall to the revolution in physics, historians and philosophers of science have long speculated about the manner in which Galileo arrived at it. Before the researches of Pierre Duhem into medieval natural philosophy, no source for Galileo appeared possible except actual measurements. Duhem's findings opened the possibility that Galileo had followed the medieval lead, going on to deduce the times-squared form of law from the fourteenth century middle-speed postulate applied to uniformly accelerated motions.[1] In that case, the novelty of Galileo's contribution would be restricted to his having cited accurate measurements as sufficient validation for a proposition in physics.

The most influential modern analyst of the Scientific Revolution, Professor Alexandre Koyré, went much further. He flatly rejected Galileo's account of the confirming measurements as fictitious and quite impossible. Koyré argued that "the decision to use mathematical language, a decision which corresponds to a change in metaphysical attitude, cannot itself be the product of experiment that is conditioned by it."[2] The Scientific Revolution in general, and Galileo's contribution to it in particular, appeared to Koyré to be capable of explanation only in terms of a philosophical shift to Platonism against the established Aristotelianism of the universities.

Professor Ludovico Greymonat, my friend of many years in whose honour I present this paper, was critical of Koyré's sweeping identification of mathematical physics with Platonism. In his valuable biography of Galileo,[3] stressing the pioneer physicist's philosophy of science, Geymonat pointed out that an instrumentalist view of mathematics fitted much better with many passages in Galileo's books, as well as with the spirit of his science generally. Writing as a mathematician and historian of mathematics, Geymonat questioned Koyré's thesis in a manner somewhat similar to that in which, writing as a philosopher and historian of science, Professor Edward Strong[4] had earlier opposed the Platonist thesis argued in E.A. Burtt's *Metaphysical Foundations of Modern Science*. Even before Koyré's influential *Etudes Galiléennes*, Strong's historical investigations (undertaken with a view to extending Burtt's conclusions) had impelled him to conclude that during the sixteenth century, especially in Italy, procedures rather than metaphysics were already shaping mathematics itself.

The conclusions of Geymonat and Strong, writing from different backgrounds about Galileo's science, are confirmed powerfully by analysis of a previously unpublished document in Galileo's own hand – f. 189v1 (Plate 15) of his working papers on motion, bound in volume 72 of the Galilean manuscripts at Florence. That was the page on which Galileo was writing at the moment he first recognized the law of fall in its simple and elegant times-squared form. Before presenting its analysis, I shall comment briefly on another page in the same volume, f. 107v (Plate 10 with transcription in Plate 11), which I long regarded as the discovery document for the law of fall.

On f. 107v, Galileo recorded a set of measurements giving almost exactly the distances that would be reached during descent from rest at the end of each of eight successive equal time intervals. They had been made by Galileo before he knew the law of fall. Other entries on the same page show that the measurements were made with a ruler divided into 60 equal parts, known from ff. 166r and 116v to have been Galileo's *punto* of 0.94 mm. The numbers he recorded could not have been calculated from any mathematical formula, though when each is divided by the first number, the quotients are very nearly the first eight square numbers. In 1975 I reconstructed Galileo's procedure in obtaining the measurements, and supposed him to have hit upon the times-squared law by simply performing those divisions.[5] One puzzle, however, remained. Entries, all in the same black ink and rather small handwriting, showed that Galileo had found the odd-number rule of increasing speeds (or distances) at once, from the measurements. But the squared numbers were in different ink, squeezed into a narrow margin and slightly smeared; evidently they had been added later. Now, simple successive addition of the odd numbers yields the square numbers, as every mathematician knew. Hence there had been two ways in which Galileo could easily have discovered the law of fall without delay. Galileo could hardly have failed to recognize its importance and interest, had he done that; yet the only entries evidencing the discovery had been added later.

This puzzle is now removed by f. 189v1 (Plate 15), as I call the first of four small pages formed by folding a folio that is now bound flat. It was recognizable as having been written about the same time as f. 107v, late in 1603 or early in 1604, but for many years I could not understand a crucial calculation on it. The measurements on f. 107v concerned motion along an inclined plane, whereas f. 189v1 related to a pendulum and a straight fall. No evident connection existed between the two folios. Because everything of later date concerned inclined planes rather than pendulums, I placed f. 189v before f. 107v when ordering all the chaotically bound notes in their probable chronological arrangement.[6] In fact, as I now know, f. 107v was begun before f. 189v1, although the last entries were made on it after the latter was completed. Galileo's pendulum studies, of which only two folios survive, had played an essential role in his discovery of the law of fall – and, simultaneously, the pendulum law. That is a new and previously unexpected fact, although in retrospect (as often happens in these matters) it should not have been unexpected.

The full story of Galileo's discovery takes us along a winding road, but I do not think that readers will find the journey tedious. What I hope will most please Professor Geymonat is the manner in which Galileo came to recognize the role of squaring in relating times and lengths, both in the case of fall and in that of the pendulum. First he had made careful measurements, not only of distances, but also of times. In that way he was led to a procedure for calculating the distance of fall in a specified time from a measured time and distance of fall. The simple and elegant times-squared relationship remained hidden from him until it emerged from his procedures in calculation. That was anything but Platonist mathematics; it was instrumentalist mathematics *par excellence*. To perceive the historical events, we must proceed step by step.

The long delay in my analysis of f. 189v1 was occasioned by the fact that one number on it, 27,834, must have been calculated by Galileo on some other working paper; yet that number was nowhere else to be found in my microfilm of volume 72. The reason was that Galileo had cut a slip of paper from a page of his pendulum studies, and discarded the remainder. On the blank side of that slip he wrote a note on a different topic, which he pasted down on f. 90. This note was numbered 90b, in pencil, at the time the volume was bound. In 1978, when compiling indexes for my chronological ordering of all the notes, I also indicated all blanks. Lacking information about the verso of 90b, I requested that it be lifted for my inspection. On it was the number 27,834, twice, and enough remaining words to identify it as a "diameter." Without the discarded diagram and calculations, it was a slender clue, but I returned to the study of f. 189v1 with some hope of successfully analyzing it. Having done so, the order of my steps is unimportant; I shall present them in an order more easily followed by the reader.

The story of f. 189v1 begins with a long letter written to Guidobaldo del Monte in November 1602.[7] He had questioned a proposition of Galileo's after making some tests by rolling a ball inside the rim of a large vertical hoop. Galileo replied that it was better to use a very long pendulum, avoiding friction and other difficulties. This is Galileo's first known mention of long-pendulum observations, from which during 1603 he was to realize that acceleration is not a brief effect at the very beginning of motion that could be neglected. That assumption, adopted a decade earlier, invalidated some of Galileo's attempted mathematical proofs. Among these was a proof mentioned in the letter, that descent along any chord to the lowest point of a vertical circle in a lower quadrant took the same time, regardless of length and slope of the chord. Galileo went on to say that descent along two conjugate chords took a shorter time, though the path was longer. Finally, he said he believed that descents along arcs to the lowest point took the same time, but had not found a proof. That is not strange, since those times are not quite equal.

When Galileo did realize that acceleration had to be taken into account, he became puzzled by his proof in which it had been omitted, of a theorem which nevertheless had been corroborated by actual tests. That seeming paradox heightened his interest in motions along inclined planes. During 1603 he found two theorems about "swiftest motion," as he called it, which we would call "least time." Time as such did not interest Galileo at first, but speed of motion did. He saw that he could go on to other and more general theorems if he could find a rule for increase of speed during descent. Three conjectures were tried, one of them based on medieval impetus theory (f. 152r), but did not yield consistent ratios among distances, speeds, and times. Finally Galileo decided to make actual measurements of increasing speeds, which he recorded on f. 107v.

By Aristotle's definitions of equal speed and greater speed, distances run in equal times, whether motion was uniform or not, had speeds proportional to the distances traversed. Galileo saw that it would suffice to equalize times, without bothering to measure them, and treat measured distances as measures of speeds. Times can be very accurately divided equally by musical beats, so I reconstructed his procedure as follows. A plane about two meters long was grooved to guide a rolling ball, and was raised 60 *punti* at one end, making an angle of about 1.7°. Positions of the ball at the end of each of eight beats, each a little over half a second, were marked, measured, and recorded.[8] Calculation shows them to have been accurate within 1/64 of a second, the precision of a good amateur musician. Since the distance from each mark to the next had been run in the same time, Galileo quickly found by subtractions that speeds in equal successive times increase from rest as the odd numbers 1, 3, 5, 7, ... and so on. He drew the "triangle of speeds," marked it with the odd numbers, and having found what he sought, stopped there.

In the same black ink, on f. 107v, there is a sketch having no evident connection with this work. Possibly it was Galileo's design of an apparatus for measuring times by flow of water. In *Two New Sciences* he spoke only of flow through a tube in the bottom of a large pail, collected during each motion and then weighed, the ratios of weights being ratios of times of motion. Whether or not he built a more elaborate apparatus, the sketch suggests a container having a funnel-shaped bottom fitted with a flap-valve that could be opened by pulling a string (the dotted line) and that would close itself on release of the string. The sketch may have been drawn when Galileo thought of a hypothesis well worth testing. For this, not just equalization, but measurement of times would be needed.

Galileo's hypothesis is illustrated in the diagram at the top of f. 189v1. If his previous chord theorem were true in general, and not just for chords in a lower quadrant, then descent along the vertical diameter (which is itself a chord to the lowest point) should take the same time as descent along a short and nearly horizontal chord. Since the short chord is sensibly indistinguishable from its arc, both in length and slope, chord and arc ought to be traversed in very nearly equal times. Then, if Galileo's conjecture about sameness of time along arcs were also true, time of fall would equal time of swing to the lowest point by a pendulum hung from the middle of the fall. Galileo marked the vertical diameter *ab*, assigned it length of 4,000, and cut off an arc marked *bd*. To the right he drew another diagram, later cancelled, that will explain itself as we go on.

The first statement on f. 189v reads "If the diameter *ab* is 4,000, arc *bd* will be run in 62 of times." Assuming the distance to be in *punti*, as on f. 107v, let us call the unit of time Galileo's *tempo*, and calculate its value in seconds.[9] A pendulum of 2,000 *punti* = 188 cm swings to its lowest point in 0.688 second. Galileo's *tempo* was then 1/90th of a second, very nearly, which seems incredible. But all units are arbitrary, and do not imply accuracy of measurement of one unit, so let us test the assumptions by applying the same units in the second statement, "In the vertical, a length of 48,143 (*punti*) is run in 280 (*tempi*)." In our units, that says that fall through 45 1/4 meters takes 3.11 seconds. At Padua, it takes 3.04 seconds, a difference of only 7/100 second. Such a coincidence is improbable, so it appears that Galileo had found a way to link half-swing of a pendulum with the distance of fall timed by it.

From the cancelled diagram it is seen that Galileo also considered use of a full swing from side to side. The first statement was so written, and then altered, the time being halved from 124 to 62. The source of 124 is seen at upper left; a column of figures was added, and the total was divided by 16 to yield 124. There are not many ways to get a time by adding up numbers; one way would be to add weights of flowing water, as described by Galileo. The smallest weight in use was the grain, often mentioned by Galileo to exemplify great precision. Supposing the numbers in question to be weights in grains, of which there are 480 to the fluid ounce, Galileo's *tempo* represented flow of 1/30th ounce of water, or one gram, at 3 ounces per second.

Because Galileo originally wrote 124 and then halved it, it is not surprising that two similar groups of numbers are seen in the column of figures, 530 + 320 + 180 and 530 + 320 + 95. Galileo had timed two half-swings, probably a downswing and an upswing. If so, the times should be equal, whereas the totals differ by 180 − 95 = 85, seemingly a very large difference. But 85 grains of weight represent only 5 *tempi*, making the difference only 0.055 second. That is not excessive, and indeed it gives us a measure of the reliability of Galileo's timing procedure. In 1961 Dr. Thomas B. Settle repeated measurements described by Galileo in *Two New Sciences*, finding that although Galileo there asserted accuracy within one-tenth of a pulsebeat, double that precision was easily obtained using his procedure.[10] One-twentieth of a pulsebeat is around 0.04 second.

In one very interesting respect, Settle modified Galileo's procedure. To save the trouble of actual weighing, he collected flows of water in a graduated cylinder, and took 1 cc. as one gram of weight. Galileo, probably to avoid overloading his delicate balance, had done the equivalent, unknown to Dr. Settle. The water collected was poured into containers marked as holding 530 and 320 grains of water, and only the remainder were weighed. Not having a graduated cylinder, Galileo made one further adjustment not needed by Dr. Settle. Some water adhered to the sides of the collecting vessel after Galileo had poured from it into the measuring and weighing containers. Apothecaries accounted for that by weighing the vessel while still damp and again when dry. The top figure in the column, 13, was such an adjustment, known to Galileo as a former medical student. It was a finicky correction in this case, being less than one *tempo* in a procedure reliable only within 5 *tempi*. But it was of great assistance to me in identifying Galileo's next weighing, which he recorded on the only other folio in volume 72 relating exclusively to pendulums.

Knowing the time of half-swing of a pendulum 2,000 *punti* long, Galileo next timed fall through 4,000 *punti*. He recorded that on f. 154v (Plate 16) as 1,000 + 107 + 107 + 107 +16 = 1,337. Divided by 16, that gives 83.5625 *tempi*, or 0.928 second. Fall through 376 cm takes 0.876 second, a difference of 0.052 second. I shall designate 1,337/16 = 83.5625 by the letter *t*, and 4,000 by the letter f (for fall), in explaining Galileo's calculation on f. 189v1.

Galileo's second measurement destroyed his hypothesis, which required a figure near 62, not a time over 80 *tempi*. Accordingly, Galileo next modified the hypothesis. Half-swing of a pendulum did not time fall through twice its length, but it did time some fall, and Galileo set out to find the correct ratio. We know, by dynamic analysis, that the ratio f/p is $\pi^2/8 = 1.2337$; Galileo did not know that π entered into the matter at all, but he arrived by measurements at 1.231067 (in effect). Of special interest is the ratio 2 $p/f = 1.62114$, and its double, 3.24228, for which Galileo's findings were, in round numbers, 1,624/1000 and 3,248/1000. I shall call 3,248 *p*, the pendulum timing fall through $f = 4,000$.

The first step Galileo took in seeking an appropriate ratio was, quite naturally, to time fall through 2,000 *punti*. Immediately below 1,337, he wrote 903, and there f. 154v ends abruptly. Galileo still needed one more ratio, which I believe he found simultaneously; that is, the length of pendulum timing a given fall in relation to that distance of fall (or vice versa). Dividing 903 by 16 gives 56.4375 *tempi* = 0.627 second for fall through 2,000 *punti*. By modern calculation the time is 0.619 second, a difference of only 0.008 second. That result is so good that I do not think it was a direct timing of fall, which requires difficult co-ordinations and judgments, as shown by the relatively poor timing of fall through 4,000 *punti*. That had been near the limit of experimental variance for Galileo's procedure, and it will be seen later that he corrected 83.5625 to 80 2/3, recognizing it as the only poor timing when he found the law of fall.

I do not assert that the following procedure was Galileo's, though it was by no means beyond his powers to have adopted it. To find the ratio of distance fallen to pendulum timing it by a half-swing, I would place a small hard ball at the edge of a support at or near eye-level; in my case, 2,000 *punti* would be a very good height. Above it I would suspend a pendulum that would dislodge the ball on reaching its lowest point. Then, keeping my eye on the bob, I would listen for the sound of the ball striking the floor. When that sound coincided with the end of the ensuing upswing, achieved by adjusting the pendulum length, I would measure the pendulum, and only then would I bother with the timing procedure. What I would time would be the half-swing of the pendulum, not the fall, because that is easier to do accurately and to repeat at leisure. But I do not assert this to have been Galileo's procedure, because it is not as easy as one might think to measure a pendulum more than 1,620 *punti* long with accuracy to 4 *punti*. Weight of the string and center of gravity of the bob come into this. Galileo was not as tall as I am, making 2,000 *punti* a less convenient height for him, and there are reasons to believe that the ratio he adopted for the calculation of f. 189vi was itself partly calculated from his geometric model relating pendulum to fall, and partly the direct result of measurements of which records were lost or discarded. Very few pages even indirectly related to his pendulum studies survive, though there are enough to evidence his pendulum-fall model.[11]

That model consisted basically of three circles, internally tangent at the lowest point. The diameter of the smallest circle was the radius of the next, while the last circle represented a ratio linking pendulum with fall. This model was drawn on f. 185r (Plate 17), and also (with elaborations I do not understand) on f. 151v (Plate 18). Later in 1604 Galileo used f. 185r in deriving a lemma needed for proper proof of the conjugate-chord theorem he had communicated to Guidobaldo in 1602. For that, Galileo needed two circles internally tangent at the highest point, so he took the page bearing only his basic geometric model for pendulum and fall, inverted, ignored the largest circle, and added lines for the common tangent and for two inclined planes, lettering these for his lemma.

When f. 185r is returned to its original position, Galileo's sketch of a pendulum is seen to the right, and that is one thing never sketched upside down.

By doubling the length of a pendulum timing a known fall, one gets the pendulum timing double that fall. Whether the timing is by half-swing, as Galileo chose, or full swing, as he considered using, or by full swing and return, as we define "period of a pendulum," does not matter, so long as one is consistent. Because we know the times-squared law of fall, we expect squaring to enter conspicuously into any valid calculation of a distance fallen from a known fall, its time, and a given time. That is why Galileo's procedure on f. 189v1 appears cryptic to us, if not simply invalid. It contains no visible trace of squaring at all. But squaring, and repeated squaring, is concealed in the number 27,834, as also doubling and repeated doubling. Only the final step in Galileo's calculations appears on f. 189v1 where he finds the distance fallen in 280 *tempi* by multiplying 6,700 by 100,000, dividing by 27,834, and doubling the quotient. Unless we knew that in his units, the result was very nearly correct, we would not bother at all with such a silly-looking calculation. To explain why it worked is complicated, not by anything in physics, but by reason of Galileo's failure to accommodate us by using algebra and decimal fractions. Mathematicians will have no difficulty in recognizing Galileo's procedure by continued proportion in whole numbers and their ratios, but it is not easy to make that clear to others.

Working in Galileo's way, numbers and ratios become very large or very small in the process of squaring. To keep them of convenient size for calculations, all but four or five significant places are ignored. That would depart from the correct order of magnitude unless a factor of the form 10^n were put into the final step, equivalent to our simply moving the decimal point as we go along. That was what Galileo did when he multiplied 6,700 by 100,000, or as we would put it, used 6.7×10^8. His number 6,700 is not hard to identify; 6.7 is $2T/t$, where $T = 280$, the time for which fall was to be found, and $t = 83.5625$, already identified as time of fall through 4,000 *punti*. The exact quotient is 6.70157, if Galileo carried it that far, but no one will blame him for rounding to 6.7. The effect on the final result is negligible. Neither is it hard to identify 27,834, once we know Galileo's units, his measurements, and his geometric model of pendulum and fall. It is $(2p^2/f)^2$ with the last three digits dropped; or rather, that is 27,823. Galileo's entire calculation is reduced to $F = T/t \, (10^6 f/p^2)^2$ with the last digit dropped; that gives 48,173 as compared with Galileo's 48,143.

These algebraic paraphrases, presented to make it evident that Galileo used only doubling, squaring, and whole numbers or their ratios, while factors of the form 10^n entered only for his convenience in calculating by continued proportion, serve also another purpose of mine. Had Galileo recognized the law of fall in its times-squared form, he would not have gone to all this trouble of calculation, especially on a page intended only for his own eyes. In working by contin- ued proportion, terms of like power are not collected as they are in algebra, so it was only as he neared the end of his calculations that Galileo recognized how much simpler it would be to collect the results of squaring (and squaring again) terms of the ratio f/p. Thus it was that the times-squared law of fall was born, and simultaneously its twin, the pendulum law relating lengths to the squares of times. Directly below the first calculation on f. 189vl, Galileo applied the times- squared law in mean-proportional form to the measured fall and the calculated fall, recognized 27,834 as very nearly the double of $(Ff)^{1/2}$, corrected t from 83.5625 to 80 2/3, and drew at lower left a diagram of the kind he used all the rest of his life in dealing with problems of this kind.

Anyone is free to try explaining all the entries on f. 189vi in some other con- sistent way. Until that happens, it cannot be reasonably doubted that Galileo's discovery of the times-squared law of fall was an event in instrumentalist math- ematics. Professor Geymonat is entirely vindicated by it in his opposition to the idea that useful physics originated in esoteric contemplation of the beauties of pure mathematics, or any other metaphysical enterprise.

Knowing how the law of fall was born, there is no difficulty in explaining consistently all entries on f. 107v, in two different inks. Careful measurements had disclosed to Galileo a simple rule that applied to freely falling bodies. It was not evident that it also applied to descents along an inclined plane. Taking up f. 107v, on which he had already recorded accurate measurements of that phenomenon, Galileo wrote in the first eight square numbers, aligned with them. Mental arithmetic sufficed to show that each, multiplied by the first such measurement, produced almost exactly the corresponding actual measurement. The balance of f. 189 was devoted to a numerical verification of the conjugate- chord theorem sent to Guidobaldo a year earlier, and to a compound-ratio rule for descents along two different planes.

When I first tried to analyze f. 189vi ten years ago, I was puzzled by the choice of 280 as a time. It bore no relation to the only other time named, 62 (or 124). Galileo seemed to have somehow linked fall with a pendulum, without his yet having discovered the law of either one. But if so, the natural procedure would seem to be for him to apply the relation to some multiple of 62, or a number easily derivable from it. It is possible to derive 280 from Galileo's measurements, using certain definitions and propositions in Euclid's *Elements*, Book V, but for the purposes of this paper his choice is more simply explained as follows.

Galileo's choice of 280 *tempi* was made because he had actually measured such a time, so his calculation on f. 189vi was one that could be compared immediately with experience. The clue is found on f. 154r, the other side of the page on which Galileo had recorded times of fall through 4,000 and 2,000 *punti*. It consists of the words *filo br[accia] 16*, "the string is 16 *braccia*." The *braccio* Galileo used when at Padua was 621 *punti*, as measured from lines drawn and labeled by him.[12] A pendulum of 16 *braccia*, or 9,936 *punti*, would be 9 1/3 meters, and would swing from one side to the other in 276 *tempi* by my calculation. The modern formula is strictly true only for very small vibrations; more time is needed for long and measurable swings. In any case, a difference of 4 *tempi* is within the range of experimental variation under Galileo's timing procedure. A pendulum of this length could be suspended from a window over the courtyard of the University of Padua, protected from wind, and I believe Galileo to have made such a timing. In calculating fall during 280 *tempi*, Galileo's procedure also gave him the length of the corresponding pendulum, which was then multiplied by *f/p* to get distance of fall. The pendulum that he timed for a full swing would be one-quarter the length of the pendulum calculated, and the calculated results agree within 1%.

Today we think of both fall and the pendulum as essentially dynamic phenomena; that is, as related to a force, the "force of gravitation." Hence it may seem surprising, or even incredible, that a useful analysis of pendulum and fall was possible without the concept of force. Galileo did not use that concept; all of his published physics was strictly kinematic; that is, was based entirely on motion alone. Because both pendulum and fall were involved, inseparably, in Galileo's geometric model, and his treatment of them was by rigorous Euclidean proportion theory, any dynamic factors would, so to speak, have "cancelled out." Only when fall and the pendulum are separately treated does π enter into the latter (and not the former), and only for dynamic reasons. Galileo's kinematic analysis is of interest and would apply where no local gravitation exists, as (say) on a space platform. A "pendulum" there would not oscillate back and forth, but would circle indefinitely if set in motion. One could call the kinematic theory more general than the dynamic theory, or disparage it as a mere approximation to the latter. Such descriptions are a matter of taste, not of physics. Some would say that by adding a local gravitation constant, we have the dynamic theory, without which the statement is general and universal; and they might consider that addition unsightly, more like engineering than like physics. Others might call the kinematic theory incomplete and primitive, regarding it as a mere first approximation to the dynamic theory. In either case, historians and philosophers of science would do well to consider it, especially those who (like Koyré) unwisely ascribe to Galileo the nonsensical notion of "circular inertia." Inertia being a dynamic concept, it simply had no place in Galileo's kinematics, which was founded entirely on relations of motions discovered by remarkably accurate measurements.

Measurements of motion, made as precisely as possible with the means available to Galileo, were what brought mathematics into physics to stay. Mathematical physics had made its first appearance in antiquity, notably with Archimedes in statics and hydrostatics. Archimedes did not appeal to actual measurements. Mathematics was again applied, this time to motions, in the fourteenth century, but again no use was made of actual measurements of motion, let alone precise measurements. Without those, Galileo could not have known what law, if any, applied to falling bodies, and without applying classical Euclidean proportion theory to them, he would not have discovered the law of fall and the law of the pendulum. As Galileo said, mathematics was an invaluable tool of discovery in physics, vastly superior to logic for that purpose. He spoke from experience, logic having led him into several errors from which measurements and mathematics later freed him. Examples are his initial belief that acceleration in fall was but of brief duration, and his error of supposing impact effects to be measures of speed, in the well-known attempted proof for Fra Paolo Sarpi of the times-squared law of fall late in 1604.

The notion popularized by Koyré that Galileo supposed space and time purely mathematical – *la géométrisation à l'outrance* – is simply mistaken. Not only that, but when the charge was brought by a philosopher of Galileo's own time (Vincenzio di Grazia), Galileo and his former pupil Benedetto Castelli replied vigorously against it. Indeed, the idea that space and time could be identified with mathematical entities, which became popular early in this century, entailed in the seventeenth century an absurdity, the confusion of a thing with its measure. Sir Isaac Newton discussed that in his scholium to the definitions in his *Principia mathematica philosophiae naturalis*: "Because parts of space cannot be seen and distinguished from one another by our senses, we use instead sensible measures of them ... and so, instead of absolute places and motions we use relative ones. ... Those men defile the purity of mathematical and philosophical truths, who confuse real quantities with their relations and sensible measures." Consciousness of that was why Professor di Grazia attacked, and it was also why Galileo indignantly replied. Only in recent times has the confusion come to be treated as a virtue, by philosophers who neglected to consider statements by their predecessors at Galileo's time, and his replies to them. To Koyré, in my opinion, Galileo would reply as he had Sagredo reply to Simplicio: *Per il primo argomento, voi riconducete in tavolo quello che ci è stato tutt'oggi, ed a pena si è levato pur ora.*[13]

Galileo's contribution to the revolution in physics was above all his pioneering of measurement with all possible precision in the study of motions undertaken spontaneously by heavy bodies near the earth. Those are what he called "natural motions," no force being required beyond simple release from constraint. Once this is known, it is easy to identify Galileo's source of inspiration in the opening pages of Ptolemy's *Almagest*.[14] Astronomy was the first exact science in the modern sense; that is, a science valid within a specified range of observational variance. Physics became the second such science, two thousand years later. Astronomers had never had anything except measurements of angles and times to work from. Those alone sufficed for construction of a science valid for prediction within the range of their measuring instruments and procedures. Galileo did the same thing for physics, beginning with measurements that could have been made two thousand years before. The sense of musical rhythm was nothing new, nor was the inclined plane, or the ball, or oscillations of a hanging weight, or flow of water as a measure of time. What was new was Galileo's idea of employing them in careful measurements of natural motions.

NOTES

1 Professor Marshall Clagett has shown that Nicole Oresme, in his *Questions on Euclid*, reached the times-squared implications of medieval mean-speed analysis, but did not apply it to the fall of bodies. The very different rule for fall implied by medieval impetus theory was formulated by Albert of Saxony and adopted by Oresme. Cf. my "Impetus Theory Reappraised," in *Journal of History of Ideas*, 36 (1975), pp. 24–46.

2 A. Koyré, *Études Galiléennes*, Parigi 1939, repr. 1966, p. 13.

3 L. Geymonat, *Galileo Galilei*, Torino 1957.

4 E. Strong, *Procedures and Metaphysics*, Berkeley 1934.

5 S. Drake, "The role of music in Galileo's experiments," in *Scientific American*, June 1975, p. 98–104.

6 S. Drake, *Galileo's Notes on Motion*, Monografia n. 3, in *Annali dell'Instituto e Museo di storia della Scienza*, 1979.

7 *Opere di Galileo Galilei*, Edizione Nazionale, vol. X, pp. 97–100.

8 See the paper cited in note 5, above. Attacks on this by R.H. Naylor and T.B. Settle appeared in *Annals of Science*, and I replied there.

9 Other documents, notably f. 116v, show that Galileo consistently used the *punto* for precise measurement; cf. my "Galileo's experimental confirmation of horizontal inertia," *Isis* 64 (1973), pp. 291–305.

10 T. Settle, "An experiment in the history of science," in *Science* 133 (1961), pp. 19–23.

11 Probably Galileo had a separate folder, now lost, of pendulum studies. A few folios were transferred to his folders *Attenenti al moto*, as was a single folio from his notes on strength of materials (f. 102v), all others being lost.

12 In his treatise on fortification Galileo specified the length of the *braccio* he used; *Opere* vol. II, pp. 31 and 102. The line identifying his *punto*, on f. 166r, measures 169 mm for 180 *punti*; cf. *Opere*, vol. VIII, p. 419, line 24, and p. 420.

13 *Opere*, vol. VII, p. 72.

14 See my "Ptolemy, Galileo, and scientific method," *Studies in the History and Philosophy of Science* 9 (1978), pp. 99–115.

17

Galileo's Physical Measurements

Galileo's two principal discoveries in mathematical physics, the laws of fall and the parabolic trajectory of horizontally launched projectiles, resulted from calculations using remarkably accurate physical measurements. A third may be added to them, the pendulum law, although historically that was not separate from Galileo's discovery of the law of fall. All three were of such importance to the rise of mathematical physics during the 17th century that close attention to their origin is justified.

Except for a description of the apparatus and procedures by which Galileo verified the times-squared law, not much was said about careful physical measurements in his published books. During the present century it became fashionable to deny that he had made any. His assertion that in a hundred trials, no departure from the times-squared law had exceeded one-tenth of a pulsebeat, was declared fictitious and impossible to achieve by the leading analyst of the Scientific Revolution some 30 years ago.[1] Not until 1973 was evidence published, from Galileo's manuscript notes on motion, that he had measured fairly accurately the lengths of projection of a ball at speeds whose ratios were known to him.[2] His calculations on the same page, for comparison with measured values, led Galileo directly to the parabolic trajectory in 1608.

Reconstruction of the work behind simultaneous discovery of the laws of fall and pendulum, in 1604, shows it to have also been the fruit of a calculation from careful physical measurements. Records of careful timings, numerically expressed, survive on folios 154 and 189 of the volume at Florence into which all Galileo's extant notes on motion from 1602 to 1636 are bound. Among the total of about 160 folio pages, those are the only two that are easily recogniz-

Reprinted from *American Journal of Physics* 54 (April 1986), 302–6, by permission.

able as dealing with pendulums. It was on f. 189v1 that Galileo was writing when he first recognized the times-squared law.

Galileo's unit of measurement for lengths was identified in 1973 as his punto = 0.94 mm. On f. 116v, recording projections, he named this unit; on f. 166r he had previously identified a line as 180 such units in length. Measurement of that line gave the c.g.s. equivalent, and calculation showed the time of fall during projections to have been just under 0.4 s. Hence it was possible not only to reconstruct the experimental work, but also to confirm its accuracy.[3] Using the same unit of length, Galileo's unit of time can be deduced from f. 189v1. I shall call this the tempo; it was less than 1/90th second, seemingly incredible, but beyond reasonable doubt when all the evidence is considered.

The critic who dismissed Galileo's published statement as fictitious likened his timing procedure to the ancient clepsydra, declaring that to have been notoriously inaccurate. Galileo's procedure was to collect water flowing through a tube in the bottom of a large pail during a motion, weighing on a sensitive balance the flows collected, and taking those weights as measures of times.[4] I would liken this not to a clock, but to a stopwatch on which small motions of a finger start and stop the uniform advance of a pointer. Small motions of Galileo's finger started and stopped uniform flows of water at low head for brief times. In 1961 Settle reported his replication of this timing procedure and found it to vary only half as much as Galileo had asserted.[5] At 1/20th of a pulsebeat, maximum variance of about 0.04 s would be expected. That is now borne out by records of Galileo's timings on the pages mentioned. Not his timing procedure, but his estimate three decades later of its reliability in fractions of a pulsebeat, had been only rough.

Any unit of measurement is arbitrary, as we know theoretically and as Galileo knew practically, living in a time when there were no standard units of actual measurement in physics. The astronomical second of time was internationally standard, but it was not a practical unit for terrestrial events. The apothecary's grain of weight was virtually standard throughout Europe, but it also was far too small (480 grains to the fluid ounce) for ordinary measurements. Customary weights such as the pound kept the same name, but meant different quantities in different countries. Units of length such as the foot varied even more greatly. In Italy, the braccio (arm) named different lengths from one city to another, and from one century to another in the same city.

Galileo did not use decimal fractions, but calculated in whole-number ratios, which made small units advantageous to him by obviating fractional calculations. His punto of about a millimeter was nearly the smallest distance accurately measurable with the naked eye and an evenly divided ruler. I think it likely that Galileo adopted it with that in mind. Likewise his tempo was near to the smallest amount of time measurable with Galileo's procedure. A striking modern confirmation of this is seen from a modification by Settle. To avoid the nuisance of weighing, he caught flows in a graduated cylinder. Finding that differences of ±1 gram did not significantly affect the ratios of times, he read times directly in cubic centimeters. Galileo, of course, recorded his raw data of timings in grains weight, but he then divided by 16 for his tempi. Flow of 16 grains was 1/30 of a fluid ounce, which happens to be almost exactly 1 g of water. It represented 0.011 s in Galileo's procedure, about ¼ of the maximum experimental variance. This happened to produce, with the punto, a very interesting system of physical units in which g' is very close to $\pi^2/8$. One will see as I go along how, although helpful in one way, that created potentially misleading alternatives in the reconstruction of Galileo's work behind the discovery of the times-squared law.

The story of f. 189 can be traced back to 1602. Galileo then still believed, as he hard argued in 1590, that acceleration in spontaneous descent of heavy bodies is a brief event only, bringing it from rest to a constant speed appropriate to its weight and the density of the medium. Along any incline, he thought, there was a constant speed depending only on the ratio of distance to vertical descent. Ignoring acceleration, Galileo deduced geometrically from these false assumptions a true conclusion, which I call Galileo's theorem: Along any chord to the low point of a vertical circle, time of descent is the same regardless of length and slope of the chord.[6] His interest awakened in vertical circles, he began observations of very long pendulums in 1602. In 1590 he had established that tendency of a body to descend on an incline, at any point, can be identified with tendency to move along the tangent circle whose center lies on the vertical through the end of the plane, whether the body is supported by the circle or is hung as a pendulum bob from the center. In 1602 he remarked that although the path was longer, the time of descent was less along two connected chords to the low point of a circle than along the single chord joining their extremes.

To his friend Guidobaldo de Monte, Galileo sent the proposition that slopes exist along which speed is so slow that a body moving down them would take as much time as fall from very great vertical heights. Guidobaldo believed this to be contradicted by experiments he was making, using a ball in a large wooden hoop. The interchange of letters is lost, but Galileo replied in a long letter dated 29 November 1602.[7] First, he recommended use of a pendulum 2 or 3 braccia long (approximately 4 or 6 ft) instead of the ball and hoop. He went on to announce the two chord theorems just mentioned, adding that he believed, but had not yet proved, that descents along arcs to the low point take equal times through different distances. Since that is not true, no proof was possible;

but in his quest for one, Galileo hit on a hypothesis, also mistaken, that was to lead on to his discovery of the laws of fall and pendulum.

Galileo's theorem implied that fall through a vertical diameter would take the same time as descent along a short and nearly horizontal chord to the low point. Both in length and slope, such a chord differed insensibly from its subtending arc, so that time along it would be nearly the same as along the arc. If his conjecture about arcs were true, it followed that fall through the diameter might be very nearly timed by swing to the vertical of a pendulum hung from the center. That was the hypothesis, which can easily be tested quite exactly. A ball resting on a support is dislodged by the bob of a pendulum whose length is half the height of the support, just as it reaches the vertical. On Galileo's hypothesis, the bob should just complete its swing as the ball is heard to hit the floor. No such luck; the bob will be well on its return swing when the ball hits. I am inclined to think that Galileo found that out.

During 1603 Galileo became aware that acceleration is not just a brief event, but continues indefinitely. Observation of a 6-ft pendulum would suggest that, and reasoning from his identification of tendency to move along an arc with tendency along the tangent probably convinced Galileo. He then became puzzled why his chord theorem, derived by assuming constant speed along a plane, held for a motion he now saw to be accelerated. He noted this query, and altered the word "speed" in an earlier document to "speeds." Toward the end of 1603, after a long illness, Galileo took up the search for a rule relating increases of speed during natural descent.

He discovered such a rule from a set of extraordinarily accurate measurements, tabulated on f. 107v, and published, with my reconstruction of the work behind it, in 1975.[8] Not realizing that Galileo's interest had been only in finding a rule of speeds, I presented f. 107v as the discovery document for the law of fall. The rule of speeds he had noted was that in successive equal times, speeds of transit increase as the odd numbers 1, 3, 5, 7 ... . In different ink from the rest of this page, squeezed into the margin of his tabulation and slightly smeared, he wrote the first eight square numbers. I supposed that he entered those simply because he knew, as did all mathematicians since antiquity, that the squares arise by cumulative addition of the odd numbers. The puzzle was why there had been any delay in recording so interesting and unexpected a discovery as the times-squared law of distances. The solution is that Galileo was not thinking of his carefully measured distances as anything but measures of certain speeds. No one would add different speeds together, so Galileo laid f. 107v aside when he found the kind of rule he had sought.

For the work behind f. 107v Galileo had used a very gently inclined plane, grooved to guide a rolling ball, set at arctan 60/2000 or about 1°0.7. The ball rolled over strings tied around this, making faint bumping sound; times between the sounds were equalized by musical beats of about a half-second each. At that rhythm anyone can detect misplacement of a beat by 1/20th of a second. With patience, a skilled musician might adjust positions of the strings to consistency within 0.01 s. Galileo managed to 1/64th of a second, so his measurements in punti left no doubt that the odd-number rule of speeds was exact. The effect on him was profound. In a postscript to his letter only a year before, Galileo had said that actual physical tests do not conform with mathematical deductions. Suddenly – and probably the first time in history that anyone had carefully measured an actual accelerated motion – a simple and exact mathematical rule had emerged from truly painstaking measurements.

Other exact rules might be found by careful measurement of times, in place of their mere equalization. At the end of f. 107v Galileo sketched first a funnel-shaped container with a kind of (impracticable) valve, then drew around it a rectangular vessel, like the simple pail-and-tube apparatus described in Ref. 4. On a fresh sheet, f. 154v, Galileo recorded raw data for his timing of fall through 4000 punti (about 12 ft) in the form 1000 + 107 + 107 + 107 + 16 = 1337. Since he did not identify this, it calls for explanatory comment.

To avoid overloading his sensitive balance, Galileo poured most of the collected water into a container marked as holding 1000 grains weight of water. Probably in a glass supposed to hold 100 grains, but found by weighing to contain 107, he measured the remainder he had collected, first marking the point at which this water stood, which on f. 189vi he took as 320 gains. The final adjustment of 16 grains was for unmeasured water adhering to the sides of the collecting beaker. That was obtained by weighting it damp and subtracting its dry weight, a practice of apothecaries known to Galileo as a former medical student. In tempi, 1337/16 is 83.5625, which at 92 tempi/s represents 0.908 s. Time of fall through 4000 punti = 376 cm is, at Padua ($g = 980.7$ cm/s^2) 0.876 s, 1/30th s less. Below 1337, Galileo wrote 903; in tempi, this is 56.4375, or 0.613 s. Time of fall through 2000 punti is 0.619 s, only 0.006 s more. The use to which Galileo put these timings is found on f. 189vi (Plate 15).

If, as I think likely, Galileo had tried his old hypothesis and found it false, he had by no means forgotten it. Now that he had means of carefully timing motions, he sought a correct rule relating pendulum and fall. On f. 189vi, at the top left, he entered two timings of swing to the vertical, by a pendulum of 2000 ±, probably for the widest swing practicable, and for the narrowest. The first appears as 13 + 530 + 320 + 180, which adds to 1043; the second, as 95 + 320 + 530, to which I believe a second adjustment in the range of 13–16 grains should have been added, for a total around 960 grains. Galileo reduced his grand total of 1988 to tempi, rounding 1988/16 = 124¼ to 124 and entering this as time of swing from side to side in his original wording of the first statement on f. 189v1. Subsequently, he halved this to 62 tempi for swing to the vertical only. In fact, the unrounded 62.125 is only 0.013 s high, being equivalent to 0.675 s. Time at Padua of swing to the vertical through a small arc for pendulum of 188 cm is 0.688 s.[9]

Galileo's three explicitly recorded timings were thus nearly exact. Now, three *exact* timings so related suffice for a valid kinematic analysis of pendulum and fall; that is, for determination of all the essential relations without the use of g'. I shall call t_1, t_2, t_3, and t_4 the times of fall through distance x, of fall $2x$, of swing to the vertical by pendulum of length x, and that swing of pendulum $2x$. From any three of these, if exact, we could get ratios equal to $\sqrt{2}$ and to π.[10] Those would suffice because the ratio of *distance* fallen to *length* of pendulum, timing it by swing to vertical, is $\pi^2/8$, whose square root $\pi/2\sqrt{2}$ is the ratio of *times* when a fixed length x serves first for a pendulum swinging to the vertical, and then as a distance of fall. I shall next briefly outline the events, and then return to comment on Galileo's calculations on f. 189v1.

It took Galileo some time to arrive at a correct analysis of pendulum and fall because he did not know that either π or $\sqrt{2}$ was involved. His analysis then led him to a long series of calculations, of which we have only the final step on f. 189v1, by which he found the distance of fall during a given time from the time and distance of a known fall (in this case, 4000 punti.) Only at the end of this calculation did he recognize in it a number of redundancies that could be eliminated by cancellations, resulting in the simple proportionality between distances and squares of times that enabled him thereafter to calculate directly from one fall to another without considering intermediate pendulums, or from one pendulum to another without considering falls. On f. 189v he went on to apply this new knowledge, amending his 83 1/2 tempi for t_2 to 80 2/3, which was virtually exact.

Next, Galileo wondered whether the times-squared law applied also to descents along inclined planes. Recalling his accurate measurements on f. 107v during such a descent, he entered in the margin of his tabulation the eight square numbers, aligned with the digits numbering the eight equal times. Each square, multiplied by the first distance (33 punti), produced almost precisely the corresponding measured distance. Returning to f. 189v2, Galileo wrote a hypothetical rule for motions along two planes differing in length and slope. Numerical trial showed it incorrect and provided the true rule. Turning the page over, Galileo applied this rule to verify his 1602 proposition that time is shorter along two conjugate chords to the low point than along the single chord, by detailed numerical calculation. Galileo's "new science of motion" was born. It was purely kinematic, derived from accurate measurements of motions accelerated from rest. It was amplified and eventually placed on a rigorous mathematical foundation, to be published in Ref. 4 in 1638.

Returning to f. 189v1, the second statement says that fall through 48 143 punti takes 280 tempi. In our units, that puts time of fall through 45¼ m at 3.043 s. The time at Padua is 3.038 s. To get 48 143, Galileo multiplied 6700 by 10^5, divided by 27 834, and doubled the quotient. No squaring is evident, although because the calculation was correct, squaring must have occurred and must be concealed in the numbers 6700 and 27 834. Galileo had calculated both numbers somewhere, but neither was to be found in my microfilm of his working papers on motion. I surmised that Galileo had discarded papers that I would need for reconstruction of his work. That was confirmed in 1979, when at my request a pasted-down slip numbered 90^b was lifted. On a blank space on one of his pendulum papers, Galileo had written a note on another topic, cut it out, and pasted it down in 1609. On the long-concealed side I found the number 27 834 with enough remaining words to identify it as a "diameter." The diagram and original calculations were gone, but not a calculation made after correction of t_2 using the law of fall. These slender clues eventually paid off.

Loss of most of Galileo's pendulum calculations renders uncertain my reconstruction and ordering of his steps, although I am confident of the essentials. Here I shall only highlight them, after some preliminary remarks that I deem necessary for understanding of Galileo's Euclidean mathematics, very different from our customary approach.

Galileo did not use algebra and never wrote an equation in his life. He calculated with whole-number ratios between magnitudes of the same kind only, according to the classical (and rigorous) theory of proportionality among mathematical continuous magnitudes, set forth in Euclid's *Elements*, Book V, in which proportion was defined as sameness of ratio. Such a procedure may seem ponderous today, but it had the rigor which algebra (and decimal fractions) lacked until little over a century ago, when Richard Dedekind created the real-number system. Every measure, and every ratio of measures used by Galileo, had to him a clear physical meaning at each step, not being just a numerical expression.

We move the decimal point as we go along to preserve order of magnitude. That Galileo used no pointing accounts for the factor of the form 10^n in the final step of his long series of calculations, as seen on f. 189v1. Had he there written 6.7×10^8 instead of 6700×10^5, the ratio character of that product would have been easier to recognize. In point of fact, 6700 was a slightly rounded figure that was more nearly 6701½ in Galileo's own calculations. Discussion of it will serve to illustrate other necessary considerations in reconstructing those.

The problem attacked on f. 189v1 was to calculate distance of fall in 280 tempi, knowing the distance and time of another fall, in this case 4000 punti. Why Galileo chose 280 as the time of the unknown distance will be explained later. This time was stated in tempi, or 1/16 the weight in grains of flow, and since it would be an unnecessary inconvenience to carry all of Galileo's whole-number ratios into this discussion, let us use decimal fractions. Writing 1337/16 = 83.5625, we then have 280/83.5625 = 3.35078, of which the double is 6.70157. Think of Galileo's 6700 as 2000 times the ratio of time of unknown fall to time of known fall.

Now, 2000 here is one term of a ratio whose terms are properly inseparable; nor were they separated in Galileo's calculation. The other term I shall take to have been 1628 and treat as having been a factor in 27 834. The ratio 2000/1628 will be called f/p, the ratio of distance of fall to length of pendulum timing it by swing to the vertical. In any units of measurement whatever, f/p is in theory $\pi^2/8$, and 1621 + is the length of pendulum timing fall of 2000; but Galileo had only his own measurements to go on. Next, let us identify the other ratio in Galileo's calculation. Writing

$T/t \times f/p = 3.5078 \times 2000/1628 = 4.11644,$

let us divide this into 24071½ (Galileo's result), getting 5848 as the factor still missing. This is found to be $Tt/4$. But,

$Tt/4 \times T/t \times f/p = f/p \times T^2/4 = 1/2$ fall in time T,

so that the cryptic-looking calculation on f. 189v1 amounts to

fall in time $T = f/p \times T^2/2$.

That was true in Galileo's units, and a very interesting fact follows – that the square of any time of fall in tempi gave the length in punti of the pendulum timing double that fall by its swing to the vertical. Galileo had not noticed this, or he would have simply squared 280 and divided by 1628/1000 = 48 157.

Two questions will immediately arise, of which the first may already have troubled the reader. In all the foregoing, I have put Galileo's tempo at 1/92 s, and although that fits perfectly with the calculation on f.189v1, the question is whether we can confirm it independently from another recorded or implied figure in Galileo's notes. The second question is where Galileo got the figure 1628 that is essential, in the ratio f/p = 2000/1628, to the whole reconstruction of the calculation just presented. In replying to these two questions, I shall preface my remarks by saying that in my oral presentation I did not use this value for the tempo, but only said that Galileo's t_1 implied 1/91 s, that his t_3 implied 1/90 s, and that 1/92 s would render his measurements virtually exact. It appeared to me impossible to be sure of the actual, physical value of g' at Padua in punti/

tempo^2. Two bits of further evidence in Galileo's extant papers had been known to me for years, but their interpretation eluded me.

On f. 151v there are tentative diagrams of Galileo's geometric model relating pendulum and fall, which in its next form survives on f. 185. On f. 151v there are also two other things that ought somehow to be relevant to this work in progress: a sketch of two gears, and the number 1590 = 53 × 30. It was usual for Galileo to measure punti in multiples of 60, using a ruler accurately divided into 60 punti; it was also usual for him to avoid fractions in multiplication or division, so it seemed to me that he had obtained 1590 by laying off that ruler 26½ times, writing 53 × 30 for 26½ × 60. What he measured was the pendulum whose swing to the vertical took the same flow of water, 903 grains, as fall through 2000 punti according to his own timing of that. That was truly an experimental measurement, requiring many adjustments and ending with remarkable accuracy, for calculation gives 902.6 grains as the time of swing to the vertical by a pendulum of 1590 punti, taking the tempo as 1/92 s. There is no other plausible explanation of the number 1590 on a page mainly occupied by diagrams relating pendulum and fall, and this account assures us that g' at Padua was very close indeed to $\pi^2/8$ punti/tempo^2. Calculation gives the constant as 1.2306 against 1.2337 for this instance.

The sketch of meshed gears on the same page is not as irrelevant to Galileo's work in hand as it may appear to be. Galileo later followed the same procedure to find the pendulum that took the same time he had recorded for fall through 4000 punti, a pendulum about 10-ft long. To make fine adjustments in length of a pendulum suspended far above his head, Galileo thought of running the string over a nail in a movable upright, anchoring it at bench height. To raise or lower the nail, through small distances, a gear affixed to the upright could be meshed with another turned by a crank at bench height. Timing of this pendulum gave Galileo the length 3478, to which the origin of his number 27 834 can be traced, and timing of the half-pendulum 1739 completed the data Galileo required. But those were subsequent steps; his first measurements, although good, led him astray for a while.

Having found the length of the pendulum that Galileo believed took the same time to swing to the vertical that a body took to fall 2000 punti, he was probably misled by this near proportionality:

1988/1337 nearly equals 1337/903,

which, in my previous terminology, would make t_2 the mean proportional between $2t_3$ and t_1. On that assumption Galileo would be led to mistake for time of fall through 8000 punti what was in fact the time to the vertical for a pendulum of 8000 punti. Details of his steps in arriving at a correct analysis by the further measurements already mentioned are omitted here.

In conclusion, I shall comment on Galileo's choice of 280 tempi for the time of fall whose distance he calculated. On f. 154v there is a calculation arriving at time 140, and on the other side of the page a notation "the string is 16 braccia." Lines drawn and labeled by Galileo at Padua indicate that his braccio was 621 punti. The pendulum of 16 braccia was accordingly 9936 punti or 9 1/3 m, say 30-ft long. Such a pendulum could be hung from a window over the courtyard at the University of Padua, and timed protected from wind. Its swing to the vertical would be 141 tempi, or its swing from side to side, 282. It is probable that Galileo timed this at 280 tempi, and that his choice shows his preference for the physically measurable world over the "world on paper" of natural philosophers up to his time.[11] Only a year after his discovery of the law of fall and the pendulum law, engaged in controversy with the ranking professor of philosophy at Padua over the location of a recent supernova, Galileo asked: "What has philosophy got to do with measuring anything?"[12] Galileo's mature physics was rooted in proportionalities among accurate physical measurements; although he said little about them in his published books, their traces are unmistakable in his working papers.

NOTES

1 A. Koyré, *Proc. Am. Philos. Soc.* 97, 60 (1953).

2 S. Drake, *Isis* 64, 291 (1973).

3 At the time, accuracy appeared to be only within 3½%. Later studies show the work to have been virtually exact and more complex than at first appeared. See S. Drake, *Annali dell'Istituto e Museo di Storia della Scienza* (Florence, Italy, 1985), pp. 3–14.

4 Galileo, *Two New Sciences*, translated by S. Drake (Univ. of Wisconsin Press, WI, 1974), p. 170.

5 T. Settle, *Science* 133, 19 (1961).

6 A reconstruction of Galileo's purported proof is given in S. Drake, *Galileo At Work* (Univ. of Chicago, IL, 1978), pp. 67 and 68.

7 Translated in Ref. 6, pp. 659–671.

8 S. Drake, *Sci. A.* 232, 98 (June 1975).

9 Galileo's calculations in whole-number ratios make it correct to use the unrounded 62.125 tempi. The appearance of 320 as a volumetric "weighing" on f. 189v1 makes it probable that the timings of falls on f. 154v were entered first.

10 Either t_2/t_1 or t_4/t_3 would give $\sqrt{2}$, and t_4/t_1 would give $\pi/2$. Galileo, of course, did not know that π was involved, and mistrusted even the good approximation to $\sqrt{2}$ of t_4/t_3 on f. 189v.

11 Galileo, *Dialogue Concerning the Two Chief World Systems*, translated by S. Drake (Univ. of California, CA, 1953), p. 113.

12 S. Drake, *Galileo Against the Philosophers* (Zeitlin & Ver Brugge, Los Angeles, CA, 1976), p. 38.

18

Galileo's Constant

Of frequent use in modern physics is what may be called the local constant of gravitation, g. It is literally a constant only at a particular latitude and altitude. Its numerical value depends on the units of length and the time employed, so that by suitable choice of units g could be given any numerical value we wish. One such value, which I shall call g', is $\pi^2/8 = 1.2337 \ldots$, literally a constant, which also has an exact physical significance. Anywhere on earth, or on the moon or on Jupiter, g' is the ratio of distance fallen from rest to length of pendulum timing the fall by swing to the vertical through a small arc. The ratio of times for such a swing and fall through a height equal to the pendulum length is $\sqrt{g'}$, anywhere and in any units. It is this ratio, $\pi/2\sqrt{2} = 1.1107 \ldots$, that I call "Galileo's constant." He arrived at it very nearly indeed by timing a pendulum 1,740 *punti* long and fall through 1,740 *punti*, getting the ratio $942/850 = 1.1082 \ldots$ The square of that ratio, $1.2282 \ldots$, played the same role in Galileo's first calculation of a distance of fall from its time that g plays in modern calculations using $S = \frac{1}{2}gt^2$.

Galileo's calculation was made on f. 189v1 (Plate 15) of his working papers on motion, just before he wrote there that the vertical distance 48,143 is run through in time 280. That statement, with an addition, was repeated on f. 115v with identification of Galileo's unit of length: "Time for the whole diameter is 280; since its length was p. 48,143, time for the diameter of length 4,000 will be 80 2/3." The abbreviation p. stood for *punti*, just as p̄ was use for the same thing on f. 116v. A line drawn on f. 166r and identified as having 180 of these units makes it certain that Galileo's *punto* was 0.94 mm in metric units.

Galileo's calculation of 80 2/3 was made on f. 189v2 (Plate 19), after a preliminary calculation at lower left on f. 189v1 immediately following the line-

Reprinted from *Nuncius* 2 (1987), 41–54, by permission.

diagram relating two distances of fall and their times, in the standard form used for the Euclidean theory of ratios and proportionality. The calculation of 80 2/3 did not use squaring, but its mean-proportional equivalent. By Galileo's law of fall, the square of a ratio of times is the ratio of distances, and from the above we have $(280/80^2/3)^2 = 12.050$ while $48,143/4,000 = 12.048$. In metric units, the fall calculated by Galileo on f. 189v1 was 45¼ meters, impossible in his day to time with accuracy; yet his result was correct within 0.03 second or less, or within a few centimeters in 45¼ meters.

Galileo's calculation of 48,143 on f. 189v1 is very condensed, and in this case his preliminary calculations no longer survive. They were probably made on the part of f. 90^bv (Plate 20) which was cut off and discarded about 1609, when Galileo used a blank part of the page for a note on a different subject and pasted it down on f. 90. The verso, when this note was lifted at my request, showed the number 27,834, used in the calculation on f. 189v1, identified as a "diameter" but lacking the diagram and related calculations. From the entries on ff. 151 (Plate 18), 154 (Plates 16 and 21), and those already named, it is possible to reconstruct the story of Galileo's discoveries of the pendulum law, the law of fall, and their immediate consequences at his hand in 1604. That has been done for the most part in previous papers, but not in precise chronological order. As I wrote:

Loss of most of Galileo's pendulum calculations renders uncertain my reconstruction and ordering of his steps, although I am confident of the essentials.[1]

Subsequent investigation has revealed clues that determine the order of entries made on the surviving working papers and has thereby removed any doubt on my part about the correctness of the numbers assumed to have existed on the lost part of f. 90^bv. It is my purpose here to set forth the steps to Galileo's law of fall from the day he devised his timing apparatus, first sketched at the end of f. 107v about the beginning of 1604. Following that, Galileo's timing procedures will be explained in detail, and finally I shall return to the matter of Galileo's constant in its historical context. In listing the steps essential to discovery of the law of fall, I shall include one (step 3) that was not truly essential, but that is very illuminating about Galileo's experimental procedures, and will omit one set of timings recorded on f. 189v1 irrelevant to the discoveries. It is hardly possible to understand those timings without the essential context, after which they will be seen to be both surprising and of extraordinary interest.

1 Timing of fall 4,000 *punti* at 1,337 grains weight of flow   (f. 154v)
2 Timing of fall 2,000 *punti* at 903 grains   (f. 154v)

3 Measurement of pendulum for 903 grains at 1,590 *punti*   (f. 151*v*)
4 Measurement of pendulum for 1,337/2 grains at 870 *punti*   assumed
5 Timing of pendulum of 2 × 870 = 1,740 *punti* at
   942 grains flow   (cf. f. 189*v*1)

I now tabulate pendulum data so obtained, extending my table to the number which Galileo refined to 27,834 on f. 90^b*v*:

Length of pendulum in *punti*	Time to the vertical in grains flow
870	668½
1,740	942
3,480	1,337
6,960	1,884
13,920	2,674
27,840	3,768

The first column is formed by successive doublings; the second, by alternate doublings. Both columns are therefore in continued proportion except for a slight discrepancy of the first two timings. To connect lengths and times, Galileo adopted his new unit of time, the *tempo* as I shall call it (= 1/92 second), by making time to the vertical the mean proportional between 2 and the length of pendulum. Because each pendulum above is double the preceding, that was the same as making each time the mean proportional between length and next-length, the latter being necessarily the "diameter" of the pendulum, as Galileo called it. When the times in *tempi* are calculated from the above lengths in this way, it is seen that the same effect is very nearly obtained simply by dividing time in grains flow by 16, as Galileo did at top left on f. 189*v*1. Before that, on f. 90^b*v*, he had made an exact calculation, which is why 27,834 appears there and on f. 189*v*1, instead of 27,840 as above.

Galileo now had the pendulum law for the case of lengths successively doubled. Whether the law was perfectly general, for any ratio of lengths, was not certain physically, so Galileo took his next step:

6 Calculation of the mean proportional of 118 and 167 = 140   (f. 154*v*)

In the new time unit, 1,884/16 = 118 *tempi* and 2,674/16 = 167 *tempi*. If the mean proportional time 140 *tempi* is indeed the time for the mean proportional length, $(6,960 × 13,920)^{1/2}$ = 9,843 *punti*, Galileo's pendulum law was perfectly general, for irrational as well as rational ratios.

7 *filo br[accia] 16*, "string 16 *braccia*" (= 9,900 ± *punti*)   (f. 154*r*)

It would be a great nuisance to measure so long a pendulum in *punti*; the common unit of length for measuring textiles was the *braccio*. In his treatise on fortification, written at Padua, Galileo drew a line to define ¼ *braccio*, and comparison of that line with the line on f. 166*r* for 180 *punti* indicates Galileo's *braccio* to have been 621 *punti*. At 615 punti per *braccio*, the string mentioned would be exactly 9,840 *punti* long, the mean-proportional length above. In either case, time to the vertical at Padua would be 141 *tempi* by modern calculation, using $g = 980.7$ cm/sec^2 at the latitude of Padua. Such a pendulum, 9¼ meters long, could be hung from a window over the courtyard of the University of Padua and timed, protected from wind. Galileo's pendulum law was confirmed as perfectly general by the next step:

8 Timing of pendulum 9,900 ± *punti* long as 140 *tempi*   (assumed)

Here I cite f. 189*v*I in evidence, though it is indirect evidence, to be explained presently. That Galileo had previously timed something at 850 grains flow is implied by the occurrence twice on f. 189*v*I of 530 + 320 (= 850), though to see this it is necessary to know his procedure in precise timings.

10 Use of the time-ratio 942/850 (Galileo's constant),
   from steps 5 and 9   (f. 90^b*v*I)

I assume that the calculations preparatory to that of distance of fall in 280 *tempi* on f. 189*v*I were on the lost part of f. 90^b*v*.

Galileo was now prepared to calculate distance of fall in a given time from his data, in which 1,337 grains flow (83.5625 *tempi*) was his time for a pendulum of specified length, and also his time for fall through 4,000 *punti*. Now, it happens that very first recorded timing by Galileo was the only one which was not remarkably precise; his timing of fall 4,000 *punti* was 1/30 second too long. That may not seem much, but its effect on fall through 45¼ meters as we now calculate it would be about 2 meters. Yet Galileo's calculation on f. 189*v*I differs from ours by only a few centimeters. The reason is that in his calculation, though he made use of 1,337, that figure cancelled out, entering once as a divisor and again as a multiplier.[2] Because he worked only in ratios, it was not until after he had completed his rather complex calculation, of which the preliminary steps are lost, that he recognized the superfluous nature of 1,337 in it, and at once saw the *relation* of times and distances to be the same for both pendulum and fall; for just as distances in fall are as the squares of the times, so the lengths of pendulums are as the squares of their periodic times.[3]

Satisfied that distances in fall from rest are as the squares of elapsed times, Galileo suspected that the same would be true of descents from rest along inclined planes. On f. 107*v* (Plate 10 with transcription on Plate 11), just before devising his timing apparatus, he had entered careful measurements of distances run by a ball on a grooved inclined plane very gently sloped (at 1.7° or arctan 60/2,000) at the ends of eight time-intervals differing by 0.55 second each.[4] Taking up that page again, Galileo entered in the margin of its tabulation the first eight square numbers. Each, multiplied by the first measured roll (33 *punti*), gave almost exactly the corresponding measured distance.

Galileo's 1602 theorem, sent to Guidobaldo del Monte, was that although the distance is shorter, the time is greater for descent along a chord to the low point of a vertical circle than along two conjugate chords joining the same endpoints. In making the calculations Galileo did not use precisely $\sqrt{2}/2$, though he calculated that and entered it to 5 significant figures near top left of the page. He used a slightly different number entered directly beneath it and calculated 2(.)003 137764 from it, correctly and in full, at lower left. That cannot fit with the view popularized by Alexandre Koyré that Galileo geometrized excessively and identified physical truth with mathematical abstraction.

Everything essential to the discovery of the law of fall and its immediate consequences for Galileo has now been listed, as well as one non-essential entry, on f. 151*v*, a page otherwise devoted to geometrical diagrams modelling relations of times and distances for pendulum and fall. Those geometric models, culminating on f. 185, once appeared to me important for the retracing of Galileo's thought and work;[5] I have neglected to mention them here because they now seem to me incidental to the main work of measuring, timing, and calculation. The non-essential entry, 53 × 30 = 1,590 (for 26½ × 60 = 1,590, Galileo's ruler having been accurately divided into 60 *punti*, and his practice having been to eliminate the fraction ½ before multiplying or dividing), was included because it throws much light on Galileo's procedures in accurate timing.

The method of timing described in *Two New Sciences* was to collect water flowing through a tube in the bottom of a large pail during a motion, weighing the collected flows on a sensitive balance, and treating the weights as proportional to (or as measures of) the times. Koyré compared this to the ancient clepsydra or water-clock, which he declared inaccurate by reason of changing head of water in the reservoir during a day, or other long period of time. That is doubtless true, but it has no bearing on an apparatus for timing intervals generally less than one second, a use which requires a stopwatch rather than a clock. Galileo's water-stopwatch was operated as we operate our stop-watches, by small motions of one finger to start and stop a pointer. In principle, his was slightly more accurate than ours, as the pointer advances by jerks, while water flows continually. Because even an inexpert person using a common stopwatch can time to the nearest tenth of a second, there is no reason to suppose that Galileo could not do that well. Thomas Settle, reporting in 1961 his experience with Galileo's procedure, remarked that double the precision claimed by Galileo (one-tenth of a pulsebeat) was easily attained without much practice. The working papers show that Galileo likewise attained such precision, though he had the wisdom not to claim it in print, being unsure that everyone would exercise the same care (and not needing to assert more than others could easily verify).

Settle, having determined that difference of ±1 gram did not affect the ratios he obtained, collected flows in a graduated cylinder and reported weights volumetrically determined. Galileo did the same with his apparatus, though not in the same way. His earliest scientific essay, in 1586, described a balance of his own making on which weighings to a fraction of one grain (1/480 ounce) were made in determining specific gravities of metals and gems. His own records of 1585–86 show those to have been extremely good, though error of one grain would have sufficed to render some notably inaccurate. Nevertheless, it is time-consuming and a source of possible unnecessary errors to weigh each collected flow with such accuracy. Galileo collected flows in a glass cylinder, on which he often marked the level of water collected before weighing it. That he used a glass vessel and marked it is shown most clearly by the non-essential entry on f. 151v.

Galileo's second known timing was 903 grains flow during fall through 2,000 *punti* (188 cm). Knowing the value of *g* at Padua and the rate of flow through Galileo's apparatus (3 fluid ounces per second), I calculate that the correct weight would be 911 or 912 grains, Galileo's finding was thus about ½ *tempo* short, or a bit more than 0,005 second, almost incredibly precise for 1604. The length of pendulum that exactly times fall of 2,000 in any units, anywhere, is 1,621 of those units. But the length of pendulum that would time flow of 903 grains through Galileo's apparatus at Padua is 1,590 *punti*; that is, the seemingly trifling shortage of time would result in some 3 cm of length for a pendulum about 1½ meters long. I conclude that Galileo, having marked the point at which 903 grains of water stood in his vessel, adjusted a pendulum until its swing to the vertical accompanied flow of water to that mark. Such patience and concern with accuracy may be surprising, but it is not impossible, and I am unable to find any other way to account for the entry on f. 151*v*. Also, it accords with the fact that Galileo's second timing, 903, is stated as if some use had been made of this flow before it was duly weighed in grains, for no detail of the weighing accompanies it, in sharp contrast with the first weighing immediately above it on f. 154*v*, stated as 1,000 + 107 + 107 + 107 + 16 = 1,337.

The round number 1,000 appears nowhere else among Galileo's records of timings, which are admittedly few, but because that first weighing was seriously defective as a timing (1/30 second too long), while all the rest are remarkably accurate, there may be a connection. The number 1,000 is conceivably due to use of a measure accepted by Galileo without checking, such as a mark on a vessel supposed to hold 1,000 grains of water, and perhaps quite accurate for some set of weights other than those used by Galileo. Certainly Galileo poured water from his collected 1,337 grains into some other vessel, for the final 16 above is Galileo's adjustment for unweighed water adhering to the sides of the collection vessel, found by weighing that while still damp and subtracting its dry weight. Also, the numbers 3 × 107 certainly represent a determination made by Galileo after he had poured off 1,000 grains (or whatever amount of actual weight in terms of Galileo's set of weights that marking represented), for he had marked the point at which the remainder stood before weighing it, and he used that mark as representing 320 grains in two recorded timings on f. 189*v*1.

Step 5 in the first group was assumed to have been timing of a pendulum 1,740 *punti* long. There was good reason to time that pendulum, double the length of the previous pendulum. Galileo appears to have considered finding the length of pendulum whose swing to the vertical accompanied 1,337 grains flow, in the tedious way he had done the same for 903 grains, for on f. 151*v* there is also a freehand sketch of two meshed gears. That seems out of place on a page otherwise devoted entirely to pendulum studies, but there is a plausible reason for it. The pendulum under consideration would be very long, probably double the length of the pendulum already measured, or some 3 meters, and hence would be suspended far above Galileo's head. Patient adjustment of its length could be facilitated by running the string over a nail in a moveable upright and anchored at bench height; the nail could then be cranked up or down conveniently by using a gear and ratchet, or two large gears, after the approximate length had been found. But instead, I believe, Galileo decided to time and measure the pendulum for *half* the flow of 1,337, and then time the pendulum double its length. Hence the measured length 1,740, which turned out to be very useful as well as convenient.

Fall through 1,740 *punti* was not timed until the pendulum law was found and confirmed for all ratios of lengths, but it was timed before Galileo began writing f. 189*v*. That is how the number 530 (grains) came to appear on this crucial page, for Galileo poured off 320 grains from the collected water, using the mark already mentioned, marked the point at which the remainder stood (530 grains), and weighed that. It was the ratio 942/850 that he used (Galileo's constant), rather than the slightly greater 945/850 which will presently appear, and which would have been a trifle high. That is evident from the calculation of distance on f. 189*v*1; only 942/850 will fit it, and I conclude that 942 had been measured earlier and probably used in the preliminary calculations for f. 189*v*1, on the lost part of f. 90^b*v*.

In the midst of that work, Galileo was distracted from completing his calculation of fall in 280 *tempi* (double the mean-proportional time 140 *tempi* that had confirmed generality of the pendulum law) into a reflection of no little interest. This had nothing to do with the law of fall, and for that reason was omitted from my listing of steps essential to that discovery. The result was that the work at the top of f. 189*v*1 concerned only a pendulum – a pendulum 2,000 *punti* long according to every indication on the page. That was not its length, but I had no way of knowing that until every other entry on these pages had been thoroughly understood and arranged in chronological order.[6] The pendulum that occupied Galileo's attention at the top of f. 189*v*1 was the pendulum of 1,740 *punti*, probably still hanging undisturbed after its timing in connection with the work leading Galileo to the pendulum law.

Indeed, there is no reason at this time that Galileo should have experimented with the pendulum of 2,000 *punti*. It was not necessary for him to measure and time such a pendulum after the pendulum law was already known, for its time in *tempi* must be simply $\sqrt{4,000}$, a very simple calculation to make. Moreover, if a pendulum of 2,000 *punti* was timed and recorded at the top of f. 189*v*1, it was a very poor timing (or rather, pair of timings), quite out of keeping with Galileo's previous work. It was in fact a pair of timings of almost incredible accuracy, and it refutes a belief that I long shared with every other specialist in the work and thought of Galileo – a belief dating back to Galileo's own time, first expressed by Marin Mersenne, one of the ablest experimentalists in physics then living.

The only pendulum timing that can be made with great precision is time of swing to the vertical. That is because, with the pendulum hanging still and plumb, a block can be accurately placed in contact with one side of the bob, and firmly fixed. The pendulum is then timed from the moment of release to the moment the bob is heard to strike the block. The fact that Galileo recorded pendulum timings to the vertical only, and recorded accurate results, would suggest that he had adopted this simple and effective procedure. But against this stood the fact that in Galileo's later *Dialogue* and *Two New Sciences*, he appeared to remain barely aware, or even reluctant to admit, that longer times are taken to the vertical through large arcs than through small arcs. Because that is easily discovered by using a block at the vertical and simultaneously releasing equal pendulum from different starting positions, Mersenne concluded that Galileo had never made the obvious test that he himself made and correctly reported. No one, so far as I know, has questioned Mersenne's conclusion, though many have been puzzled that Galileo should overlook so simple an investigation.

I now think that Galileo knew the fact no later than November 1602, when he sent two theorems and a conjecture to Guidobaldo. One theorem was that along any chord to the low point of a vertical circle, time of descent is the same regardless of length and slope of the chord. That is true, and Galileo proved it correctly after he had the law of fall. (His 1602 proof had reasoned correctly from false assumptions to a true conclusion.)[7] His conjecture was that descent along arcs of different lengths likewise took equal times to the low point, which is not true, and Galileo said that he had sought a proof in vain. The reason for which he did not despair of finding a proof was not ignorance of the facts, but is found in the final passage of his letter. When matters come to actual test, he wrote, the conclusions reached absolutely by mathematicians are not borne out. There are material impediments to exact agreement, as Galileo had written in 1590, and continued to point out to the end of his life, but from 1604 on his attitude toward this situation was very different from the Aristotelian (and the different Platonic) traditional position. The drastic change was undoubtedly brought about by the measurements recorded on f. 107v, and thoroughly confirmed by the measurements discussed above.[8]

Supposing, then, that Galileo was quite aware that in material tests a pendulum takes a bit longer through a long than through a short arc to the vertical, the reflections that diverted him from his calculation of distance of fall from its time alone will be quite understandable. What inaugurated those reflections was, I believe, something else. In his laborious and redundant calculations for adapting his pendulum law to the phenomenon of vertical fall, the necessity of doubling a length in deriving a time, or doubling a doubled length to double a time, awakened the idea of doubling time to the vertical and calculating from times of swing side to side, rather than just to the vertical. The cancelled diagram at top right on f. 189v1, and the first statement as originally worded, show Galileo to have been considering that, while his rewording and return to the central diagram show that he abandoned the idea. The pendulum timings at top left show why he did. Yet the pendulum of radius 2,000 indicated in the statement and diagrams seemed to show that the timings were of such a pendulum.

The data, in grains weight of flows, are for two timings, added together. That is in accord with the notion of using swing from side to side, each timing being to the vertical because that is the only way of timing accurately. Separately the totals are

$$13 + 530 + 320 + 180 \ (= 1{,}043) \text{ and } 95 + 320 + 530 \ (= 945)$$

The large numerical difference in two times for swing of the same pendulum arose from timing of a very wide arc (a full quadrant is shown in the cancelled diagram) and a very small arc, probably not more than 10°. Modern dynamic analysis shows that the latter takes only 1½% more time than swing through a vanishingly small arc, while the former takes 12¼% longer. The difference in times above is about 10 1/3%, in close agreement with that, and the smaller arc took virtually the same time as Galileo's original timing for the pendulum of 1,740 *punti* (942 grains). Galileo was doubtless surprised to get as long a time as 1,043 grains, and timed again the swing through a small arc, though he did not alter the original figure used in getting the ratio 942/850. The total, 1,988 grains, was divided by 16, giving 124 *tempi* for the statement he wrote, later halved to 62 *tempi* for swing to the vertical of a pendulum of 2,000 *punti*. In fact, $\sqrt{4{,}000}$ is 63¼, very nearly the same as the time in *tempi* for what I shall call a "representative" swing for the pendulum of 1,740 *punti*, between a maximum and a minimum arc to the vertical.

Galileo's reflections during this gambit, which was then abandoned, were probably along the following line. Difference in time for large and small arc, though small in amount – about 6 *tempi*, or 0.066 second, little more than one-tenth of a pulse-beat – was very considerable proportionately. The time of each swing must become continually less as a pendulum "died down" and came to rest. The only evident reason, for Galileo, was that greater speed acquired during wider swing encountered greater resistance from the air. If a pendulum could swing in the absence or air, no reason appeared to him why it should not continue forever, rising always to the height from which it descended. Now, in the process of "dying down" from maximum swing of 90° in air, there would be one swing in which the time from side to side would be equally divided at the vertical position, as it would always be for the ideal pendulum not impeded by air. This swing from side to side would be "representative" of the perpetual pendulum, and in the case of the pendulum of 1,740 *punti* that Galileo had timed for maximum and minimum arcs, the times combined very nearly equaled time for minimal swing of the pendulum of 2,000 *punti* from side to side. Thus there would be a way of measuring the effect of air resistance on pendulum times, had Galileo's belief been correct that only speed and air resistance were involved. In fact, it is the shape of the circular path that makes the time greater for large arcs than small ones, and Huygens later discovered that a cycloidal path divides time equally at the vertical.

Any interested reader who credits Galileo with like reflections can understand why he mentioned only insensible differences of times of swing in his later *Dialogue*, where he added another reason for which swings would never continue perpetually, and why in *Two New Sciences* he appeared to neglect entirely any difference in times, at least for arcs less than 20°. Full discussion of his later statements about pendulums would be out of place here, his early work on pendulums being the focus of attention.[9]

The manner in which Galileo selected his *tempo* as the unit of time explains how it came about that of all the possible numerical values of g, Galileo's units of length and time implied $g' = \pi^2/8$. In one way it may seem incredible that he selected units that implied a universal rather than a local constant of gravitation, while in another way it is obvious that he could not have reached any other kind of gravitational constant. He simply did not know enough physics to have done that. Galileo's mature physics was entirely kinematic, without any dynamic principles whatever, and it is only because we regard gravitation as a force that we think in terms of the local gravitational constant g. Ratios and proportionalities of measures of distances and times do not distinguish one place from another; Galileo thought and worked entirely within the rigorous Euclidean theory of ratios and proportions among mathematically continuous magnitudes set forth in Book V of the *Elements*. Galileo's measurements were of distances and times alone, not of forces or even of speeds as such. Proportionalities among those measurements were necessarily universal rather than local.

I have called g' Galileo's constant because he measured that, very nearly, and used it in the calculation that led him to recognition of the law of fall. I did not call it "Galileo's gravitational constant" because that phrase has a modern connotation very different from the simple sameness of ratio by which Euclid defined proportionality. What Galileo had in mind, and used, was sameness of ratio alone, not a supposed "force of gravitation," when he found first a proportionality between pendulum lengths and the squares of the periodic times, and then a proportionality between distances of fall from rest and the squares of elapsed times. Until that day, the notion of squaring a time, or squaring a measure of time, had never entered his mind, because no physical meaning, or indeed any meaning at all, could be given to such an operation. Squaring a length had a meaning in terms of area, and units already existed analogous to our "square meter" or "square mile." No unit of "square second" or "square year" exists even today, and though those could be given meanings in terms of operations, they seem not to have counterparts in nature. But squaring a *ratio* of times, or any other ratio, was legitimate, and was recognized by Euclid (who named this the "doubled ratio"). It arose when a ratio was taken twice, or compounded with itself, and that is how Galileo's constant came into his calculation of distance of fall from its time alone, giving the same ratio as the ratio of related lengths.

I would not say that Galileo had a gravitational constant, but only that in his calculation on f. 189v1 there was a constant which we can factor out and recognize as g', or as $2/g'$, or in other ways, with a small error of measurement. Galileo anticipated in 1604, at the outset of his mature physics, what was to become the gravitational constant in later physics, much as William Heytesbury arithmetically, and Nicole Oresme geometrically, anticipated the mathematical analysis of uniformly accelerated motion in the 14th century that Galileo applied to physics in the 17th century. To say that Galileo anticipated the dynamics inaugurated by Newton is misleading in a serious way, and so it is misleading in another serious way to say that medieval writers anticipated the mature physics of Galileo. Galileo worked from careful measurements of actual physical phenomena of motion, which no one had made before, and Newton worked from concepts of mass and inertia, clarified in terms of actual measurements as no one had clarified them before.

NOTES

1 S. Drake, "Galileo's Physical Measurements," *Amer. Journ. of Physics*, 54, 4, 1986, p. 304.
2 If we write T for 280, the time in *tempi* for which distance was sought, and t for 83.5625, Galileo's timing in *tempi* for fall of 4,000 *punti*, the calculation on f. 189v1 can be factored as $Tt/4 \times T/t \ (942/850)^2 = \frac{1}{2}$ distance of fall in time T. Fall in 280 *tempi* is therefore $\frac{1}{2}g'(280)^2$ or $\frac{1}{2}g'T^2$ in the usual modern form.
3 It is an interesting fact that in Galileo's units, the square of any time of *fall* is the length of *pendulum* timing double the distance of that fall by its swing to the vertical. Galileo did not mention this because in his published books he did not use his own units of *punti* and *tempo*; in ordinary units the proposition would be false.

4 See S. Drake, *Galileo At Work*, Chicago 1979, pp. 86–90.

5 S. Drake, "Galileo and Mathematical Physics," in *Scienza e Filosofia, Saggi in onore di Ludovico Geymonat*, ed. C. Mangione, Milano 1985, pp. 636–637.

6 More than a decade of study of Galileo's notes on motion has shown me that only when every entry on a group of folios relating to a particular topic has been identified and fits with all the others in a single chronology is it possible to be sure that Galileo's work is fully understood. Selected entries arranged at will can be very misleading.

7 See op. cit., note 4, pp. 67–68.

8 Only in optics and in statics had physical laws been discovered which conformed exactly, or very nearly, with actual measurements. On f. 107v Galileo had found that speeds from rest along an inclined plane increased during equal increments of time as the odd numbers 1, 3, 5, 7 ... exactly, despite any material impediments. His recorded measurements were accurate within 1/64 second variance in equalizing the times.

9 One obstacle to understanding of pendulum passages in *Two New Sciences* has been neglect of the fact that when writing of comparison of swings through large and small arcs, Galileo meant swings allowed to continue undisturbed, not swings kept at their original amplitudes. Another has been his use of the word *punto* for "count," or "point in a game or contest." For the length of pendulum and amplitudes that Galileo named, the count would differ by one after about 30 swings, but would never increase to a difference of two, as both pendulums would thereafter swing through small arcs until they came to rest. Cf. *Opere*, Ed. Naz. VIII, p. 277.

19

Galileo's Gravitational Units

Galileo is remembered in physics chiefly for his discoveries of the pendulum law, the law of fall, and the parabolic trajectory of horizontally launched projectiles. Those consequences of his early studies of motion were set forth in his last book, *Two New Sciences*, in 1638. There he derived the times-squared law for distances in fall mathematically, from the definition of uniformly accelerated motion, but Galileo did not explain how he had first discovered the law.

The measurements and timings that led Galileo to the pendulum law, and from that to the law of fall, have now been found in his working papers of 1604, from which his experimental procedures and mathematical reasoning can be reconstructed. Until it was known that the pendulum law had been a necessary preliminary to the law of fall in the procedure followed by Galileo, historians generally believed that he must somehow have followed the lead of certain 14th-century writers on accelerated motion who had devised the so-called Merton rule, a mean-speed postulate. But none of these earlier writers had ever applied that postulate to the fall of heavy bodies or carried out any measurements of accelerated motions, something obviously required in order to determine what law (if any) might govern free fall. The introduction of careful measurements into physics was, in a way, Galileo's principal contribution to the birth of modern physics. But the units he devised are of even greater theoretical interest, because they imply a universal constant of gravitational acceleration that has been strangely neglected.

In 1604 there were, of course, no standard units of length. Every nation, and in Italy each province, had its own length-unit. For time, the astronomical second was available, but that was not a practical unit of measurement until Huygens invented the pendulum clock and fitted it with cycloidal checks. The units

Reprinted from *The Physics Teacher* 27 (September 1989), 432–6, by permission.

Numerical Example of Galilean Units

(Editor's Addition)

For a fall of 1.00 m (g = 9.80), starting from rest, h = 1.00 m/(0.942 × 10^{-3}m/λ) = 1060 λ

$$t_{Fall,h} = \sqrt{\frac{2h}{g}} = 0.452\,s = (0.452\,s)(91.88\;^1/_s) = 41.5\,\tau$$

$$t^2_{Fall,h} = 1720\;\tau^2$$

$$t_{Fall,2h} = \sqrt{\frac{2(2h)}{g}} = 0.639\,s$$

$$T_{\frac{1}{4}} = \frac{2\pi}{4}\sqrt{\frac{L}{g}}$$

$$0.639 = \frac{\pi}{2}\sqrt{\frac{L}{9.80}} \rightarrow L = 1.62\,m = 1720\,\lambda$$

Hence, in G.U., the square of the time of fall through 1 m is numerically equal to the length of the pendulum whose quarter period is equal to the time of fall through 2 m.

Galileo used in his discoveries of the laws of pendulum and fall, and again in 1608 when he identified the paths of projectiles to be parabolic in shape, were of his own creation. His punto, 0.94 mm in length, was entirely arbitrary, but useful because the millimeter is a precision measure that can also be quite accurately bisected with the naked eye.

Galileo's tempo, 1/92 s, was not arbitrary for it was gravitationally linked to the punto. How Galileo managed that, and why it is still of interest today, will be explained here. It may seem that all units of length and time are necessarily arbitrary, and that in principle no pair can have any advantage over another in physics, though some units are more convenient than others for making measurements and calculations. In modern physics, for instance, units for distance and time are related by the invariant speed of light. For convenience in some calculations, this speed is given the arbitrary magnitude of one.

Mathematically speaking, space and time are incommensurable magnitudes. But that does not mean that a measurement of length cannot be numerically identical with the measurement of some related time, and, in fact, for gravitational phenomena measured lengths are associated with measured times. What will here be called "Galilean units," or G.U. for short, are λ = 0.9422119 mm and τ = 1/91.88025 s (to seven significant figures). In relating these units, we will use the quarter period of a pendulum, the time for the bob to swing (in a vanishingly small arc) through any small amplitude to the vertical. Call this time, $T_{1/4}$. Then in G.U. the square of the *time* of free fall, starting from rest through *any* height, h, is numerically equal to the *length* of a pendulum whose quarter period is equal to the time taken for a fall from 2h. (See example in box.) This relationship holds anywhere in the universe where heavy bodies fall and pendulums oscillate. It does not depend on the strength, g, of the local gravitational field. The rate of acceleration in G.U. is $g\lambda/\tau^2$, where $g = \pi^2/2^3 =$ 1.23370055 ... This follows from the ordinary equations for distance fallen from rest in a given time, and for the time of a quarter period of a pendulum.

It is no mere coincidence that Galileo's punto and tempo were nearly λ and τ, or that the implied rate of acceleration $g\lambda/\tau^2$ is a universal physical constant – unlike standard $g = 9.80665$ m/s^2, which holds only at sea-level at 45° latitude (or the gravitational equivalent elsewhere). Galileo did not know enough physics to hit on a dynamic constant of acceleration, varying with the strength of gravitational field. Modern physical constants arise from physical equations, and such constants would have cancelled out in the older notations using the classical mathematics of proportion. Galileo never used algebra, even in his working papers; the first modern mathematical physicist never wrote a physical equation in his life.

What Galileo did use was the Euclidean theory of ratios and proportionality for mathematically continuous magnitudes (such as lengths and times) set forth in Book V of the *Elements* around the year 300 B.C. That was enough for Galileo's physical discoveries. But for his proof of the law of fall he needed something that the ancient Greek mathematicians did not supply. Euclid excluded the infinite from mathematics by his axiom that "the whole is greater than the part." Galileo needed to prove that there are as many different speeds as there are instants of time during any fall from rest. For that, he introduced the concept of one-to-one correspondence between members of an infinite set and its half. That concept was no small contribution to pure mathematics, and it was quite essential to the birth of rigorous mathematical physics.

About the beginning of 1604 Galileo set out to find, if he could, a rule for increases of speed during "natural descent," as any spontaneous motion of a heavy body was called. That would be hard to do if he had had to rely on direct measurements of free fall. But the motion of a ball rolling down an inclined plane was just as surely a "natural descent" as was straight fall, and rolls could be made slow enough to time by musical beats. On folio 107 of his working papers (as they are now numbered), Galileo recorded the results of some measurements that led him on to the discovery of the law that distances in fall from rest are as the squares of the elapsed times. For about ten years I believed this page to have been the discovery document for Galileo's law of fall.

The calculations down the middle show how Galileo arrived at the eight figures he tabulated at upper left. A number was multiplied by 60, and a number less than 60 was added. Clearly Galileo had a ruler accurately divided into 60 equal parts which he used when making measurements of lengths. In fact, he used the same ruler for drawing and measuring several diagrams on other pages of these notes. Those divisions were 0.94 mm each, or a punto as Galileo called this. What he had measured were places of a ball at each of eight successive equal times when it rolled from rest down a plane grooved to guide it. The plane was something over 2,100 punti long, raised 60 punti from the horizontal at one end, to an angle of about 1.7°. Full roll would take about four seconds, permitting eight half-second marks. Calculation shows that Galileo was accurate within 1/64 s for every mark except the last, when the ball was moving about 1,000 punti per second. That was the only length that Galileo changed later, though he put a positive or negative sign on four of the other marks.

Galileo's father and brother were professional musicians, and he was a good amateur performer on the lute. At beats of half a second, anyone can detect an error of 1/25 or 1/30 of a second. A good amateur musician can do twice as well, and a professional could detect an error of 1/100 s or less. I assume that Galileo tied strings around his grooved plane, as frets were tied around the neck of his lute. His bronze ball would then make a bumping sound as it struck the plane after passing over a fret. Galileo had only to adjust the strings until every bump agreed with a beat of a crisply sung march, and measure distances from the resting point to the lower side of each string. That takes patience to do, but the apparatus described would permit anyone to adjust the strings to the limit of accuracy of his own sense of rhythm.

The rule Galileo found was that speeds during successive equal times increase as do the odd numbers 1, 3, 5, 7, ... as seen from the diagrams he drew with the page turned sideways. That was the kind of rule he sought, and he put this page aside. Later, in different ink and a bit smeared, he squeezed into the left-hand margin the first eight square numbers, which is why I took *f.107v* (Plate 10 with transcription in Plate 11) to be the discovery document for the law of fall. But it puzzled me why, in that case, he had delayed entering so important a result. Any mathematician knew that adding the odd numbers together gives the square numbers. Yet there was no point in adding speeds together, and Galileo was not thinking of these lengths as distances, but only as measures of successive speeds in a series of equal times. It had not yet occurred to him that some rule might exist the linked the distance directly to the time of descent.

What did occur to him before he laid *f.107* aside was that if a simple rule for increase of speed in descent could be found by merely equalizing time, still more might be found by *measuring* times of motions. At the bottom of the page he sketched a kind of water stop-watch. In *Two New Sciences* he described his device as simply a bucket of water with a tube through its bottom; the flow of water during a motion was collected and weighed on a sensitive balance – the weights being taken as measures of times. His sketch suggests a more elaborate device. Analysis of the data recorded on two other pages shows that water

Table I. Summary of Galileo's Procedure

Length of pendulum in punti	Time to the vertical in grains flow	(see text below)	
		$\sqrt{(2L)}$	T/16
870	668.5	41.7	41.8
1,740	942	59	58.9
3,480	1,337	83.4	83.6
6,960	1,884	118	117.8
13,920	2,674	166.9	167.1
27,840	3,768	236	235.5

flowed at very nearly three fluid ounces per second. Galileo recorded his timings in grains of weight, each equal to 1/480 ounce of flow.

Galileo measured fall time of 4,000 punti = 3.76 m at 1,337 grains of flow (about 1/30 s too high). He found the pendulum taking half that time, or 668 1/2 grains of flow, to be 870 punti long. He then doubled that pendulum to 1,740 punti, and found its swing to take 942 grains of flow while reaching the vertical through a small arc. In all this work Galileo timed pendulum swings to the vertical only, not the full period. He used the quarter-period because the only swing he could time with precision was from the instant of releasing the bob to its sound of impact with a block that was fixed in advance against a side of the bob when hanging plumb.

From the pendulum measurements of lengths and times, a table of the kind shown above could be compiled. Although Galileo may not have actually compiled a table, this will serve to show the way in which he arrived at his discoveries simply by applying the general theory of ratio and proportionality set forth in Euclid's *Elements*, Book V. The table has been extended far enough to show the source of a very important number found in two of Galileo's surviving working papers, on one of which he was writing when he first recognized the law of fall in times-squared form from its mathematically equivalent mean-proportional form.

The third figure in the second column, 1,337 grains, was the measure of time Galileo had found for free fall of 4,000 punti. The first column above was formed by successive doublings, and the second by alternate doublings. Except for one slight discrepancy, each of the first two columns taken separately is in continued proportion. Galileo's new time unit, the tempo, came into being when he related the two columns horizontally, so to speak. His original measure of time in grains of flow through a particular device having been completely arbitrary, he was free to alter it in any ratio he pleased. Taking each time to be the mean proportional between 2 and the length of pendulum, the data in each column become related line by line. The same numbers also result, almost exactly, from division of each time in grains by 16, and 16 grains ($\cong$ 1 cc) became 1 tempo – the new unit adopted by Galileo as a result of these investigations relating the times of pendulums to their lengths.

Because division by 16 of the times in grains weight of flow did not *exactly* produce that mean-proportional relation between the two columns, Galileo made a calculation that resulted in his changing 27,840, as shown above, to 27,834. That work was done on one of the working papers of which only a part survives. On the blank side he wrote, in 1609, a note on another topic, cut it out, and pasted it on *f.90*. That was lifted at my request, and on the hidden side I saw the number 27,834 twice, with enough words to identify it as a "diameter." Galileo's diagram and calculations were thrown away with the part of the page he cut off, but the work was done before Galileo discovered the law of fall, because 27,834 played a crucial part in the calculation on *f.189v1* (Plate 15) which put that discovery into Galileo's hands.

With his adoption of the tempo as the unit of time, Galileo had the pendulum law in a restricted form; that is, for any set of pendulums successively doubled in length. It would have been a difficult task to test it for successively tripled pendulums, let alone for any other integral multiples, and quite impossible to establish it by induction in its complete generality. For that purpose, Galileo next calculated the mean proportional of 118 and 167 – the times to the vertical, in tempi, for the two pendulums of lengths 6,960 and 13,920 punti – and got 140 tempi. The mean proportional of those two lengths is 9,843, so if his restricted pendulum law were perfectly general, a pendulum 9,843 punti long would swing to the vertical in 140 tempi.

On the other side of the same page (*f.154* – see Plate 21), Galileo wrote the note *filo br. 16* – "the string is 16 braccia long." From two lines drawn and labeled by Galileo at Padua, the braccio that he used in 1604 was about 620 punti. At 615 punti per braccio, length of that pendulum would be 9,840 punti, or about 30 feet. Such a pendulum could be hung from a window over the courtyard of the University of Padua and timed, protected from wind. Its time to the vertical at Padua would be 141 tempi, by calculation.

In the foregoing table, Galileo's timing of fall through 4,000 punti, 1,337 grains of flow (or 83.5625 tempi) was shown as the time for the pendulum of length 3,480 punti, whereas that fall is in fact timed by the pendulum of 4,000/g = 3,242.28 punti. The error arose from his having timed the fall as 1/30 s too long. The correct time was 80.527 tempi, not 83.5625, which Galileo adjusted to 80 2/3 immediately upon his discovery of the law of fall.

That discovery occurred when Galileo calculated, on *f.189v1* (Plate 15), the distance a body would fall vertically in time 280 tempi, double the time he had just measured for swing to the vertical by his 30-foot pendulum. When he made that calculation, he supposed some distance of fall to be needed, and for that he used fall of 4,000 punti. In fact, it did not matter what fall he used, because that factor was cancelled by its presence as the numerator of one ratio and as the denominator of another. The result he got was very near to the modern result for the latitude of Padua, his 48,143 punti as compared with our 48,317, a difference of 174 punti or 15 cm in 45 1/4 meters (about 0.36% low).

As you see, Galileo at once took the mean proportional of his 48,143 punti and the assumed base, 4,000 punti, saw this 13,863 to be nearly the half of 27,834, and thus realized that his roundabout calculation of a distance of fall through an intermediate pendulum relation had been unnecessary; the distance depended only upon the time. At lower left he drew the conventional proportionality diagram for magnitudes "commensurable in the square," which he used always thereafter for relating distances and times in fall.

Strictly speaking, Galileo never made use of any "constant of gravitation" in the modern sense. The acceleration implied by his units of measurement was, of course, very nearly $g = \pi^2/8$, valid anywhere that pendulums oscillate and heavy bodies spontaneously fall when released from restraint. But the constant Galileo used was about $\sqrt{g} = \pi/(2\sqrt{2})$, and that is what I call "Galileo's constant." In theory it is 1.110720735 ... , whereas Galileo used the ratio 942/850 = 1.108, his ratio of time to the vertical for a pendulum of length 1,740 punti, to time of fall 1,740 punti from rest. He did not know that π had anything to do with the matter but relied entirely on the most accurate measurements he could make and the rigorous Euclidean theory of proportionality for mathematically continuous magnitudes.

G.U. provide a method of experimental measurement of π that physics students should be invited to test in the laboratory. No better way exists of showing them the nature of units of length and time based not on the distance from pole to equator through Paris and 1/86,400 of one axial rotation of Earth, but on actual phenomena of vertical fall and of horizontal motion spontaneously produced by gravitation. That physical relationship was built into Galileo's units as accurately as he could measure it, and of course we can build into G.U. the exact relationship, involving a transcendental number, π, that cannot be exactly measured in any actual phenomena of nature.

In conclusion, I add that in Galileo's day there existed no measurements of planetary distances in terrestrial units such as millions of miles. There are now, and it is easy to convert those into G.U. Anyone who does that will find some very interesting relations among the places and periodic times of planets over and above those deducible from Kepler's well-known planetary laws.

Index

Fabri, Honoré, 183, 274

fall, 81, 259, 334; along circular arcs,
303–4; and contrary proportion, 202–5,
263, 289–90; development of law of,
179, 195, 201–2, 250–2, 257, 262–4,
282–3, 284–5, 335–41, 358–61, 372–3;
in *Dialogue*, 60, 61–3, 64, 65–6; and
inclined planes, 301; law of, 66–7, 122,
149, 161, 251, 261, 288–9, 303, 341,
376; Mersenne's law of, 60–1, 64–5,
66; and the Merton rule, 199, 200; and
pendulum, 300, 339–40, 342, 350–2,
354; proof of the law of, 258; semicircular, 77–8; and the tower argument, 87

Favaro, Antonio, 148

Fermat, Pierre, 58; and fall, 61, 64–5, 66,
67

Feyerabend, Paul: and tower argument,
86, 94

Finnocchiaro, Maurice, 266, 269, 270–2;
and tower argument, 86, 94–5

force: impressed, 48, 126. *See also* inertia

Galilean units, 371–2, 376

Galilei, Galileo: and Archimedes, 78; and
Aristotelianism, 39; on Aristotle, 71;
and Copernicanism, 30, 31–2, 50, 52,
54, 131; and cosmogony, 79–80; and
cosmology, 79–80, 83; and Euclid,
184, 352; and fall, 66–7, 179, 188–91;
and Kepler, 54, 97, 104nn1, 2; and
kinematics, 48, 50, 342; and mathematics, 293, 304–5; and metaphysics,
35, 50; and motion, 79; and physics,
14, 48, 79, 89, 229, 355, 367; and planetary motion, 50–1, 80–1; and Platonism, 38, 64, 79–80, 158, 197; and
projection argument, 88; and Sarpi,
191; and scientific method, 121, 123,
172–3, 178, 179, 185, 195; and teach-

ing, 122; and tidal theory, 8, 29–30,
31–2, 33, 34, 49, 51–2, 53, 97–8, 99,
100–3, 108–9, 110, 112; and the tower
argument, 86–9; trial of, 18–19, 50
– Works: *Assayer (Il Saggatore)*, 63, 74,
220; La Bilancetta, 139; *De Motu (On
Motion)*, 70–1, 72, 73, 142, 174, 176,
180; *Dialogue Concerning the Two
Chief World Systems*: — and Copernicanism, 9, 11, 14, 17, 18, 21nn16, 17,
28, 30, 39–40; — and fall, 60, 61–3, 64,
65–6; — and motion, 239; — preface
of, 11–16, 27–8, 41, 43–4; — reasons
for writing, 5; — and Theorem I, 238,
242; — and tides, 16, 26, 28, 41–2, 44,
46, 49; — title page of, 6–8, 28, 29, 40–
1, 42–3, 55–6n6; — tower argument in,
88, 90–1; *Letters on Sunspots*, 142–3;
On Mechanics, 71, 176; *On Uniform
Motion*, 276; — and Theorem I, 275;
— and Theorem II, 265, 266–70, 273;
Two New Sciences (Discorsi), 75, 216,
225; — and accelerated motion, 233;
— and double distance rule, 240; —
and Theorem I, 233, 238, 240, 242–3

Galileo's constant, 357, 367–8, 376

Galileo's theorem, 348, 349

Gassendi, Pierre, 181

Geminus, 32

Grazia, Vincenzio di, 232, 343

Greymonat, Ludovico, 333–4

Guidobaldo del Monte. *See* Monte,
Guidobaldo del

Heytesbury, William of, 241–2, 243

Hipparchus, 32, 124, 135

Holste, Luca, 36

impetus, 48, 72, 124–5, 126–7, 135, 180–
1, 274; relation to inertia, 135–7

projection argument: and Copernicus, 88,
89; and Galileo, 88; and Ptolemy, 88
Ptolemy, Claudius: and motion of the
earth, 87–8, 94

Riccardi, Niccolò, 23, 24, 27, 42–3, 44,
47, 90

Sarpi, Paolo, 53–4, 187–8, 301; and
Copernicanism, 99; and tides, 98–9,
100
Settle, Thomas, 147, 362
Soto, Domingo de, 225
space-proportionality, 210, 211, 214, 220,
224, 226, 240
speed, 179–80, 193–4, 260, 262–4, 285;
equal, definition of, 277; measurement
of, 193; representation of, 286–7
Strauss, Emil, 38, 101
Strong, Edward, 333–4

Tartaglia, Niccolo, 274

tides, 112–13; and Galileo, 28, 29–30,
31–2, 33, 34, 51–2, 53, 97–8, 99, 100–
3, 108–9, 110; and Kepler, 33, 103,
107, 111–12, 113
time: measurement of, 307–8, 336, 350,
359, 362, 365–6, 373–4
time-proportionality, 213, 214, 224, 225,
226, 240, 260–1
tower argument, 88, 90–1; definition, 86;
and motion of the earth, 87–8

Urban VIII, 8, 10, 16–17, 45. *See also*
Barberini, Maffeo Cardinal

Valerio, Luca, 209
Varro, Michael, 182, 198

Wallis, John, 101
Wohlwill, Emil, 101

Zollern, Cardinal, 10, 11